# Empire and Environment in the Making of Manchuria

**Contemporary Chinese Studies**

This series provides new scholarship and perspectives on modern and contemporary China, including China's contested borderlands and minority peoples; ongoing social, cultural, and political changes; and the varied histories that animate China today.

A list of titles in this series appears at the end of this book.

# Empire and Environment in the Making of Manchuria

EDITED BY NORMAN SMITH

UBC Press • Vancouver • Toronto

25 24 23 22 21 20 19 18 17     5 4 3 2 1

Printed in Canada on FSC-certified ancient-forest-free paper (100% post-consumer recycled) that is processed chlorine- and acid-free.

**Library and Archives Canada Cataloguing in Publication**

Empire and environment in the making of Manchuria / edited by Norman Smith.

(Contemporary Chinese studies)
Includes bibliographical references and index.
Issued in print and electronic formats.
ISBN 978-0-7748-3289-2 (hardback). – ISBN 978-0-7748-3291-5 (pdf).
ISBN 978-0-7748-3292-2 (epub). – ISBN 978-0-7748-3293-9 (mobi)

1. Human geography – China – Manchuria – History. 2. Manchuria (China) – Environmental conditions – History. I. Smith, Norman (Norman Dennis), author, editor II. Series: Contemporary Chinese studies

GF657.M35E46 2017     304.20951'8     C2016-906294-5
C2016-906295-3

Canada

UBC Press gratefully acknowledges the financial support for our publishing program of the Government of Canada (through the Canada Book Fund), the Canada Council for the Arts, and the British Columbia Arts Council.

This book has been published with funds provided by the Social Sciences and Humanities Research Council of Canada and with the help of the University of British Columbia through the K.D. Srivastava Fund.

UBC Press
The University of British Columbia
2029 West Mall
Vancouver, BC V6T 1Z2
www.ubcpress.ca

# Contents

# Illustrations

## Maps



## Figures

# Acknowledgments

This volume is the result of a workshop organized by Diana Lary and me at the Institute of Asian Research at the University of British Columbia, Canada, May 16–18, 2013: "The Manchurian Environment: Natural Resources, Climate and Disease." The workshop was generously funded by the Social Sciences and Humanities Research Council of Canada (SSHRC), the University of British Columbia (UBC), and the University of Guelph (UG). It built on a 2011 conference organized by Mark Elliott at Harvard University, "Empire and Beyond: Manchuria in the Modern World," and featured many of the same scholars and graduate students. I thank Victor Seow for his kind extension of an invitation to that conference at the 2011 Association for Asian Studies Annual Conference in Honolulu.

"The Manchurian Environment" workshop gathered together more than twenty senior, mid-career, and emerging scholars, as well as graduate students, from Canada, China, Great Britain, and the United States. This volume is the product of some of the material presented at that workshop. I am grateful to all of the participants in the workshop: Olga Bakich, David A. Bello, Blaine Chiasson, Sakura Christmas, Annika A. Culver, Colin Green, Jack Patrick Hayes, C.J. Huang, Loretta Kim, Diana Lary, David Luesink, Christos Lynteris, Kathryn Meyer, Robert Perrins, Jonathan Schlesinger, Victor Seow, Song Nianshen, Ronald Suleski, and Sun Xiaoping. Most of all, I am grateful to Diana Lary for her never-ending inspiration and support. Our initial planning of the workshop took place while *yum cha*-ing with Victor Zatsepine at the Summer Palace Restaurant in the Island Shangri-La Hotel Hong Kong, a most propitious start.

A considerable amount of work went into organizing the workshop. At UG, I thank Donald Bruce, Stuart McCook, Sandra Sabatini, and Ornella McCarron. The organizing and execution of the workshop would not have been possible without the skills of Jacqueline McIsaac. And I especially thank Lori Irvine and Karen Bygden for their handling of the financing. At UBC, I thank Timothy Cheek and Karen Jew for their support and provision of the inspiring setting of the Institute of Asian Research. At the workshop in Vancouver, Wang Xian, Zhang Hongbing, and Wu Yang worked seamlessly to ensure the workshop's success. I thank Jocelyn Smith, too, for her timely, post-workshop contribution.

At UBC Press, Emily Andrew has proven, again, to be an endlessly supportive editor; where she gets her energy from, I just don't know. Megan Brand marshalled us through the editing process with care and finesse. Jillian Shoichet was a very thorough copy editor. Holly Keller was supportive with image reproduction. The cartographer, Eric Leinberger, was a great help. And, once again, David Drummond has outdone himself with what I consider to be a beautiful and evocative cover.

I thank the external readers, who generously shared their knowledge and expertise. I especially thank Timothy Brook for his incisive suggestions and encouragement. Stevan Harrell proved very helpful early in the publication process. David A. Bello and Ronald Suleski went above and beyond the call of duty with the glossary.

I thank all of the above for their contributions to the workshop as well as this volume, which, at every stage of publication, made it to the Press on time. Their enthusiasm and cooperation has made assembling this collection a joy.

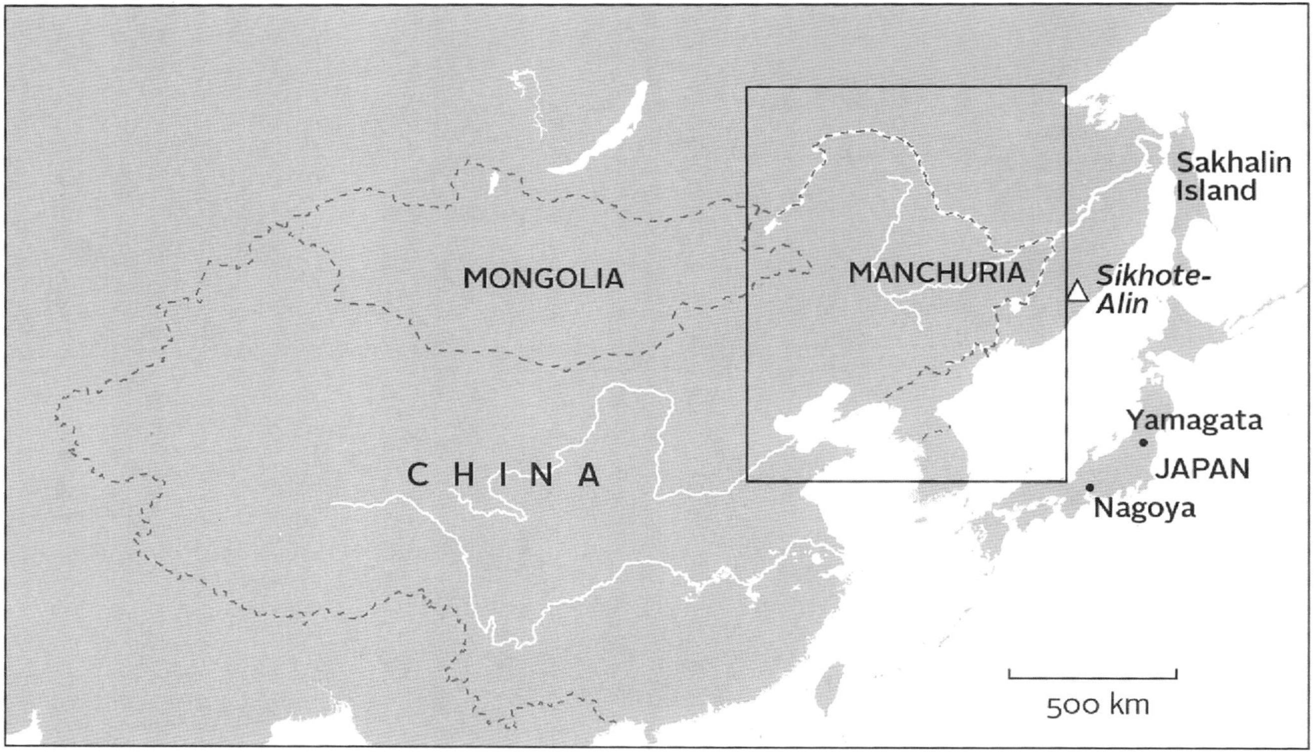


Map 1 Manchuria in Northeast Asia. *Courtesy of Eric Leinberger.*

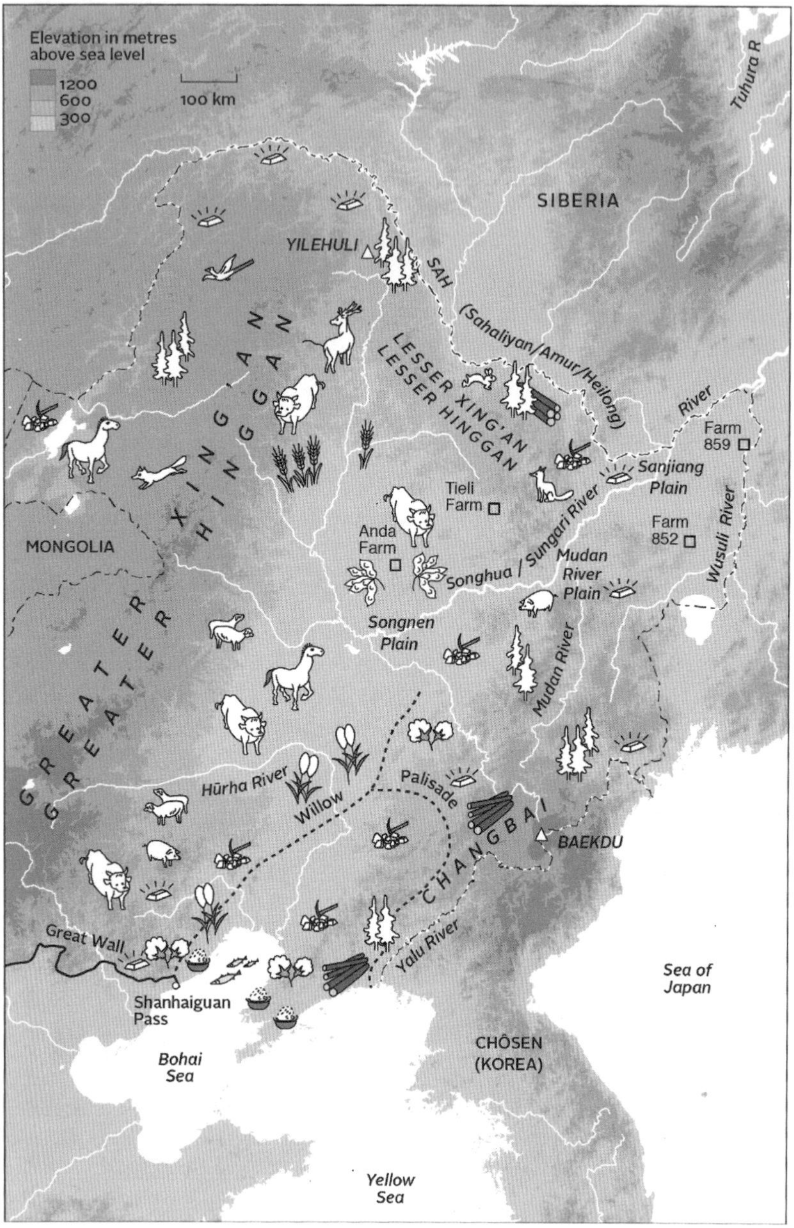


Map 2 Manchuria's natural wealth, early to mid-twentieth century. *Courtesy of Eric Leinberger.*

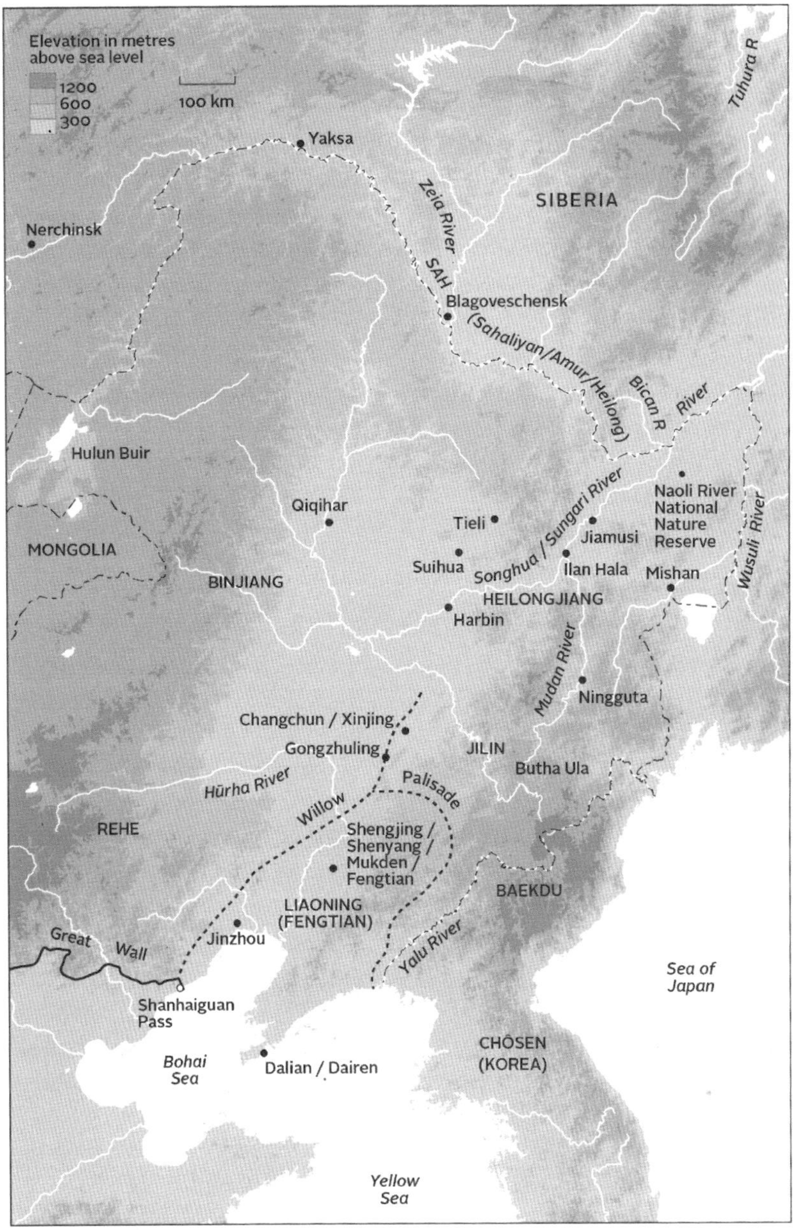


Map 3 Manchurian Settlements, early to mid-twentieth century. *Courtesy of Eric Leinberger.*

# Empire and Environment in the Making of Manchuria

# Introduction

*Norman Smith*

In the early decades of the twentieth century, Japanese- and English-language commentators characterized the vast lands of Northeast China, then known as Manchuria, as a "Land of Opportunities," a "Promised Land," and the "Cockpit of Asia."[1] Within China, it was more commonly known as the land "beyond the passes" or the Three Eastern Provinces.[2] "Manchuria" conjured up images of far-away adventure, romance, and wealth for the foreigners who most widely used the term, while Chinese nationalists came to view the term as an embodiment of foreign imperialism and "Manchus" as an ethnonym linking them to their ancestral home.[3] The number of names applied to the region attests to a centuries-long contestation over its sovereignty. From the 1600s until the mid-twentieth century, sovereignty over Manchuria was challenged by Manchu, Russian, and Japanese empires, Chinese warlords, Soviet invaders, and finally by Chinese nationalists and communists. Mass migrations of people from across Eurasia, lured by Manchuria's rich resources, vast territory, and relatively sparse population, transformed the region from a Manchu homeland to a quixotic imperial frontier or borderland and, now, one of China's "Four Economic Engines."[4]

This volume advances new understandings of Manchurian history through two unifying themes: empire building and the environment. While the former has been the focus of previous volumes on Manchuria, the latter has been more rarely, so the chapters in this book direct the reader's attention to a multiplicity of colonizing agendas *as well as* human interactions with the environment to highlight significances over a period of four centuries.[5] The region's history has been forged at the interstices of Chinese, Japanese, Korean, Manchu, Mongolian, and Russian states within a diverse array of regimes, including colonial, agricultural, economic, military, moral, and penal, as represented in chapters of this volume.[6] Each regime sought control over the region's resources and engaged in a range of conflicting sovereignty claims, in what Jonathan Schlesinger has termed a "commercial dynamism of the frontiers."[7] For historians, Manchuria offers a unique space to study Manchu and Russian imperialism, Japanese fascism, and Chinese communism. A study of Manchuria also allows for research into farming, resource extraction, northern identity, and the interconnections between them. Such research reveals regional significances that have been obscured for decades. Today, even with over 100 million residents (almost one-tenth of China's population), Manchuria continues to be denigrated as a cultural backwater, if not the imagined "empty space" of earlier times. *Empire and Environment in the Making of Manchuria* questions the ways that control over Manchuria was linked to its distinctive environmental conditions. How did these conditions shape the rivalries of those who sought to claim Manchuria for their own? The answers to this question posit Manchuria as a major factor in the modern history of China and East Asia from a perspective that goes beyond the immediately political.

## Imperial Approaches to Manchurian Environment

"Manchuria" (*Manzhou*) is a controversial name given to the homeland of the Jurchen peoples whom Hong Taiji (1592–1643) banded together under the name "Manchu" (*Manzu*) to form the last imperial dynasty in China, the Qing (1644–1912).[8] Once the dynasty was firmly established, the Kangxi emperor (1654–1722) and his successors attempted to seal off the area to protect "Manchu" heritage and prevent the Sinicization of local peoples such as the Daur, Ewenk, and Oroqen; Evelyn Rawski notes that these latter groups were "culturally 'Manchuized'" during Qing rule.[9] For two hundred years, efforts were made to keep Han Chinese migrants from the multi-ethnic region that the Manchus claimed as their own. The Great Wall and Willow Palisade, Qing statutes, and the region's famously ferocious winters initially succeeded, to an extent, in limiting Chinese migration, especially to the northernmost regions of Jilin and Heilongjiang.

During those years, the land marked as the Manchu people's homeland – an area far greater in extent than their foraging grounds – acquired the name "Manchuria," which Mark Elliott has termed a "troublesome toponym,"[10] a product of Manchu-Jesuit cartographic projects, designed to distinguish the region from "China proper."[11] The most egregious application of the term was by Japanese and Russian imperialists, through their efforts to permanently sever the region from Chinese control. Perhaps the least egregious use was by Manchu royals, who in the early 1930s saw in the term a chance to regain mastery over their namesake land from a Chinese warlord. For each group, "Manchuria" was a toponym reflecting questionable sovereignty claims. Over the past several decades, the term "Manchuria," once widely used, has fallen into disfavour in China as a token of the region's imperialist past – yet it is suggestive of a regional identity that continues to exist.[12] In the early 1960s, in a Chinese Communist Party (CCP) internal document (*neibu wenjian*), historian Sun Wenliang argued that the term *Manzhou* dated to Nurhaci (1559–1626), was prevalent by the Qianlong emperor's era (r. 1735–96), and was improperly used from at least then to the time of his writing. Sun stressed that *Man* could no more denote a country name (*guohao*) than that of an area (*diyu*).[13] He stressed that the term was properly only the name of an ethnic group (*zucheng*), and that ongoing improper usage was due to mistaken understandings of the term in magazines, books, and common speech, as well as ongoing political manipulation by Japanese and American imperialists, which forced him to clarify that the proper usage of the term was *Manzu* for a people and not *Manzhou* for a region. Clearly, CCP members required instruction that the term, though widely used, was to be deemed politically incorrect.

"Manchuria" is used in this volume to reflect the term's historical applications and the distinct status of the region before it came to be treated as a constituent part of the Chinese nation. How successful were the Manchus at securing the land that came to bear their name? Their control over it, and their coordination of the local peoples, enabled them to replace the moribund Ming Dynasty (1368–1644) and establish a regime that lasted for nearly three hundred years as the Qing Dynasty. But in the eighteenth century, as ecological and cultural pressures arising from the region's successful incorporation increased, waves of Han Chinese migrants began to surge in and out of Manchuria, while Russian and Japanese expansionism threatened the region's borders.[14] Faced with few choices as the long-term ambitions of the Russians and the Japanese became clearer, Qing rulers reluctantly allowed Han Chinese migration, heightening tensions in the region in the Age of Imperialism, when – unlike recent decades – Manchuria attracted a great deal of international attention.[15]

In 1888, H.E.M. James (1846–1923) remarked in *The Long White Mountain* that Manchuria was "the scene of a great conflict in the past, for supremacy over Eastern Asia, and perhaps of a still greater [one] in the future. Manchuria merits alike the attention of historical students and contemporary statesmen."[16] His observation almost certainly inspired Owen Lattimore (1900–89), who, having travelled extensively through the region in the 1920s and 1930s, formulated his important "reservoir theory," in which he conceptualizes Manchuria as a historic source for the Inner Asian invasion of China. This theory would establish Lattimore as a leading figure of frontier studies.[17] Thomas Barfield argued that "Manchuria, for political and ecological reasons, was the breeding ground for foreign dynasties when native dynasties collapsed in the face of internal rebellions."[18] Even in 1929, the significance of Manchuria was lauded by Frederick Simpich (1878–1950) in *National Geographic Magazine:* "Here, through turbulent years, three ancient empires met – the Bear, the Dragon, and the Rising Sun. Their struggles shook the earth. Korea succumbed, absorbed by the Rising Sun; the Dragon mothered Manchuria. War mangled the Bear, and to the north rose an evanescent Far Eastern Republic."[19]

These men argued, in no uncertain terms, that great changes with worldly significance were being wrought in Manchuria, "The Wild West of the Far East"[20] – a land that was simultaneously a "frontier" or "borderland" and a "contact zone" for peoples from China, Japan, Korea, Russia, and beyond. Mark Elliott, in his recent analysis of words approximating "frontier" and "borderland" in Chinese and Manchu, argues the continued relevance of Lattimore's description of Manchuria as a "zone of contestation."[21] Mary Louise Pratt's concept of the contact zone – "social spaces where disparate cultures meet, clash, and grapple with each other, often in highly asymmetrical relations of domination and subordination" – is also pertinent to the study of Manchuria.[22] Pratt argues that such "peripheral" zones can produce effects unanticipated in the imperial metropolis.[23] The chapters in this volume demonstrate how successive waves of disparate cultures were attracted to Manchuria's environment, over which and within which they grappled with one another (and the environment itself) in a heavily contested zone, often with dramatic ramifications beyond the region's borders.[24]

In the seventeenth and eighteenth centuries, two expansionist empires – the Qing and the Romanov – clashed over the settling of their borders. Pamela Kyle Crossley has argued that the Qing "was an empire, one of the largest, most powerful and influential of the early modern period."[25] Ho

Ping-ti has called it "the most successful dynasty of conquest in Chinese history."[26] According to Evelyn S. Rawski, the Qing owed its rise to power to "its ability to use its cultural links with the non-Han peoples of Inner Asia and to differentiate the administration of the non-Han regions from the administration of the former Ming provinces."[27] A powerful multi-ethnic Qing empire then faced its Romanov counterpart; both empires expanded with the settling – and sharing – of Manchuria's borders. The empires grew in what Li Narangoa and Robert Cribb have labelled a "twin story," incorporating peoples and lands at a remarkable pace.[28] Manchuria held a special place in both regimes. The Qing treated the region as a Manchu imperial reserve *par excellence* (notwithstanding the presence of twenty to thirty ethnic groups and subgroups).[29] For the Romanov, the region was a fount of considerable wealth, initially in furs, and later a vital link to a year-round, ice-free Asian seaport.[30] Qing policies, while effective for the dynasty's early modern ambitions, led in time to unanticipated imperialist contention over Manchuria's sovereignty.

Early Qing-Romanov treaties, notably Nerchinsk in 1689 and Kiakhta in 1727, permitted a supplicant Russia special, limited rights. European powers from farther afield and, later, Japan, would subsequently wage much larger wars in order to exploit these rights as victors.[31] Qing and Romanov histories entwined, but not always as peacefully as the treaties might suggest. Subsequent treaties in 1858 and 1860, following humiliating Chinese defeats, transferred vast tracts of land to Russia, demonstrating just how weak the Qing dynasty had become. In *International Rivalries in Manchuria (1689–1922),* Paul Hibbert Clyde describes Chinese and Russian tensions in Manchuria over the *longue durée.* Clyde is especially critical of the vicious treatment of local peoples by the Russians, which, he argues, undermined Russian efforts to bring the local population peaceably into the tsarist empire. Clyde's view receives qualified support from David A. Bello in this volume.[32] R.K.I. Quested, too, argues that the Chinese and Russian empires shared uneasy borders that necessitated treaties of reconciliation because ongoing quests for land and resources pitted them against one another.[33] S.C.M. Paine stresses how Russia repeatedly violated these treaties as Qing power declined.[34] More recently, Rui Hua has argued that the Russian occupation of south Manchuria in the aftermath of the Boxer Rebellion was characterized by more collaborationism between Russians and local populations than was the case earlier in northern and western Manchuria and, in fact, paved the way for later collaboration with the Japanese.[35]

Scholarship on the Russian presence in Manchuria highlights the significance of railways and the culture of Harbin, one of the largest Manchurian cities and an important railway hub in Heilongjiang. To celebrate the twenty-fifth anniversary of Russia's founding of the Chinese Eastern Railway (CER) (est. 1896), E.Kh. Nilus published a historical survey that described the CER as an economic, civilizing, and beneficial force in the development of the region.[36] Felix Patrikeeff and Harold Shukman's *Railways and the Russo-Japanese War* emphasizes the "railway imperialism" through which both Russia and Japan sought to settle Manchuria as their own; this is also the focus of Bruce Elleman and Stephen Kotkin's *Manchurian Railways and the Opening of China*.[37] Patrikeeff and Shukman argue that the Russo-Japanese War (1904–5) was a "watershed event" as decisive as Waterloo, though now mostly forgotten.[38] The long-term Russian presence in Manchuria has perhaps best been illustrated in monographs on the Russo-Chinese culture of Harbin, which stress the roles played by Russians in Manchuria and by Manchurians in Russia in the decade before and after the revolution of 1917, as Harbin became a centre for tsarist loyalists and a truly international city.[39] There, the increasingly common sight of Russians working for the Chinese as labourers or sex workers led many to view the city as the "grave of the white man's prestige."[40] Harbin, the "Paris of the East," thus named for its European, cosmopolitan flavour, contributed to the creation of a "distinct regional identity" that still exists today.[41]

Ultimately, the Russian state's ambitions in Manchuria cost Russia dearly, as Japan decisively defeated Russia in 1905, in the first modern war in which an Asian country defeated what was then widely perceived as a European one. This war was notable in three other ways as well. First, it gave birth to a new terror: war in the trenches, with machine guns and barbed wire that presaged the calamitous Great War in Europe (1914–18). Second, the defeat in Manchuria, distant though it may have been to most, inspired revolutionaries around the world and set Manchuria firmly on the world's radar. Vladimir Lenin (1870–1924) noted: "The catastrophe of our vilest enemy signifies not only the approach of Russian freedom. It also portends a new revolutionary upsurge of the European proletariat ... Progressive, advanced Asia has struck a blow at retarded, reactionary Europe."[42] Lenin explicitly linked the Russian loss in Manchuria with the revolution that would replace tsarist rule with a communist regime. Third, the war spurred the Qing court to permit large-scale Chinese migration to Manchuria in order to prevent further attempts at external appropriation, in what has been called "a migration perhaps without parallel in recorded history."[43] The Russian and Japanese contestation over Manchuria attracted international attention. To date, however, the early- to mid-twentieth-century Russian presence in Manchuria has not attracted the degree of critical attention it deserves, overshadowed as it has been by the harsh Soviet invasion of Manchuria in 1945 and the subsequent cooling of Sino-Soviet relations. Soviet and Russian historians have generally ignored the topic,[44] while historians in China have tended to downplay the Russians' presence in Manchuria. Two of the chapters in this volume (by David A. Bello and Blaine Chiasson) highlight Russian involvement in the region's history and the varied agendas of tsarist and Soviet regimes, as demonstrated in their studies of tribute collection and farming practices.

The collapse of the Qing Dynasty in 1912 ushered in a period of Chinese warlord rule that lasted for two decades, until the Japanese invaded in 1931. The following year, Harry L. Kingman further outlined the development of "railway diplomacy" over the control of Manchuria's rail system, which had fuelled Chinese nationalism and created the context for a Japanese invasion to protect Japanese regional investments.[45] Gavan McCormack and Ronald Suleski have demonstrated how ruler Zhang Zuolin (1878–1928) secured a respectable level of stability in the early 1920s while overseeing the expansion of the railways, the creation of the largest munitions factory in China, and, for a while, a flourishing economy.[46] As the chapters in this volume make clear, Chinese rule over Manchuria during this period was made even more complex by continuing Russian and Japanese interests. In 1922, as a harbinger of events to come, Japanese officials of the South Manchuria Railway (SMR / Ja., *Mantetsu*) (est. 1907) lauded Manchuria as "The Garden of China," citing it as "the most favored [*sic*] spot for agriculture in the Far East."[47]

The Japanese invasion of Manchuria in 1931 capped a decades-long struggle for dominance in the region.[48] Japan's "incremental" empire building, especially through the SMR, is the focus of Yoshihisa Tak Matsusaka's *The Making of Japanese Manchuria.*[49] The SMR is also the subject of John Young's *Research Activities of the South Manchurian Railway Company, 1907–1945,* which examines the conglomerate's wide-ranging activities. It operated rail lines, hospitals, schools, and research facilities that produced massive collections of statistics tarnished by their imperialist provenance.[50] Louise Young has detailed the impact in Japan of such huge Japanese investments in, and romanticized perceptions of, Manchuria and the puppet state of Manchukuo (1932–45). Manchuria served as a "lifeline" for Japan, inspiring the most romantic of Japanese "visions of empire."[51] Prasenjit Duara has termed these visions that manifested in Manchukuo as "tradition within modernity," stressing the role that modern concepts such as Asianism, citizenship, and ideal womanhood played in the Confucian rhetoric–based regime that the Japanese named "Country of the Manchus."[52] Janis Mimura has cogently argued that the regime was "techno-fascist."[53]

Received interpretations of widespread, immediate Chinese condemnation of the Japanese establishment of Manchukuo are the focus of Rana Mitter's *Manchurian Myth,* in which he carefully traces "the development of the narrative of resistance to the occupation" that played a central role in the creation of Chinese regional and nationalist identities.[54] Mitter argues that while the Japanese occupation was negative, over the long term more positive outcomes included the strengthening of a Northeast Chinese self-identity to counter Japanese attempts to sever the region from the republic. Mitter underlines the contentious nature of sovereign claims to the region, as does Dan Shao, who traces shifting local self-identities of Manchus and the impact on them of multiple colonial contestations as the region transformed from a Manchu homeland to a Qing borderland and then to an inalienable part of China.[55] The long-term ramifications of Japanese rule even extended into neighbouring Korea, as Suk-jung Han argues: "Manchukuo incubated the leadership of both Koreas."[56] Although fourteen years under Japanese rule did little to empower the Manchus for whom the state was named, those years have structured much understanding of the region's history.

Under Japanese occupation, Manchuria's industrial sector grew to be the largest in mainland China; there, multinational automobile manufacturer Nissan had its start.[57] A modern, urban environment grew in Changchun as it was renamed the "New Capital" (Ch., Xinjing; Ja., Shinkyō) and became home to one of the largest movie-making enterprises in Asia, the Manchukuo Film Association, which ultimately became one of the leading mainland Chinese film production companies, the Changchun Film Group Corporation. Japanese consumer goods spread through local marketplaces. Japanese-sponsored, Chinese-language newspapers, journals, and books reflected a growing literary community and an increasingly identifiable regional identity. The population of Manchukuo grew by a third, mostly Chinese migrants fleeing war and poverty for the relative security of the region. Officials called their new state a "Kingly Way paradise land" (Ch., *Wangdao letu;* Ja., *Ōdō rakudo*).[58] Despite potentially positive attributes, however, the state of Manchukuo failed to gain recognition from most of the outside world as it devolved into a racist, colonial regime that engaged in crimes against humanity, including Harbin's infamous Unit 731, a biological warfare unit operated by the Japanese until 1945 and which carried out experiments on human beings. Sheldon H. Harris's *Factories of Death*, which recounts horrendous Japanese war crimes, reflects mainstream Chinese understandings of Japanese involvement in the history of the region, if not the nation.[59] A related legacy was the enormous drug market that serviced users and addicts within and well beyond Manchuria's borders, as reflected in Kathryn Meyer's contribution to this volume.[60] Few who lived through Manchukuo, and stayed afterward, publicly mourned the state's disappearance in 1945.[61]

The contestation of sovereignty over Manchuria was one of the most significant influences of the region's history.[62] Russian and Japanese colonial ambitions, empowered by the relative weakness of Manchu and Chinese regimes, ensured that Manchuria was a feature of China's "century of humiliation" (*bainian guochi*), while bestowing on the region an unprecedented prestige. Yet for decades after Russia and Japan had been driven out, there was only modest international interest in the region, which was transformed into a cornerstone of national economic development in the People's Republic of China (PRC) after communist victory in the civil war (the crucible of which was in Manchuria).[63] As in the late Qing-Republican, Romanov-Soviet, and Japanese regimes, development became a major focus of PRC state policy, which sought to maximize the potential of the region's resources and peoples. As this volume demonstrates, though the PRC implemented different policies in Manchuria, it faced many of the same rewards and challenges (albeit within larger regional ecological conditions of continuity and change) as its predecessors.

## Environmental Approaches to Manchurian Empire

Manchuria's environmental context is the second major theme of this volume. The latitude, topographical features, and wind currents from Siberia create an environment that has long attracted the attention of governments, scholars, missionaries, and journalists. Owen Lattimore argued that "historical processes of real importance are possible within the forest, river, and mountain world that overlaps from Manchuria into Siberia and Korea and sweeps to the edges of both the Mongol steppe and the Chinese Pale of lower Manchuria."[64] Lattimore took Manchuria's environment into dynamic analytical account, recognizing that certain factors acted to inhibit human habitation, at least in the northernmost regions. He was not alone in this view. Diplomat Alexander Hosie (1853–1925) noted that it was "so cold that we rarely summoned up courage to ride"; he wrote of winter travel challenged by snow, special footwear, and his frozen moustache.[65] But while travel was difficult in the winter, according to famed traveller Sir Francis Younghusband (1863–1942) it was nearly impossible in the summer, when marshy land, nonexistent or lacklustre roads, and insect infestations made getting around even more treacherous.[66] But these conditions seem not to have dampened the ambitions of Manchus, Chinese, Russians, or Japanese in their attempts to settle the region, especially in the south. *Empire and Environment in the Making of Manchuria* analyzes the environment's role in Manchuria's history during times of great change. This analysis illustrates that imperial rivalries and socio-economic upheaval are significantly conditioned by existing ecological parameters, as embodied in the unprecedented levels of immigration and growing individual and state demands on the region's resources that constituted the main human responses and adaptations to Manchurian ecosystems.

Manchuria's rich, vast "emptiness" was a powerful lure in the Age of Imperialism. Once officially closed to outsiders, the Imperial Age saw Manchuria pried open by massive international railway projects that further exposed the region's riches and attracted increasing numbers of visitors as well as those with more permanent ambitions. In 1888, H.E.M. James argued that it was "intolerable that a great continental highway like the Amur [River] should be closed for ever to commerce, and vast regions to colonists, simply because of China's crass obstructiveness."[67] To James, it was unacceptable that the Qing court sought to isolate such wealth and abundance, preserving the region for itself. But "China's crass obstructiveness" did not arise solely from the state's greed or its preservation efforts. The very practical difficulties of settlement and resource extraction are the focus of Loretta Kim's chapter in this volume. Railways provided new, more reliable and effective transportation, importing increasing numbers of migrants to harvest – and extract – the region's resources. Manchuria's farmers soon cornered the world market in soya beans. Highly valued Manchurian furs were sold internationally. Manchuria's forests were harvested by the Japanese, who, according to Ma Zongshen, "plundered 70 million cubic meters of timber [from Manchuria], or one-tenth of China's entire timber reserves at that time."[68] In short order, the "opening" of Manchuria transformed relations between the people and the environment by states eager to take control of both, and use them to their advantage.

Scholars such as Ted Steinberg have shown how the environment can be an "active, shaping force" in human agency and historical change.[69] Over the past two decades, notions of development, modernity, and northern-ness have been linked explicitly with the environment.[70] In Asia, recognition of the environment's significance to human life has been articulated in diverse philosophical traditions that often depict the human world as one part of an interactive natural system. Yet environmental history per se is relatively new in China. Bao Maohong has dated the emerging field in China to the late 1990s, noting that the term "environmental history" first appeared only in 2000.[71] Bao argues the need for a "theory of environmental history with Chinese characteristics," to be created through regional and broader studies.[72] While Bao makes an important distinction between regionally specific and more general forms of environmental history, J.R. McNeill has identified three main branches in the field, namely material (changes in the environment and how they affect human society), cultural intellectual (representations and images), and political (law and state policy).[73] This volume features Manchuria-specific essays that engage each of the branches identified by McNeill to underline the utility of what Mark Elvin has described as an "almost disconcertingly diverse"[74] range of environmental history.

One of the most important recent general studies of China's environment is Elvin's *Retreat of the Elephants,* which centres its analysis on China proper, setting aside a detailed analysis of Manchuria.[75] Elvin's work provides a useful, broad-ranging analysis of the environmental costs in China of an intensifying human presence, noting that "human systems ... cannot be fully understood over time without reference to their environment."[76] This truism is highlighted by Robert Marks's *China: Its Environment and History,* in which he demonstrates how "wild" environments in China were viewed historically as needing to be tamed and civilized – a process achieved through a "particularly successful combination of family farming with the strategic interests of the state."[77] Marks reveals how the expansion of agriculture led in many areas to resource exhaustion, environmental change, and deforestation.[78] These consequences have also left their mark in Manchuria, though later here than in other parts of China, and often as a result of different processes. Sun Xiaoping highlights in this volume how state deployment of human energies transformed the region, at great cost, from the Great Northern Wilderness (*Beidahuang*) into the Great Northern Granary (*Beidacang*), echoing the title of Judith Shapiro's text *Mao's War against Nature.*[79]

Marks's *China* engages in a limited discussion of Manchuria, outlining its rich resources, the failure of Qing policies aimed at keeping the region separate from China proper, and the region's utility as a refuge for famine-induced migration.[80] Importantly, Marks argues the need to examine "Chinese" as well as "non-Chinese" peoples' relationships with their environment. The studies collected here have consulted sources in Chinese, English, French, Japanese, Manchu, Polish, and Russian, in order to show how various peoples and states sought to come to terms with each other and their environment in Manchuria. In *The Unending Frontier*, John F. Richards argues that "continuing internal colonization, followed by continuing agricultural intensification, dramatically transformed the Chinese landscape and the human relationship to nature."[81] Manchurian colonization, undertaken by Manchu, Chinese, Russian, and Japanese regimes, had dramatic consequences for the region. Chinese agricultural practices spread. Massive migration was coupled with railway imperialism in the race to secure, and profit from, the regime's resources. Citing studies by Chinese researchers, including Xiong Datong, Eduard B. Vermeer argues that the "Russian and Japanese exploitation of the Manchurian forests early in [the twentieth] century was the most extensive and best organized ever undertaken."[82] But Vermeer stresses that the environmental exploitation of Manchuria "did not just further the 'capitalist,' 'colonialist,' and 'imperialist' interests of these foreign powers; it also furthered China's strategic aims and interests."[83] In this volume, we point to the many layered ways in which a diverse array of regimes interacted with the Manchurian environment – and each other – in their efforts to implement sovereignty.

All of the studies mentioned above stress that, despite considerable environmental change and degradation, China "continues to be one of the most biologically diverse places on planet," whose environmental history deserves more critical scholarly attention.[84] J.R. McNeill argues that it is precisely the territory's ecological diversity that "helps explain the resilience of the Chinese state."[85] For almost four centuries, Manchuria has been a key element of that resilience. Since the first stirrings of the Qing Dynasty at least, Manchuria's rich, northern environment has ensured that the region was a locus of multiple state ambitions. The studies here focus on the intersections of Manchuria's environmental and colonial history. But many important areas, including forests, mines, and rivers, as well as the region's Korean population, remain beyond the scope of this volume. *Empire and Environment in the Making of Manchuria* thus seeks to encourage much-needed further study of the multiplicity of human encounters and encounters with the environment in Manchuria, especially through these centuries of intense change.

## *Making Manchuria in Environmental and Imperial Detail*

This volume begins with Diana Lary's "Manchuria: History and Environment," which describes the natural and human environment of Manchuria in the *longue durée,* highlighting comparisons with Canada, which have been made regularly by observers over the past two hundred years. Lary points to significant interactions between the northern environment and its peoples, Indigenous as well as Han Chinese, Japanese, and Russian. The chapter considers the region's riches as well as seemingly less positive climatic attributes to provide a wide-ranging backdrop for the chapters that follow.

The next two chapters, by David A. Bello and Loretta Kim, detail the tribute systems aimed at structuring and extracting the resources of expanding empires. Bello's "Rival Empires on the Hunt for Sable and People in Seventeenth-Century Manchuria" demonstrates that both Qing and Romanov empire building was informed by Manchurian ecology as a habitat for sable tribute to incorporate local peoples. The Manchu and Russian projects, however, were not equally sustainable in cultural and ecological terms. Manchu incorporation generally put less stress on local resources and peoples, who were less amenable to more immature and intense Russian exactions. Qing responses to Romanov incursion, nevertheless, substantially alienated these peoples from their previous foraging lifestyle. Manchu identity was, thus, consistently under construction by human cultural elements, which were mediated by their surrounding ecology. In "Inclement Weather and Human Error," Kim analyzes the challenges of harvesting the tribute described by Bello, and how delivery of this tribute to the Qing court was complicated by environmental and human factors. Kim posits that there were three main types of problem: "human errors, natural causes, and environmental abnormalities resulting from human actions." Over time, as the dynasty matured and the local population grew, various factors combined to lessen the quantity and quality of the resources that could be extracted successfully from the region. To make matters worse, at any time "inclement weather" could interrupt the gathering and transportation of resources. Although the Manchu rulers left their homeland behind as they moved into China proper, their continued appetite for the region's riches produced deleterious effects on the very resources that had once empowered them to take the Chinese throne.[86]

Blaine Chiasson's chapter, "Producing Full-Fat Controversy," reveals the Sino-Russian contestation in Manchuria over agricultural development by detailing the rising tensions between Chinese and Russians that led to Chinese farmers in the 1920s fighting back against what they perceived to be the unfair and illegitimate seizure of property. Local Chinese rejected Russian as well as Japanese claims that the land from which they had been evicted was "empty space." Further, the Chinese objected to the dairy industry that the Russians were establishing on what the Chinese viewed as their fertile agricultural land. Chiasson situates dairy production as one element of the Russian program to establish legitimacy in the region, using notions of "modernity" and "progress" to frame their farming practices. The Chinese criticized these actions as neo-colonialist meddling in a land to which the Russians had no legitimate claim. Together, Bello, Kim, and Chiasson bring to the fore state agendas and environmental circumstances that the Russians, Manchus, Chinese, and Japanese tried to coordinate in Manchuria, problematizing their sovereignty claims.

The following four chapters, by Norman Smith, Annika A. Culver, Ronald Suleski, and Kathryn Meyer, deal with the period of Japanese occupation, one that has to date shaped significantly our understanding of Manchuria's modern history. Smith's and Culver's chapters consider the ideological campaigns to structure personal behaviour and perceptions of the northern region. "'Hibernate No More!': Winter, Health, and the Great Outdoors" evaluates Chinese-language discourses on winter health, housing, and sports, stressing the environment's omnipresent influence over those who sought to claim the northern, "pioneering" region as their own. Locals were criticized by physicians and officials for what was characterized as an unhealthy tradition of lying "dormant," like animals, through the winter. Russian and Japanese migrants, especially, promoted the development of winter sports such as skating and skiing, which came to be seen as markers of civilized, modern behaviour, making the winters more suited to human life. Despite contestation over other Russian and Japanese political and economic activities in the region, these winter narratives were promoted as models of modern behaviour and, to some extent, these models stand to this day. But winter did not always feature in the materials generated to entice migrants to Manchuria. In "Constructing a Rural Utopia," Culver analyzes Japanese propaganda depicting Manchukuo as a "paradise" for Japanese migrants. Culver focuses on the *Manshū gurafu* (*Manchuria Graph*) (1935–44), a journal that portrayed Manchuria as a rich territory, ripe for Japanese migration. Images in the journal and gracing its covers provided visuals that reinforced the positive, laudatory tone of the articles. Culver stresses that although the images do not reflect the violence of Japan's wartime aggression, they accurately depict the "right-wing proletarian" utopia idealized by Manchukuo officials. Not surprisingly, the harsher aspects of the region's winter, which dominate other chapters in this volume, were consistently minimized by the journal. Significantly, rural winter scenes that would have most approximated the migrants' actual experiences are not to be found on the journal's covers. While winter propaganda may have appealed to locals, it was thought less capable of compelling migration.

The next two chapters, by Meyer and Suleski, detail the failings of Japan's empire in Manchuria. The idealized lifestyles described in *Manshū gurafu* were not part of the experience of the impoverished, who are the focus of Meyer's chapter, "The Garden of Grand Vision." Meyer analyzes Japanese descriptions of the lives of Chinese in a Harbin slum, a decrepit urban environment in which prostitution and drug use proliferated as the poor collected (or stole) and sold animal carcasses, rags, cigarette butts, and anything else of any value in order to support themselves. Horrendous conditions prevailed in the slum throughout the year, in stifling summer heat and in freezing winter cold. The inhabitants suffered immensely and were not participants in the debates on housing, food, and clothing discussed in the earlier chapter "'Hibernate No More!'" Japanese observers, both horrified and fascinated by the slum community, provide an unambiguous portrait of the lives of the poorest in Manchuria. The poverty and hopelessness that dominate Meyer's chapter presages Suleski's "Salvaging Memories," which centres on the escape of a young Japanese migrant after the collapse of Manchukuo. Suleski describes the experiences of colonists forced to flee for their lives and relates the terror experienced by Kubota Isamu, who was only fifteen years of age. His experiences were the polar opposite of the utopia described by Culver. Japanese farmers often endured a hard life of outdoor work with few material comforts, living among – yet, like the Russians described by Chiasson, for the most part separate from – the Chinese whose land had been expropriated for the Japanese. In August 1945, when Soviet troops invaded, many Japanese agricultural colonists were killed by the Russians or took their own lives. One scene of horror followed another, as righteous indignation was voiced and vengeance was exacted. Overall, colonial contestation would take an enormous personal toll on the hapless

Japanese farmers who went to Manchuria, voluntarily or involuntarily, and were essentially abandoned there.

The final two chapters, by Wang Ning and Sun Xiaoping, examine the Chinese who, voluntarily or not, migrated to Manchuria. Wang's "Exile to Manchuria" compares the experiences of exiles who were banished to Manchuria in the early Qing period with those of Maoist exiles to the Northeast in the PRC of the 1950s and early 1960s. Wang demonstrates how "unfavourable physical and ecological conditions" combined with state persecution resulted in hardships for both groups. However, Wang stresses that Qing exiles tended to fare much better than their Maoist counterparts. Qing exiles were treated by locals as valued Chinese gentry, whose lifestyle brought the exiles a considerable degree of respect and informal power. Maoist exiles, on the other hand, were viewed as politically if not morally questionable, and their punishment was compounded by ideological remoulding that had not been applied to their Qing-era predecessors. For both groups, the environment of Manchuria necessitated adaptation to more extreme weather patterns than the exiles had been accustomed to. Sun's "'War against the Earth'" analyzes the post-1949 development of Manchuria into one of China's most important agricultural areas. Sun recounts how millions of migrants from across China were sent to the wetlands of Heilongjiang to transform the region. Maoist-era hyperbole about the wilderness echoed earlier Japanese and Russian rhetoric, stressing Manchuria's vast emptiness and inexhaustible resources. In this rhetorical model, the hardships faced by the exiles became challenges that only "extraordinary heroes" could overcome. Farming tools became weapons and, as Sun suggests, the extensive burning of vegetation for land clearance "literally opened fire on nature." Sun argues that this making of Northeastern heroes was not only a product of top-down state mobilization but involved the participation of the migrants as well, in campaigns that swept across China's northernmost region far more aggressively than any earlier Manchu, Russian, Chinese, or Japanese precedents, firmly positioning the region as an inalienable part of China.

Together, the essays in this volume reveal the degree to which Manchuria's modern history has been structured by expanding empires within the region's environment. Manchuria's vast lands and seemingly inexhaustible resources drove imperial aspirations and rivalries. The region's bounty nourished the ambitions of the Manchus and then neighbouring states, bringing the states' ambitions into close and, sometimes, deadly contestation – whether contestation over land use, resource extraction, health discourses, or sovereignty claims. *Empire and Environment in the Making of Manchuria* delineates the connectedness of empire building and an environment that tempted and defied colonists who dared to claim the region for their own. The "opening" of Manchuria, coupled with incessant state demands for land and resources, pushed Manchuria onto the world stage, with enormous consequences for the region, the environment, and the peoples who lived there. Over the period covered in this volume, ever-intensifying state presence transformed peoples' lives and how the environment was perceived and used. The research here demonstrates that those transformations still have signal contributions to make to our understandings of empire and environment in China, East Asia, and beyond.

In Russia, despite enormous human and financial investments, neither Manchuria (and the war that contributed to the toppling of the tsar) nor Manchukuo has inspired much scholarship.[87] Since *perestroika* in the mid-1980s, the large Russian émigré populations of Harbin and Shanghai have attracted more popular interest and praise for their perceived protection of traditional Russian culture.[88] In Japan, Manchuria is a focus of great attention, as many Japanese scholars from across the political spectrum have devoted their careers to its study.[89] Japanese public study groups such as the Manchukuo Literature Research Group (*Manshūkoku bungaku kenkyūkai,* established in 2001) engage in informal research and activities, many of which focus on the positive aspects of Manchukuo and Japanese memories of it.[90] The formation of such a group in China today would be unthinkable, as much of the history recounted there is either considered best forgotten or officially condemned for its imperialist provenance. Some Chinese scholars have researched Manchukuo (especially its Chinese literary world),[91] but more scholars study the Qing or the region's ancient past. A notable exception is the 2010 publication of a joint Sino-Japanese volume on Manchukuo, *Wei Manzhouguo de Zhenxiang: Zhong Ri xuezhe gongtong yanjiu* (*The Truth of False Manchukuo: Sino-Japanese Scholars' Joint Research*), which – like this volume – urges less politicized approaches to the study of the region's history.[92] Yet in April 2015, Manchuria's contentious nature was demonstrated anew by a scathing denunciation from Professor Li Zhiting (China's National Qing Dynasty History Compilation Committee), of "New Qing History," which treats the Qing as a Manchu (not a Chinese) dynastic state. He decries New Qing History as "pseudo-academic" and politically suspect.[93] Li echoes an assertion made by Owen Lattimore in 1931: "The final Manchu conquest of China was less an alien invasion than the triumph of the strongest regional faction in a colossal Chinese civil war."[94] But unlike Lattimore's reasoned thesis, Li engages in a starkly nationalistic diatribe, aimed at leading US-based scholars in the field, Evelyn Rawski, Pamela Crossley, Mark Elliott, and James Millward.[95] In this sense, Manchuria retains a remarkable, continuing capacity to act as a "cradle of conflict." *Empire and Environment in the Making of Manchuria* likewise demonstrates that, far from being an "empty" frontier or borderland, Manchuria has been a highly contested zone at the heart of the modern history, and historiography, of China and East Asia.

## Notes

1 South Manchuria Railway Company, *Manchuria: Land of Opportunities* (New York: South Manchuria Railway, 1922); Frederick Simpich, "Manchuria: Promised Land of Asia," *National Geographic Magazine* 56, no. 4 (1929): 379–428; and P.T. Etherton and H. Hessell Tiltman, *Manchuria: The Cockpit of Asia* (London: Jarrolds, 1932).

2 Other names include "East of the Wall/Barrier," "Country of the Eight Banners," and, for the Japanese occupation era (1932–45), *Manzhouguo* (Country of the Manchus; alternative English-language Romanizations: "Manchukuo" and "Manchoukuo"), and, from 1934, *Manzhoudiguo* (Empire of the Manchus; English Romanization: "Manchoutikuo").

3 Mark Elliott, "The Limits of Tartary: Manchuria in Imperial and National Geographies," *Journal of Asian Studies* 59, no. 3 (2000): 603–46, 607.

4 Cheng Li, "China's Northeast: From Largest Rust Belt to Fourth Economic Engine?" *China Leadership Monitor* 9 (2004): 1.

5 For example, empire-building books include Prasenjit Duara, *Sovereignty and Authenticity: Manchukuo and the East Asian Modern* (Lanham, MD: Rowman and Littlefield, 2003); and Alexander Hosie, *Manchuria: Its People, Resources and Recent History* (London: Methuen, 1901). *Crossed Histories,* edited by Mariko Asano Tamanoi, focuses on three main themes: Pan-Asianism, nationalism, and memory. Tamanoi stresses how Manchuria was consistently, since the seventeenth century, depicted as an "empty space," an assertion contradicted by the diversity of research in that volume and this. Tamanoi, "Introduction," in *Crossed Histories: Manchuria in the Age of Empire,* ed. Mariko Asano Tamanoi (Honolulu: University of Hawai'i Press, 2005), 2. For a volume on the Manchurian environment, see David A. Bello, *Across Forest, Steppe, and Mountain: Environment, Identity, and Empire in Qing China's Borderlands* (Cambridge, UK: Cambridge University Press, 2016).

6 Asada Masafumi has argued the need for "a triangular comparative point of view" to interpret early-twentieth-century military relations among China, Japan, and Russia. See Asada Masafumi, "The China-Russia-Japan Military Balance in Manchuria, 1906–1918," *Modern Asian Studies* 44, no. 6 (2010): 1283–1311. Asada's bibliography of Russian and Japanese works is noteworthy.

7 See Jonathan Schlesinger, "The Qing Invention of Nature: Environment and Identity in Northeast China and Mongolia, 1750–1850" (PhD diss., Harvard University, 2012), 322.

8 For an incisive analysis of the origins and uses of the term "Manchuria," see Mark Elliott, "The Limits of Tartary." Elliott's other main studies include *Emperor Qianlong: Son of*

*Heaven, Man of the World* (London: Longman Pearson, 2009) and *The Manchu Way: The Eight Banners and Ethnic Identity in Late Imperial China* (Stanford, CA: Stanford University Press, 2001). See also Pei Huang, "New Light on the Origins of the Manchus," *Harvard Journal of Asiatic Studies* 50, no. 1 (1990): 239–82.

9 The degree to which Manchus were "Sinicized" is a matter of great debate in the school of New Qing Studies, as discussed by Evelyn S. Rawski in "Re-envisioning the Qing: The Significance of the Qing Period in Chinese History," *Journal of Asian Studies* 55, no. 4 (1996): 829–50, 836.

10 Elliott, "Limits," 605–7.

11 "China proper," also a controversial term, refers to the territory that comprised much of imperial China, with the exception of "frontier" regions, such as the Northeast.

12 Such an identity is underscored by the publication of Ma Sizhou and Jiang Guanghui, eds., *Dongbei fangyan cidian* [Northeast dialect dictionary] (Changchun: Jilin wenshi chubanshe, 2005).

13 Sun Wenliang, "'Manzhou' yu 'Dongbei' de mingcheng jiqi yu Manzu de fazhan guanxi" [The development of the relationship between the names "Manchuria" and the "Northeast," and Manchus] (CCP: Neibu wenjian, n.d.).

14 Serge Cipko provides an important reminder of the ethnic divisions among those identified as "Russians." See Serge Cipko, "Ukrainians in Manchuria, China: A Concise Historical Survey," *Past Imperfect* 1 (1992): 155–73. See also Thomas Lahusen, ed., *Harbin and Manchuria: Place, Space, and Identity, special issue of the South Atlantic Quarterly* 99, no. 1 (2000).

15 For a concise discussion of Manchurian history and the status of Japan's colony of Dalian, see Robert John Perrins, "Great Connections: The Creation of a City, Dalian, 1905–1931" (PhD diss., York University, 1997).

16 H.E.M. James, *The Long White Mountain* (London: Longmans, Green, 1888), vi.

17 Lattimore's major works include *Manchuria, Cradle of Conflict* (New York: Macmillan, 1932) and *The Gold Tribe, "Fishskin Tatars" of the Lower Sungari* (New York: Kraus Reprint, 1933).

18 Thomas J. Barfield, *The Perilous Frontier: Nomadic Empires and China* (Cambridge, MA: Basil Blackwell, 1989), 11.

19 Simpich, "Manchuria," 379. A Far Eastern Republic (*Dal'nevostochnaia Respublika*) (April 1920 to November 1922) was established with Soviet political and financial support as a buffer state between the Soviet Union, which then was the Russian Soviet Federative Socialist Republic (RSFSR), and the areas occupied by the Japanese. With the Soviet victory in the Far East, the Far Eastern Republic joined the RSFSR. The Far Eastern Republic claimed jurisdiction over the Russian China Eastern Railway, established an office in Harbin, and announced that residents should register with it. "The Far Eastern Republic," *Russian Information and Review* 1, no. 10 (1922): 232–33.

20 Simpich, "Manchuria," 393.

21 Mark Elliott, "Frontier Stories: Periphery as Center in Qing History," *Frontiers of History in China* 9, no. 3 (2014): 340.

22 Mary Louise Pratt, *Imperial Eyes: Travel Writing and Transculturation* (New York: Routledge, 1992), 4.

23 Ibid., 6.

24 For discussion of "layered" migrations over a longer period, see Tsukase Susumu, *Manchuriashi kenkyū: "Manshū" roppyakunen no shakai hen'yō* [Research on history of Manchuria: 600 years of social transformation in "Manchuria"] (Tokyo: Yoshikawa kōbunkan, 2014).

25 Pamela Kyle Crossley, *The Manchus* (Cambridge, MA: Blackwell Publishers, 1997), 8. Crossley's other main studies include *A Translucent Mirror: History and Identity in Qing Imperial Ideology* (Berkeley: University of California Press, 1999) and *Orphan Warriors: Three Manchu Generations and the End of the Qing World* (Princeton, NJ: Princeton University Press, 1990).

26 Ho Ping-ti, "The Significance of the Ch'ing Period in Chinese History," *Journal of Asian Studies* 26, no. 2 (1967): 191.

27 Rawski, "Re-envisioning," 831. This essay offers a state-of-the-field discussion of the Qing period. A major contribution is Ronald Suleski, *The Modernization of Manchuria: An Annotated Bibliography* (Hong Kong: Chinese University Press, 1994). One of Rawski's main monographs is *The Last Emperors: A Social History of Qing Imperial Institutions* (Berkeley: University of California Press, 1998).

28 Li Narangoa and Robert Cribb, eds., *Historical Atlas of Northeast Asia, 1590-2010: Korea, Manchuria, Mongolia, Eastern Siberia* (New York: Columbia University Press, 2014), 3.

29 Carl Skutsch, *Encyclopedia of the World's Minorities* (New York: Routledge, 2005), 291.

30 For example, see B.A. Romanov, *Russia in Manchuria (1892–1906)* (Ann Arbor, MI: Edwards Brothers, 1952).

31 Sergei M. Shirokogoriff, *Social Organization of the Manchus* (New York: Ams Press, 1973), 104. The Qing and Romanov settling of borders has attracted more scholarly attention than that between the Chosŏn and Qing states. For discussion of the latter, see Adam Bohnet, "'On Either Side the River': The Rise of the Manchu State and Chosŏn's Jurchen Subjects," in *The Exploitation of the Landscape of Central and Inner Asia: Past, Present and Future,* ed. Michael Gervers, Uradyn E. Bulag, and Gillian Long (Toronto: University of Toronto Asian Institute, 2008), 111–25.

32 Paul Hibbert Clyde, *International Rivalries in Manchuria, 1689–1922* (Columbus: Ohio University Press, 1926).

33 See R.K.I. Quested, *Sino-Russian Relations: A Short History* (Sydney: George Allen and Unwin, 1984). Quested underlines the relative weaknesses of the Russian business and military classes as well as the inability of the Russians to compete with or train the Chinese. See also R.K.I. Quested, *Matey Imperialists? The Tsarist Russians in Manchuria, 1895–1917* (Hong Kong: Centre of Asian Studies, University of Hong Kong, 1982), 156.

34 S.C.M. Paine, *Imperial Rivals: China, Russia and Their Disputed Frontier* (Armonk, NY: M.E. Sharpe, 1996), 348.

35 Rui Hua, "Creating Meaning in Manchukuo: The Old Chinese Kingly Way in a New Multi-National 'Nation'" (paper presented at the Chicago Association of Asian Studies Conference, March 27, 2015). For a study of the plague, see Carl F. Nathan, *Plague, Prevention and Politics in Manchuria, 1910—1931* (Cambridge, MA: Harvard University Press, 1967). International cooperation and competition in the region has been underlined by Liu Yishi and Wang Xinying, who argue that "the city of Changchun before 1932 was a collage of territories divided according to ethnic sovereignty: the Russian railway town at Changchun, the Japanese South Manchuria Railway settlement, the Chinese commercial district, and the old walled city." See Liu Yishi and Wang Xinying, "A Pictorial History of Changchun, 1898–1962," *Cross-Currents: East Asian History and Culture Review* 5 (2012): 193.

36 See E.Kh. Nilus, *Istoricheskii ocherk Kitaiskoi Vostochnoi zheleznoi dorogi, 1896-1923 gg.* [Historical survey of the Chinese Eastern Railway, 1896–1923] (Harbin: Tipografiia Kit. Vost. zhel. dor., 1923). See also *Severnaia Man'chzhuriia i Kitaiskaia Vostochnaia zheleznaia doroga* [North Manchuria and the Chinese Eastern Railway] (Harbin: CER Printing

Press, 1922). For an environmental assessment by the CER, see V.I. Surin, *Man'chzhuriia i ee perspektivy* [Manchuria and its prospects] (Harbin: CER Economic Bureau, 1930) and P.L. Glushakov, *Man'chzhuriia. Ekonomichesko-geograficheskoe opisanie* [Manchuria. Economic and geographical description] (Moscow: Gos. izd-vo geogr. lit-ry, 1948).

37 Felix Patrikeeff and Harold Shukman, *Railways and the Russo-Japanese War: Transporting War* (London: Routledge, 2007); Bruce A. Elleman and Stephen Kotkin, *Manchurian Railways and the Opening of China: An International History* (Armonk, NY: M.E. Sharpe, 2010).

38 Patrikeeff and Shukman, ibid., 104.

39 For example, Søren Clausen and Stig Thøgersen, *The Making of a Chinese City: History and Historiography in Harbin* (Armonk, NY: M.E. Sharpe, 1997); David Wolff, *To the Harbin Station: The Liberal Alternative in Russian Manchuria, 1898–1914* (Stanford, CA: Stanford University Press, 1999); James Carter, *Creating a Chinese Harbin: Nationalism in an International City, 1916–1932* (Ithaca, NY: Cornell University Press, 2002); Blaine Chiasson, *Administering the Colonizer: Manchuria's Russians under Chinese rule, 1918–29* (Vancouver: UBC Press, 2010); Olga Bakich, *Valerii Pereleshin: The Life of a Silkworm* (Toronto: University of Toronto Press, 2015); Olga Bakich, "Émigré Identity: The Case of Harbin," in *Harbin and Manchuria,* ed. Lahusen, 51–73; and Olga Bakich, "Russian Émigrés in Harbin's Multinational Past: Census and Identity," in *Entangled Histories: The Transcultural Past of Northeast China,* ed. Dan Ben-Canaan, Frank Grüner, and Ines Prodöhl (Cham, Switzerland: Springer, 2014). For a history of Mennonites fleeing Russian persecution for the freedom of 1920s and 1930s Manchuria, see Abram Friesen and Abram Loewen, *Escape across the River* (Winnipeg: Echo Historical Series, 2001).

40 Chiasson, *Administering the Colonizer,* 4.

41 Elliott, "Limits," 635.

42 Patrikeeff and Shukman, *Railways,* 112.

43 Simpich, "Manchuria," 379. See also Thomas R. Gottschang and Diana Lary, *Swallows and Settlers: The Great Migration from North China to Manchuria* (Ann Arbor: Center for Chinese Studies, University of Michigan, 2000).

44 Patrikeeff and Shukman stress that Soviet and Russian historians have thought it best to ignore the topic: *Railways,* 121–22. A notable exception is a two-volume book published in the US: Petr Balakshin, *Final v Kitae. Vozniknovenie, razvitie i ischeznovenie Beloi emigratsii na Dal'nem Vostoke* [Finale in China. Emergence, development, and disappearance of white emigration in the Far East] (San Francisco: Sirius, 1958–59). See also Felix Patrikeeff, *Russian Politics in Exile: The Northeast Asian Balance of Power, 1924–1931* (New York: Palgrave Macmillan, 2002).

45 Harry L. Kingman, *The Effects of Chinese Nationalism upon Manchurian Railway Developments, 1925–1931* (Berkeley: University of California Press, 1932), 4, 57.

46 Gavan McCormack, *Chang Tso-lin in Northeast China, 1911–1928: China, Japan, and the Manchurian Idea* (Stanford, CA: Stanford University Press, 1977); and Ronald Suleski, *Civil Government in Warlord China: Tradition, Modernization and Manchuria* (New York: Peter Lang, 2002).

47 South Manchuria Railway Company, *Manchuria,* 13.

48 For a Japanese encyclopedia of Manchurian history, see Kishi Toshihiko, Matsushige Mitsuhiro, and Matsumura Fuminori, *Nijū seiki manshū rekishi jiten* [Twentieth-century Manchurian history encyclopedia] (Tokyo: Yoshikawa kōbunkan, 2012). For contemporary Russian discussion of Japanese-controlled Manchuria, see *Man'chzhuriia.*

*Ekonomichesko-geograficheskoe opisanie* [Manchuria. Economic and geographical description]; and *Velikaia Man'chzhurskaia imperiia. K desiatiletnemu iubileiu, 1932–1942* [Great Manchu empire: On the tenth anniversary, 1932–1942] (Harbin: Kiowakai, 1942).

49 Yoshihisa Tak Matsusaka, *The Making of Japanese Manchuria, 1904–1932* (Cambridge, MA: Harvard University Press, 2001).

50 John Young, *The Research Activities of the South Manchuria Railway Company, 1907–1945: A History and Bibliography* (New York: Columbia University, 1966). For three recent evaluations of the company, see, in English, Maruya Akihiko, *The South Manchuria Railway Company as an Intelligence Organization* (Washington, DC: CSIS, 2012); in Chinese, Chen Dingyin, *Cong Wangdao letu dao Zhongguo Yanjiu de Ziliaoku: Chaoyue diguozhuyi de Mantie* [From Kingly Way to Chinese research storehouse: South Manchuria Railway transcending imperialism] (Taipei: Guoli Taiwan daxue zhengzhi xuexi Zhongguo dalu ji liang'an guanxi jiaoxue yu yanjiu zhongxin, 2014); and, in Japanese, Kobayashi Hideo, *Mantetsu chōsabu* [South Manchuria Railway research department] (Tokyo: Kōdansha, 2015).

51 Louise Young, *Japan's Total Empire: Manchuria and the Culture of Wartime Imperialism* (Berkeley: University of California Press, 1998). For Japanese ideals of empire, see Bill Sewell, "Reconsidering the Modern in Japanese History: Modernity in the Service of the Prewar Japanese Empire," *Japan Review* 16 (2004): 213–58. For further discussion of Japanese migration to Manchuria, see Tsukase Susumu, *Manshū no Nihonjin* [Manchuria's Japanese] (Tokyo: Yoshikawa kōbunkan, 2004).

52 Duara, *Sovereignty and Authenticity*. A recent discussion of women's lives in Manchuria can be found in Michiko Ikuta, ed., *Onnatachi no manshū: Taminzoku kūkan o ikite* [Women of Manchuria: Life in a multi-ethnic space] (Ōsaka: Ōsaka Daigaku shuppankai, 2015).

53 Janis Mimura, *Planning for Empire: Reform Bureaucrats and the Japanese Wartime State* (Ithaca, NY: Cornell University Press, 2011), 70.

54 Rana Mitter, *The Manchurian Myth: Nationalism, Resistance and Collaboration in Modern China* (Berkeley: University of California Press, 2000), 228.

55 Dan Shao, *Remote Homeland, Recovered Borderland: Manchus, Manchoukuo, and Manchuria, 1907–1985* (Honolulu: University of Hawai'i Press, 2011).

56 Suk-jung Han, "Those Who Imitated the Colonizers: The Legacy of the Disciplining State from Manchukuo to South Korea," in *Crossed Histories*, ed. Tamanoi, 165–83, 180.

57 Nissan "played a major role in turning Manchukuo into one of the top ten industrial bases in the world"; see Harold M. Tanner, *The Battle for Manchuria and the Fate of China: Siping, 1946* (Bloomington: Indiana University Press, 2013), 27.

58 For discussion of the Kingly Way paradise land, see Norman Smith, *Resisting Manchukuo: Chinese Women Writers and the Japanese Occupation* (Vancouver: UBC Press, 2007), 22. For contemporary media, see Takashi Toshihiko, *Manshūkoku no bijuaru media – posutā ehagaki kitte* [Manchukuo visual media: Posters, postcards and stamps] (Tokyo: Yoshikawa kōbunkan, 2010). Among positive evaluations of the regime, D.M.B. Collier and Lt.-Col. C.L'E. Malone's *Manchoukuo: Jewel of Asia* stands out for its effusive praise for the education system and improvements to urban life. Harbin is called "Elysium on the Manchurian Plains" (170) and Xinjing is described as "the most amazing and arouses every traveller's admiration" (172). See D.M.B. Collier and Lt.-Col. C. L'E. Malone, *Manchoukuo: Jewel of Asia* (London: George Allen and Unwin, 1936).

59 Sheldon H. Harris, *Factories of Death: Japanese Biological Warfare, 1932–1945, and the American Cover-up* (New York: Routledge, 1994). For the treatment of women, see Liu

Jinghui, *Minzu, xingbie yu jieceng* [Nation, gender, and social stratum] (Beijing: Shehui kexue wenxian chubanshe, 2004).

60 See also Miriam Kingsberg, *Moral Nation: Modern Japan and Narcotics in Global History* (Berkeley: University of California Press, 2013); Mark Driscoll, *Absolute Erotic, Absolute Grotesque: The Living, Dead, and Undead in Japanese Imperialism, 1895–1945* (Durham, NC: Duke University Press, 2010); and Norman Smith, *Intoxicating Manchuria: Alcohol, Opium and Culture in China's Northeast* (Vancouver: UBC Press, 2012).

61 Contemporary books with lurid titles such as George E. Sokolsky's *The Tinder Box of Asia,* Amleto Vespa's *Secret Agent of Japan,* and Alexandre Pernikoff's *Bushido: The Anatomy of Terror* outline most clearly for English readers the dangers of a Japanese-sponsored Manchukuo, ally of Germany's Third Reich. George E. Sokolsky, *The Tinder Box of Asia* (Garden City, NY: Doubleday, Doran and Company, 1932); Amleto Vespa, *Secret Agent of Japan* (Garden City, NY: Garden City Publishing, 1938); and Alexandre Pernikoff, *Bushido: The Anatomy of Terror* (New York: Liveright, 1943).

62 Manchuria left long influences in Japan. For example, see Mariko Asano Tamanoi, *Memory Maps: The State and Manchuria in Post-War Japan* (Honolulu: University of Hawai'i Press, 2008).

63 Major studies of the civil war include Steven I. Levine, *Anvil of Victory: The Communist Revolution in Manchuria 1945–1948* (New York: Columbia University Press, 1987); and Donald G. Gillin and Ramon H. Myers, eds., *Last Chance in Manchuria: The Diary of Chang Kia-ngau,* trans. Dolores Zen (Stanford, CA: Hoover Institution Press, 1989).

64 Owen Lattimore, *Inner Asian Frontiers of China* (New York: Oxford University Press, 1940), 114.

65 Hosie was in charge of the British Consulate at Newchwang in Manchuria, 1894–97 and 1899–1900. Alexander Hosie, *Manchuria: Its People, Resources and Recent History* (London: Methuen, 1901), 1.

66 Sir Francis E. Younghusband, *Heart of a Continent: A Narrative of Travels in Manchuria, across the Gobi Desert, through the Himalayas, the Pamirs, and Chitral, 1884–1894* (London: John Murray, 1896).

67 James, *The Long White Mountain,* 104.

68 Eduard B. Vermeer, "Population and Ecology," in *Sediments of Time: Environment and Society in Chinese History,* ed. Mark Elvin and Liu Ts'ui-jung, 235–79 (Cambridge, UK: Cambridge University Press, 1998), 252–54. Vermeer cites Ma Zongshen, "Woguo lishishangdi shuitu baochi" [Water and soil conservation in our country's history], *Nongshi yanjiu* [*Agricultural History Studies*] 3 (1983): 61–74.

69 Ted Steinberg, "Down to Earth: Nature, Agency, and Power in History," *American Historical Review* 107, no. 3 (2002): 798–820.

70 For example, see Kenneth S. Coates and William R. Morrison, "The New North in Canadian History and Historiography," *History Compass* 6, no. 2 (2008): 639–58, and Phil Macnaughten and John Urry, *Contested Natures* (London: Sage, 1998).

71 Bao Maohong, "Environmental History in China," *Environment and History* 10, no. 4 (2004): 476. Important books on China's environment include Vaclav Smil, *The Bad Earth: Environmental Degradation in China* (New York: M.E. Sharpe, 1984); Peter C. Perdue, *Exhausting the Earth: State and Peasant in Hunan, 1500–1850* (Cambridge, MA: Harvard University Asia Center, 1987); Kenneth Pomeranz, *The Making of a Hinterland: State, Society, and the Economy in Inland North China, 1853–1937* (Berkeley: University of California Press, 1993); Mark Elvin, Nishioka Hiroaki, Tamura Keiko, and Joan Kwek, eds., *Japanese Studies on the History of Water Control in China: A Selected Bibliography*

(Canberra: Institute of Advanced Studies, ANU, 1994); and Lillian M. Li, *Fighting Famine in North China: State, Market, and Environmental Decline, 1690s–1990s* (Stanford, CA: Stanford University Press, 2007).

72 Bao, "Environmental," 491–92.

73 J.R. McNeill, "Observations on the Nature and Culture of Environmental History," *History and Theory* 42 (December 2003): 6.

74 Mark Elvin, "Introduction," in *Sediments of Time*, ed. Elvin and Liu, 2.

75 Mark Elvin, *The Retreat of the Elephants: An Environmental History of China* (New Haven, CT: Yale University Press, 2004).

76 Ibid., xx.

77 Robert B. Marks, *China: Its Environment and History* (Lanham, MD: Rowman and Littlefield, 2012), 3, 5. In a regional study, Marks further analyzes how the south Chinese economy and environment were interrelated. Robert B. Marks, *Tigers, Rice, Silk and Silt: Environment and Economy in Late Imperial South China* (Cambridge, UK: Cambridge University Press, 1998).

78 Marks, *China,* 331–33.

79 Judith Shapiro, *Mao's War against Nature: Politics and the Environment in Revolutionary China* (Cambridge, UK: Cambridge University Press, 2001).

80 Marks, *China,* 190, 256–59.

81 John F. Richards, *The Unending Frontier: An Environmental History of the Early Modern World* (Berkeley: University of California Press, 2003), 114.

82 Eduard B. Vermeer, "Population and Ecology," 252. Vermeer cites Xiong Datong, *Zhongguo jindai linye shi* [A modern history of forestry in China] (Beijing: Zhongguo linye chubanshe, 1989), 56–78.

83 Vermeer, "Population and Ecology," 54.

84 Marks, *China,* 6–7.

85 J.R. McNeill, "China's Environmental History in World Perspective" in *Sediments of Time,* ed. Elvin and Liu, 34.

86 See also David A. Bello, "The Cultured Nature of Imperial Foraging in Manchuria," *Late Imperial China* 31, no. 2 (2010): 1–33.

87 For reasons behind this, see Victor Zatsepine, "An Uneasy Balancing Act: The Russian Émigré Community and Utopian Ideals of Manzhouguo," *Journal of Northeast Asian History* 10, no. 1 (2013): 130–53. A notable exception is G.F. Zakharova, *Politika Iaponii v Man'chzhurii, 1932–1945* [Policies of Japan in Manchuria, 1932–1945] (Moscow: Nauka, 1990). For a Japanese discussion of Russia in Manchuria, see Ikuta Michiko, ed., *Manshū no naka no Roshia: kyōkai no ryūdōsei to ninteki nettowaku* [Russia in Manchuria: Border movements and personal networks] (Yokohama: Seibunsha, 2012).

88 For example, see G.V. Melikhov, *Man'chzhuriia dalekaia i blizkaia* [Manchuria distant and close] (Moscow: Nauka, 1991); G.V. Melikhov, *Rossiiskaia emigratsiia v mezhdunarodnykh otnosheniiakh na Dal'nem Vostoke, 1925–1932* [Russian emigration in the international relations in the Far East, 1925–1932] (Moscow: Russkii put', 2007); Elena Taskina, *Neizvestnyi Kharbin* [Unknown Harbin] (Moscow: Prometei, 1994); N.E. Ablova, *KVZHD i rossiiskaia emigratsiia v Kitae. Mezhdunarodnye i politicheskie aspekty istorii (pervaia polovina XX veka)* [CER and Russian emigration in China. International and political aspects of history (first half of twentieth century)] (Moscow: Russkaia panorama, 2005).

89 For example, Shin'ichi Yamamuro, whose exceptional book *Manchuria under Japanese Dominion* has been translated into English by Joshua Fogel.

90 For Japanese writings on Manchukuo literature, see Nihon shokuminchi bunka kenkyū [Japanese Colonial Research Society], ed. *"Mansūkoku" bunka saimoku* [Detailed bibliography of Manchukuo literature] (Tokyo: Fuji shuppan, 2005). Prominent examples detailing the experiences of the Japanese who fled Manchukuo at the end of empire include Yeeshan Chan, *Abandoned Japanese in Postwar Manchuria: The Lives of War Orphans and Wives in Two Countries* (New York: Routledge, 2011); and Mayumi Itoh, *Japanese War Orphans in Manchuria: Forgotten Victims of World War II* (New York: Palgrave Macmillan, 2010). One of the most popular TV series ever produced in Japan is *Daichi no Ko* [A son of the good earth] (1995), about Japanese in Manchuria suffering after the collapse of Manchukuo.

91 Interest in the Manchurian literary world is perhaps best illustrated by the 2015 founding in Shanghai at East China Normal University of the "Manzhouguo wenxue yanjiu zhongxin" (Manchukuo Literary Research Centre), directed by Professor Liu Xiaoli. A 1990s state-of-the-field collection of essays can be found in Feng Weiqun, Wang Jianzhong, Li Chunyan, and Li Shuquan, eds., *Dongbei lunxian shiqi wenxue guoji xueshu yantaohui lunwenji* [Collection of papers from the International Symposium on Literature of the enemy-occupied Northeast] (Shenyang: Shenyang chubanshe, 1992). See also Karen Laura Thornber's magisterial study, *Empire of Texts in Motion: Chinese, Korean, and Taiwanese Transculturations of Japanese Literature* (Cambridge, MA: Harvard University Asia Center, 2009).

92 Dongbei lunxian shisi nianshi zong bian shi (The Chief Editing Office of the Fourteen Years' Northeast Occupation) and Nihon shokuminchi bunka kenkyū (Japanese Colonial Research Society), eds., *Wei Manzhouguo de Zhenxiang: Zhong Ri xuezhe gongtong yanjiu* [The truth of false Manchukuo: Sino-Japanese scholars' joint research] (Beijing: Shehui kexue wenxian chubanshe, 2010).

93 See "New Qing History Targeted," *Modern Chinese Literature and Culture Resource Center*, April 25, 2015, http://u.osu.edu/mclc/2015/04/25/new-qing-history-targeted/.

For a more scholarly version of the issues raised by Li, see Zhong Han, "Beimei 'Xin Qingshi' yanjiu de jishi hezai?" [Where does the cornerstone of the new Qing history lie?], *Zhongguo bianjiang minzu yanjiu* [*Studies of China's Frontier Regions and Nationalities*] 7 (2014): 156–213.

94 Lattimore, *Manchuria*, 46.

95 Millward's main studies include *Beyond the Pass: Economy, Ethnicity and Empire in Qing Xinjiang, 1759–1864* (Stanford, CA: Stanford University Press, 1998); and *Eurasian Crossroads: A History of Xinjiang* (New York: Columbia University Press, 2007).

# 1
# Manchuria: History and Environment

*Diana Lary*

Blinding blizzards, howling gales, vicious cold: winter in Manchuria. Torrid heat, terrifying thunderstorms, huge mosquitoes: summer in Manchuria. In between the two long seasons came two very short ones, a brief spring and a brief fall.

This was the standard traditional Chinese view of Manchuria; it was how those few Chinese who even thought about the remote region saw the vast northeastern parts of the country.[1] The few who did think about Manchuria thought about it with dread; they were the unfortunate officials posted there and the even more unfortunate people who were sent there in exile. Manchuria had a reputation in the rest of China as a dangerous place that bred people who periodically turned greedy eyes towards the south. In the twelfth century, the Jurchens poured out of Manchuria and took over northern China, forcing the Song Dynasty to move its capital to the south. Five centuries later, the Manchus came south out of Manchuria, as the Ming Dynasty collapsed in the face of peasant risings, and established the Qing Dynasty, which ruled China until 1911.

The northernmost regions of the globe are a demanding zone of habitation. Small numbers of Indigenous people found ways to live in these areas into modern times, but the northern lands are rich in resources and potentially rewarding enough to lure outsiders to settle there as well. Two huge northern regions underwent almost parallel patterns of development in the late nineteenth and early twentieth centuries: Manchuria and the three Canadian Prairie provinces – Manitoba, Saskatchewan, and Alberta. The regions are similar in climate, topography, natural resources, flora, and fauna – including mosquitoes. The development of the two vast regions was based on similar concerns: sovereignty, migration, railway building, and urban development. The parallels have been given some recognition over the past decades, in the twinning of cities (Harbin/Edmonton, Calgary/Daqing) and in investment in mining.

There were also major dissimilarities between the regions, to do with attitudes towards winter, the foundation of communities, and systems of governance. In looking at the history and environment of Manchuria, it is helpful to make comparisons with the Canadian Prairies, to see how very different political and social systems affected the course of development.

## The Setting

The scale of Manchuria is hard to grasp. The region covers over a million and a half square kilometres. It has two great mountain chains, the Xing'anling and the Changbaishan, each with many peaks above two thousand metres. Its longest river, the Heilongjiang (Amur) is well over four thousand kilometres long. Its largest tributary, the Songhuajiang (Sungari) flows for fourteen hundred kilometres before it joins the Heilongjiang.[2] These two rivers create an enormous plain, highly fertile but so cold and windswept in winter that it has been hostile to human settlement. In southern Manchuria, the Liao River runs for more than thirteen hundred kilometres through an equally rich but slightly less cold plain. To the north and east of the plains are vast forests, home to great numbers of fur-bearing animals. The largest animal of the forests is the tiger, an animal that has a potent symbolism in Chinese culture, admired for its fierceness but also having high-minded characteristics of justice and compassion ascribed to it.[3]

## Early Outsiders

Manchuria was isolated and sparsely populated into the nineteenth century. The Qing did not want Chinese in Manchuria. The Willow Palisade, a system of ditches planted with willow trees, was constructed in the late seventeenth century, soon after the establishment of the dynasty, to keep the region free from Chinese settlement. Some bold and hardy people still settled there, without benefit of state sanction. They farmed the rich alluvial soil of southern Manchuria, and, in a lawless world, established their own forms of rough justice. Other, even hardier men went into the northern forests and mountains to find the treasures of the natural world: furs of sable, ermine, and other animals, as well as tiger skins, ginseng, fungi, pearls, and gold (see Chapter 2 of this volume, by David A. Bello, and Chapter 3 by Loretta Kim). There were even greater hidden riches, waiting to be exploited once modern means to extract them were available: minerals and oil, in enormous quantities. The harvesting of Manchuria's forests, which supplied the timber critical for China's economic development, had to wait for modern tree-cutting equipment and new means of transport to ship the timber.

The natural riches of the wild north of Manchuria paralleled the treasures that first lured Europeans westward across the great plains of North America. From the late seventeenth century, the first non-natives to penetrate into what became the Canadian West were fur traders. They were men of fearsome courage, endurance, and entrepreneurship, the *coureurs de bois,* from what is now the province of Quebec. They travelled in canoes. The furs – otter and beaver – they bought from Indigenous people and carried back to Montreal were less glamorous than the Manchurian sable and ermine but still valuable enough to establish a significant trade. The trade was enhanced as trading posts were set up by the Hudson's Bay Company and the Northwest Company; the posts were staffed by another breed of tough men – dour, hardy men often recruited by the trading companies in the Western Isles off Scotland, the Hebrides and the Orkneys. These two types of tough men pushed west against formidable obstacles of climate and terrain.[4] Almost all the men were single when they arrived. Many married women from the large Indigenous populations of the "empty" plains. With these marriages came a mixing of cultures, *métissage,* in which the Indigenous peoples taught the incomers how to live in the harsh environment. Their descendants are the Métis, with a current population of more than 400,000. These early families of mixed heritage founded in the wilderness were the forerunners of the agricultural settlers who flooded the Prairies from the late nineteenth century on.

In Manchuria there were few Indigenous peoples. There seems also to have been little *métissage*, little intermingling of blood and culture, between the Indigenous peoples and the incomers from China. Most members of the major Indigenous group, the Manchu, had already moved south into China proper (south of the Great Wall). The closest that most members of the Manchu imperial family got to Manchuria was the lovely summer palaces at Chengde. Manchu commoners lived in communities and military settlements throughout China. There were other Indigenous peoples in the north of the region, small groups such as the Nanai (Fish Skin Tartars), who lived in almost complete isolation along the banks of the great rivers. There are few traces today of the Neolithic Hongshan culture, a people who lived in widespread areas of Manchuria, including the south of the region. They left behind beautiful, tiny jade figures, including the mysterious "pig-dragon" (*zhulong*) (now thought to be a bear).

FIGURE 1.1 "Hongshan Pig-Dragon – now thought to be a bear" postcard.
*Source:* Collection of N. Smith

The isolation of Manchuria did not last into the modern age. The opening up of the region was a consequence of a combination of several coincidental factors. In the latter half of the nineteenth century, pressure from Russia, and later Japan, created a political determination in China's government to settle lands that the Qing had always claimed as its own. The spread of new means of transportation – railways and steamships – allowed for the export of natural resources and gave economic justification for settling Manchuria. The existence of a vast resource of potential settlers just to the south, in north China, made the increase in the settler population rapid. These changes mirrored those in the Canadian West, where the threat of foreign encroachment from the south, the coming of the railways, and the pressure for emigration from Europe to the New World, now easily accessible by steamship, created a drive to open up the West. Manchuria and the Canadian West, both once seen as hostile to outside settlement, went through, almost simultaneously, a seismic process of change in which their economies and societies were transformed.

## Sovereignty

In the late nineteenth century, China was beset by foreign pressures from the south and the east, from Britain, France, and Germany. But the Qing government was more worried about incursions in the north. The Russians had their eyes on the riches of the Russian Far East, and on the milder regions that lay to the south of Eastern Siberia: Manchuria. The intrepid Russian explorer Count Nikolai Nikolaievich Muraviov (1809–81; later Muraviev-Amursky), governor of Eastern Siberia from 1847 to 1861, foresaw a rapid expansion of Russia towards the east, which would counter the expansion of other European countries in Africa and more southern parts of Asia. In 1858, Muraviov negotiated the Treaty of Aigun; it established a frontier between Russia and the Qing Empire along the Heilongjiang and gave Russia the right to explore Manchuria.

A little later, the Japanese, transformed by the Meiji Restoration, developed their own vision for Manchuria, which included developing mineral extraction, growing agricultural products for export, and building factories. Along with these economic visions went the desire to create living space for the people of the crowded Japanese islands, an expansion not unlike the early-twentieth-century German vision of *Lebensraum,* the push of population into eastern Europe, which predated the Nazis but reached its most extreme under the Nazi regime.

In response to Russian and Japanese pressure, the Qing government decided it had to assert control over the ancestral region of its own people and its treasure house of natural resources. The threat to Manchu sovereignty became even more serious with the competition between Russia and Japan for the domination of Northeast Asia, much of which was played out on the soil of Manchuria. By the last years of the Qing Dynasty, two wars – one between China and Japan in 1895 (the *Jiawu zhanzheng*) and one between Japan and Russia in 1905 – had given Japan an almost impregnable position to develop industry, railways, and agricultural settlements in Manchuria. The 1905 war set Russia on the path to revolution, which eventually came about in 1917, and led to a major decline in Russian influence in Asia. All that Japan lacked in Manchuria was actual political sovereignty.

These developments in Manchuria came at the same time that Canadians were opening up the western plains and mountains. The government of Canada had a major political concern – the urgent need to cement sovereignty over the vast Prairies. There was a deep-seated anxiety in the Canadian polity about the northern expansion and encroachment of the United States, and the dire possibility that Canada might be swallowed up by the behemoth to the south.

The sovereignty situation played out differently in Manchuria and Canada. Canadian sovereignty was gradually strengthened on the Prairies through political and administrative means, and through settlement. The "southern threat" faded. The real threat to sovereignty was to come from within, from the separatist movement in Quebec in the mid to late twentieth century. Chinese sovereignty over Manchuria was lost to Japan in 1931. The Chinese military and civilian authorities were ousted.[5] The League of Nations found against Japan after its investigation of Chinese protests. Japan responded by withdrawing from the League and setting up a puppet state, Manchukuo (Manzhouguo), nominally headed by the last emperor of the Qing Dynasty. The new state was flooded with Japanese and Korean settlers, as discussed by Annika Culver in this volume (Chapter 6). For fourteen years, until the defeat of Japan in the Second World War, Manchuria was severed from China.

## Migrants

Behind the large sovereignty concerns of Manchuria and Canada was a more practical problem: finding the people to settle the land. In both states, the same lure enticed rural settlers: the longing for land, the dream of people who were landless in their own farms. The idea of vast, almost empty space – as the Chinese saying goes, "few people much land" (*renshao diduo*) – waiting for those who wanted to work the land, was a magnet for ambitious peasants who wanted what they knew they could not have at home, their own land. Millions were willing to brave any hardship to have their own land.

Manchuria needed several categories of migrant. The first category was the labourers (usually peasants) who would build the railways and the infrastructure of the new world; almost all of them were young, single men. The second category was the peasants who would settle the land; these people would migrate as families and create their own rural settlements. The third category included labourers, miners, and factory workers; again these were mostly single men, and most would live in the new cities. The fourth category, critical for the development of trade, was made up of merchants, shopkeepers, and other professionals, a burgeoning middle class.

There was a vast reserve of potential migrants in each category, just across the Gulf of Bohai from southern Manchuria, in Shandong. Shandong was one of the most populous and poorest provinces in China; Shandong people made up a huge proportion of the migrants to Manchuria.[6] They came in one of the largest migrations in world history. Between the late Qing and the Second World War, roughly twenty-five million Chinese migrated to Manchuria.

Shandong people were renowned for their ability to bear hardship. In spite of the poverty of the province, men from Shandong tended to be husky and were usually taller and stronger than southern Chinese. The provincial self-image was that Shandong men were competent and good-looking, the stereotype borne out by the fact that a large number of China's film and opera stars have come from the province. Shandong people also had a reputation for being quarrelsome, in the *haohan* tradition of tough but decent men, given to fighting and drinking but also to looking out for the downtrodden. The *haohan* were immortalized in the immensely popular novel, attributed to Shi Nai'an, the *Shuihu zhuan* (*Outlaws of the Marsh*); it was set in the western part of the province. Shandong people were involved in frequent protests against authority. The *Yihetuan* (the Boxer Movement), which brought terror to foreigners in China in 1900, started in Shandong; most Boxers were from the province.

In the late nineteenth century, extensive migration networks began to develop in Shandong. These networks were essential for large-scale migration. Some networks were set up by commercial agencies, such as the ones that recruited miners for the Fushun Mines (Liaoning). Most of the migrants, however, were recruited through personal networks, which relied on ties of family and locality (*laoxiang*).[7] A local labour contractor (*batou*), or a migrant already settled in Manchuria at home on a visit, would recruit other locals as migrants. As an incentive to the potential migrant, a payment – actually an advance on wages – was made to the family left behind in the village. The *anjiafei* (money to settle the family) was critical in fostering migration; it was one of the few significant sums a poor family might ever touch. Most of the migrants recruited in Shandong were destined for Manchuria, though some were recruited to work in Siberia, and a large number (140,000) was recruited during the First World War to serve as labourers on the Western Front in France, digging trenches and cleaning up the battlefields.[8]

Much of the migration to Manchuria was destined to be seasonal. Most outdoor work was impossible in the Manchurian winter. Opencast mines and construction sites closed down as the winter came on. Migrant "swallows" who worked in these industries flew back south of the Bohai during the cold months. Forestry, on the other hand, was largely a winter occupation; it was impossible to move cut timber except when the ground was frozen.

Permanent settlement on the great plains of Manchuria was haphazard; however much the state may have wanted the land settled and turned over to agriculture, it did little to smooth the settlement process. The state was minimally involved with the recruitment and settlement of migrants. It did expand the system of local government, but individual officials were often ineffective, before and after the 1911 Xinhai Revolution. Local officials had a great deal of difficulty coping with banditry and other forms of lawlessness, not least because the men sent to administer local government were often themselves in some form of disgrace. The Qing had traditionally dispatched officials in trouble to posts in Manchuria, as a type of punishment. These officials were only slightly better off than the families of executed officials who were banished permanently to the region. The perception that being sent to Manchuria is a form of punishment has never completely faded. Wang Ning recounts that in the 1950s thousands of Beijing intellectuals designated as Rightists were sent to labour camps in Xinkaihu in the furthest reaches of northern Manchuria.[9]

Neither the Qing Dynasty nor the early Republican governments of China set up effective means of rural community building. Instead, to a great extent, the process of community building depended on the actions of the migrants themselves. People from one place in Shandong would congregate in the same place in Manchuria, in effect recreating their original village. These communities were given the same name as the original village, with the addition of *haiwai cun* ("village across the sea," referring to the Bohai). The village name was often the only element of community; there were few other community-based institutions such as schools, shops, temples, or graveyards. The notion of "settlement" still had a large element of the temporary about it. Many of these migrants intended to return home in old age or even after their death. "Home" was always Shandong, even if only for their burial.

After 1932, the status of the Chinese in Manchuria was ambiguous. The new state, Manchukuo, was clearly a puppet state of Japan, but its existence seemed to imply that Manchuria had never been fully incorporated into China. Lending strength to the doubt over the "Chinese-ness" of Manchuria, the migrants, even the settlers, seldom thought of themselves as belonging anywhere but their native Shandong; the concept of China was remote. But they suffered because of their Chinese nationality; the Japanese occupiers treated them as inferior, colonial subjects. This ambiguousness of the identity of the Chinese in Manchuria went against the nationalist current in China itself. After the annexation of Manchuria by Japan, Chinese nationalists saw the loss of Manchuria as a great disaster; they considered Manchuria to be an inalienable part of China and saw its loss as a national humiliation (*guochi*). The lyrics for the "March of the Volunteers" (later the Chinese national anthem) start with the line "Rise up all those who refuse to be slaves." The lyrics were written by Tian Han and the music composed by Nie Er as a rallying call to patriotism after the loss of Manchuria.

One group of people was unquestionably Manchurian. The Manchus, in disgrace and without power in China since 1911, were quite willing to work with the Japanese. The last Qing emperor, Pu Yi (1906–67), was enthroned as the first (and last) emperor of Manchukuo. By Chinese standards, he had committed treachery, a crime for which he was later imprisoned.

On the Canadian Prairies, from the late nineteenth century on, waves of migrants came first from eastern Canada, soon to be followed by people from northern and eastern Europe. Some were recruited directly at the initiative of the visionary federal minister Clifford Sifton (1861–1929), whose agents looked for migrants in eastern Europe, especially Ukraine. Sifton's concern was not for the nationality or religion of migrants but for their ability to adapt to a harsh environment. He looked for "a stalwart peasant in a sheep-skin coat, born on the soil, whose forefathers have been farmers for ten generations, with a stout wife and a half-dozen children,"[10] to open up and cultivate the land, and to create communities.

Prairie communities grew around two key institutions brought in from eastern Canada: schools – which were state-run and either non-religious or run by the Roman Catholic Church – and parish churches of many denominations. The combination of state and religious organizations was a highly effective means of community creation. The schools forced the learning of English and led to a certain level of community integration. School was compulsory for all children; winter weather was no excuse for non-attendance. Churches allowed for close-knit congregations, of a range of Christian denominations. Quite small communities often supported several churches – Anglican, Roman Catholic, Eastern Orthodox, Presbyterian, Methodist (the latter two were combined as the United Church of Canada in 1925). Church membership was not compulsory but almost universal, for reasons of faith and of community. Regular attendance at Sunday service was expected, whatever the weather. After the service, the congregation met in the church basement, to eat and to share a sense of community; marriages were contracted, business agreements made, social contacts nurtured – all in the warmth of the church basement. Political movements grew out of the Prairie churches, notably the Co-operative Commonwealth Federation (later the New Democratic Party [NDP]). Charismatic clerics led their congregations in movements for social change. Tommy Douglas (1904–86), a Baptist minister from Manitoba and the father of the Canadian health care system, was voted by popular ballot in 2004 as the Greatest Canadian of all time in a poll run by the Canadian Broadcasting Corporation.

## Winter

One of the central issues for those who moved to Manchuria or to the Canadian Prairies was how to cope with the long, harsh winters. Manchurian and Prairie winters are fearsome.[11] They may last as long as half the year. Agriculture flourishes in the hot, moist summers, but seeding is often threatened by late frosts, and the harvest is often hit by early frosts. Frost can occur in almost any month.

Those moving to Manchuria had to build secure housing before they could endure the winter. Northern Chinese peasant housing, single-storey structures of adobe brick, was quite well suited to the northern lands. The

FIGURE 1.2 Manchuria winter scene postcard.
*Source:* Collection of N. Smith

brilliance of the Chinese *kang*, a heated platform that often occupied much of a room and was used for sleeping and sitting, ensured at least some warmth. It bore some similarity to the European tile stove, which radiated heat outwards. Europeans settling on the Prairies abandoned the tiled stove in favour of forced-air heating produced by furnaces, a much less efficient way of heating a room. The Chinese houses had two weaknesses: the windows and roofs. The windows used paper rather than glass; paper was no match for a blizzard. The roofs were flat and could not withstand a great weight of snow. However, few adaptations were made to adjust to the conditions.

Chinese garments, padded with cotton (for the poor) or silk floss (for the affluent), were well adapted to the cold. Layers of clothing were added as it became colder. Furs and sheepskin were used as they were in the Arctic, with the fur worn on the inside rather than on the outside. Manchuria did not have one fur that was essential for opening the Prairies – the buffalo hide. Buffalo robes were commonly worn on the Prairies in the depth of winter, one of the many Indigenous practices adopted by newcomers.

Adaptation to the Manchurian winter did not cure settlers from the south of their fear of winter. Winter was hostile, brutal, and even deadly, not just from hypothermia or frostbite, but also from a mysterious condition, *hanbing* (disease of the cold). This may have been a version of SAD (seasonal affective disorder) or of the vague but real condition known as cabin fever. The fear of winter led to virtual hibernation in much of Manchuria; settlers hunkered down indoors to wait for the return of warmth. But the disease that was the scourge of Manchuria in the late nineteenth and early twentieth centuries was bubonic plague, which spread easily when humans, fleas, and rats were in close proximity, as they were in winter.[12]

Many of the Chinese migrants in Manchuria coped with winter by avoiding it. A large population of migrants, known as "swallows," went to Manchuria in the spring and left for home in China at the end of the fall; their jobs in construction and in the open-cast mines dried up as winter came on, and they were able to get home in time for the lunar New Year (in January or February). It was essential to take some money home; home communities depended on remittances from those who worked in Manchuria, much of which was brought home on annual visits. The embarrassment of coming home without money was so overwhelming that those who did not have money to bring home simply stayed away, often for years.

Russians in Manchuria did not share the gloomy attitude of the Chinese towards winter. The Russian population included employees of the Chinese Eastern Railway, and, after the 1917 Revolution, a large number of White Russians – refugees from the Red Revolution. The Russians enjoyed the winter, especially in Harbin, and were out in force skating and sledding. They marked the course of winter with festivals, religious and secular, many of which included processions out onto the river ice. The most deeply devout would bathe in holes cut in the thick ice. The greatest festival came at Easter, which fell towards the end of winter.[13]

The attitudes towards winter in Canada contrasted with those of the Chinese and the Russians. Settlers on the Prairies embraced winter, less out of enthusiasm than necessity. Avoiding winter was out of the question until the arrival of cheap vacation packages to warm destinations in the 1980s.[14] Very few settlers on the Prairies could afford to leave at all, either for the winter or for good. All their assets were bound up in the family farm. Hibernation was out of the question. Settlers were forced by law and by custom to keep moving through the winter; their children had to go to school, by law, and the whole family went to church on Sundays, by custom. Settlers lived out the line of the national anthem that describes Canada as the "True North strong and free," and many of them did it very well.[15]

From the start of non-native settlement on the Prairies, there was an official emphasis on the healthiness of the cold: communicable diseases did not survive the cold. The brilliance of winter, blue skies, and gleaming white snow were stressed as positive factors. The city of Edmonton, founded in 1904, was named "Canada's Sunshine City." The city's winter temperatures went as low as −30°C, but the inhabitants learned early to insist that they loved the winter, and were exhilarated and strengthened by it.

Winter sports evolved rapidly in Canada; they have remained dominant, the sports at which Canada excels. Hockey, skating, and sleigh riding all became part of Prairie winters. Hockey has been the national obsession, from the late nineteenth century on. It was virtually obligatory for Canadian boys to play. For a gilded few, hockey was the way out of Prairie villages and on to the prestige and riches of the National Hockey League. Gordie "Mr. Hockey" Howe (b. 1928), the pride of the unincorporated village of Floral, Saskatchewan, is one of the most celebrated members of the Canadian hockey aristocracy.

Whether one was in Manchuria or the Prairies, one factor common to surviving the winter was strong drink. In Manchuria, the drink of choice was ferocious homebrew. Early descriptions of the region show men lying in a stupor on their *kangs*. Women were seldom mentioned as drinkers, partly because men outnumbered women in the settler populations, but also perhaps because women did not drink.[16]

In Canada, alcohol was an essential feature of the country's development and of the spread of the population from east to west. Several of the wealthiest Canadian families made their fortune on alcohol, including the Labatts, the Molsons, and the Bronfmans (the owners of Seagram's). The Molson Company, founded in the late eighteenth century, is the second oldest company in Canada, after the Hudson's Bay Company. Alcohol was consumed in vast quantities, often in taverns that excluded all women except those with dubious reputations. Alcohol was also viewed by many a shameful cause of social ills. The campaign against alcohol was led by women; the first Canadian feminist, Nellie McClung (1873–1951), was heavily involved in temperance. One of the first women's organizations in Canada was the Woman's Christian Temperance Union, founded in the late nineteenth century.

Manchurian attitudes towards winter have evolved over time (see Chapter 5 by Norman Smith). In recent decades, they have become very positive. Figure skating is strongly developed and promoted. Many national and world figure-skating champions have come from Manchuria. The luminous Chen Lu (b. 1976), from Changchun, was the first Chinese skater to become a world champion, in 1995. Hockey is also played by an increasing number of people in Manchuria, and all major Manchurian cities have professional teams. Every year, the Harbin Ice and Snow Festival brings in great numbers of tourists, who come to see the remarkable collection of buildings and sculptures carved from ice.

The beauty of Manchurian winter is also celebrated in film. Two of the most celebrated Manchurian films, *Along the Sungari River* (1947) and *The Day the Sun Turned Cold* (1995), include breathtaking winter scenery.

## Railways

Railways were essential for opening up Manchuria, as they were for opening up the Canadian West. In both regions, railway building was quite easy. The land was flat, no tunnels needed to be dug, and there were few rivers to bridge. There were also few problems with land title. In Europe, on the other hand, one of the greatest problems in building the railways was dealing with land owners. The railways were also a vast improvement over roads. In Manchuria there were very few roads, and these were impassable for much of the year. Rivers could take the place of highways, especially in the winter when they froze, but there were long periods – during freeze-up and break-up – when rivers could not be used for transportation.

The opening up of Manchuria and of the Prairies coincided with the global age of railways, when the romance of steam was at its peak in the late nineteenth century. Around the world people were captivated by railways. Government and private investors poured money into railway building, driven more by dreams of profit than by real profit possibilities. Grand railway stations were built, almost as temples for the new age. Railway building was dangerous work; countless lives were sacrificed during construction. Railways provided opportunities to ship goods and people, and the possibility of penetrating remote regions in comfort. Railways satisfied the needs of the rich and the poor, incorporating an explicit class system: passenger trains were made up of luxurious first-class carriages, with bedrooms and dining cars for the affluent traveller, and virtual cattle cars for the poor, who travelled in discomfort but at relative speed.

In Manchuria, a vast rail network stretching east-west and north-south spread across the region. Much of the network was controlled by

FIGURE 1.3 South Manchuria Railway train, South Manchuria Railway leaflet, no date. *Source:* Author's collection

foreign interests: the Chinese Eastern Railway was built by Russia in the late 1890s, and the South Manchuria Railway was built by Japan and opened in 1906. In Canada, the great east-west transcontinental lines took migrants westward and the products of their labour eastward. Rail lines connected to the Great Lakes ports; cargo was transferred to large seagoing vessels that transported goods through the Welland Canal and the St. Lawrence to the Atlantic Ocean en route to Europe. The construction of railways through the mountains to the Pacific Ocean, which saw dramatic feats of engineering, came slightly later than the eastern lines, in the late 1880s.

In both Manchuria and the Prairies, the history of the cities and the railways is intertwined. In each region, cities and railways evolved in response to each other, and in both regions they continued to be dependent on one another for decades.

Two of the great cities established in late-nineteenth-century Manchuria were the products of the Russian expansionist vision. The railway city Harbin and the port city Dalian formed a geographical triangle along with Russia's most easterly port, Vladivostock, the "ruler of the East," founded in 1860. Harbin and Dalian were built with extraordinary speed, in less than a decade. Harbin, founded in 1898, grew at the junction of the Chinese Eastern Railway (the line across northern Manchuria that cut off the long northern loop of the Trans-Siberian Railway within Russia) and the line coming north from Dalian. Known as the "Ice City," Harbin was sunny year-round, its climate bracing, with an average January temperature of -19°C. The city was designed on an ambitious scale to be "the Paris of the East."[17] It was dominated by fine churches, schools, department stores, and houses. It remains one of the greatest collections of Russian architecture in the world, since it did not suffer the ravages of war and modernization to the extent that cities in Russia did.

Dalian was principally a port. Its beautiful location on the southern coast of Manchuria also helped establish it as a resort city and a retirement destination for many Chinese politicians and warlords. The Russian naval base was located at Luxun, far enough away from Dalian to ensure that the city remained quite separate from the military installation. The Russian city, Dalny, was founded in 1898. It was lost to Japan less than a decade later, but the elegant layout was preserved, though later development reflected the Japanese preference for neoclassical architecture.

The third great Manchurian city, Shenyang, was not a new city; it had been the capital of the Manchus before they moved south. It incorporates a small, imperial city, and there are imperial tombs in the suburbs. In the late Qing Dynasty and the early years of the republic, it emerged as a commercial and industrial hub. In addition to having its own factories, the city acted as a centre for mines and other factories in Liaoning.

The fourth Manchurian city was never fully built. Xinjing, the "new capital" of Manchukuo, was designed on a grandiose scale. Parts of it were built, but construction stopped when the Japanese were defeated in 1945. The city, renamed Changchun, subsequently suffered one of the worst disasters imaginable, the siege of 1947–48 during the civil war, in which as many as 200,000 people starved to death. This tragedy, which the communist state purposely hid for many decades, has made Changchun a city haunted by ghosts and grief.[18]

The Prairie cities were founded at around the same time as the Manchurian cities. Winnipeg, Regina, Saskatoon, Edmonton, and Calgary were all railway cities, dominated by the railways and the commercial interests they supported. Each city had a splendid railway hotel, most of them built in the Baronial architectural style. Banks and department stores built equally splendid structures, but none could rival the grandiloquent administrative buildings, city halls, and provincial legislatures.

The Prairie cities are smaller than the Manchurian cities, but they function in the same way as the Manchurian ones, as a combination of administration and commerce. This is a departure from the separation between the two functions found in other parts of China and Canada – namely, Beijing and Shanghai, or Ottawa and Toronto.

FIGURE 1.4 Saskatchewan Provincial Legislature (completed in 1912), postcard.
*Source:* Collection of N. Smith

FIGURE 1.5 Manchukuo Board of General Affairs, photo, *Manchoukuo Illustrated,* 1941.
*Source:* Hsinking: Manchoukuo Photo Service, 1941, p. 58

## INDUSTRY

Manchuria's first industries were related to food. From antiquity, salt pans on the coast of southern Manchuria produced salt for local use and for export to other parts of China. In the late nineteenth century, soya beans and soya bean products became a major export industry. Manchuria was also destined for heavy industry; abundant coal and iron ore deposits were exploited by both the Russians and the Japanese. Under Japanese rule, Manchuria became the industrial heartland of China. The Japanese government and major corporations invested heavily in industrial plants and mines. The Fushun Coal Mine became the largest open-cast mine in the world. Steel plants were established at Anshan and Benxi. These plants and factories all demanded a huge supply of labour.

There were a number of factors that encouraged industrial development. One was the prevalence of raw materials. Another was the need to provide materials and equipment for Japan's military build-up in Asia; the Guandong Army, based in Manchuria, was the spearhead of Japanese expansionism. These military investments never paid off for Japan. In 1945 these industrial projects were abandoned. Soviet troops entered Manchuria in the last weeks of the Second World War; over the next months, they expropriated much of Manchuria's industrial plant as war booty. That, combined with the start of the civil war between the Guomindang and the Chinese Communist Party, meant that Manchuria's industrial recovery was severely delayed.

There was no industrial or mining development in the Prairies in the early stages of settlement. The Prairies were expected to consume the industrial products of Ontario, which produced the machinery needed in extensive farming. Industrial-scale extraction of mineral resources came late, long after the Second World War. Even when the tar sands in Alberta were developed in the 1980s, and potash was mined on a large scale in Saskatchewan, the products were intended for export, not to serve local industry. Today, one of the key issues is how resource products are transported to what is referred to as "tide-water" – that is, the Pacific Ocean. Whether it is shipped by rail or in pipelines, oil is a difficult product to move, unlike grain, the first product to be moved in large quantities by rail.

## Agriculture and Animal Husbandry

The great plains of southern and central Manchuria have some of the most fertile soils in the world – rich, black alluvial earth. The extent of the land is vast. The enormous southern plain, drained by the Liao River, is itself dwarfed by the more northerly Songhua Plain, which extends over 1.4 million square kilometres. Both plains were ideal for growing sorghum (*gaoliang*), also known as tall millet, which became the staple food for many of the inhabitants. While not the most exciting food in the world – Alexander Hosie described the taste as "insipid"– it was highly nutritious, almost a complete food. A man could eat up to four pounds a day.[19] Sorghum could also be used to produce fierce, home-brewed liquor. Sorghum was to Manchuria what the potato was to Ireland, a crop that could fulfill almost all human and animal needs.

The Manchurian plains came under cultivation at the time when new crops, for export as well as for local consumption, came into production. These crops included wheat and barley, but the most prominent was the humble soya bean, which demanded good soil, warmth, moisture, and skilled cultivators. By the 1930s, Manchuria was producing 60 percent of the world's crop of soya beans. Maize was also grown widely. Like sorghum, maize is highly nutritious and has many uses – as food for humans and animals; as fuel; as thatch for roofs; and as the base for home-brew. Other crops were also cultivated on a large scale. Opium and tobacco were widely grown, more for home consumption than for export. And ginseng farms multiplied, though cultivated ginseng was never as valuable as the wild variety.

One major hindrance to the development of Manchurian agriculture was the region's chaotic system of landownership. The acquisition of land by peasant cultivators was complicated, and title was often insecure. In many parts of the region, large rural estates, often owned by absentee landlords, evolved instead.

Under the Japanese, there was a serious effort to develop hardy crops. Crop research stations, run by agencies such as the South Manchuria Railway Company, bred new strands of agricultural crops. Part of this work was predicated on the assumption that those who would most benefit from the research would be the incoming Japanese and Korean settlers.

Under Japanese rule, too, there was an emphasis on developing hardy livestock breeds. This included some less-than-successful efforts to breed hardy cattle and sheep, as discussed by Blaine Chiasson in Chapter 4. Efforts to breed the Heihe horse, in the most northern part of Manchuria, were more successful. This breed of draft horse could stay outside in temperatures as low as -30°C. There was little interest in breeding riding horses in Manchuria, where there was no real cattle-raising industry. Dogs and goats were bred for their furs and skins.

The pattern of agricultural settlement was more orderly on the Prairies. New arrivals who planned to settle there were promised land grants. The provincial governments took the responsibility of surveying land, putting in roads, and marking out future villages and towns. Settler families were given the right to exploit a minimum of a quarter-section (160 acres), which became theirs after five years if they cleared the land and opened it up for cultivation. They were given free advice on cultivation, and could sell their crops through regulated markets. The rural communities that grew on the Prairies were made up of owner-cultivators, who had at least the chance of making a decent living. Official and commercial involvement was critical. Governments ran agricultural research stations, which helped to introduce hardy crops such as durum wheat. Private companies, notably Massey-Harris, provided the huge pieces of agricultural machinery necessary for farm families to cultivate hundreds of acres on their own. A passion for agricultural machinery swept the West as soon as the hard years of the Great Depression were over. In the Prairies, flotillas of gigantic tractors and combine harvesters became as much a symbol of man's progress as railways and dams were in many parts of the world.

After 1949, Manchuria took up mechanized agriculture. Large-scale farming was modelled on the Soviet system and also on the North American system, though this was not recognized publicly. The major difference was that in China's war against the earth (Sun Xiaoping, Chapter 10 of this volume), the equipment was owned by the people; on the Prairies, it was owned by individual farmers.

## The Political Environment

The largest difference between Manchuria and the Canadian Prairies has to do with the political environment. Manchuria was fought over constantly from 1895 to 1949, in both international and civil wars in which the local population had no role. The region suffered foreign incursion – in 1895 and in 1931 – in two of the most critical wars in modern Chinese history. Then came the long, eight-year war with Japan, with Manchuria far behind the front lines. The civil war between the Guomindang government and the Chinese Communist Party started in Manchuria in 1946, very shortly after the occupation of the region by the USSR in 1945 at the end of the Second World War. Cities in Manchuria changed hands four times in three years, from the Japanese to the Soviets to the Guomindang to the Chinese Communist Party.

Beyond actual involvement in war, the region also lived in a constant, grim state of insecurity. Banditry was rampant from late in the Qing Dynasty. Manchuria's greatest leader, Zhang Zuolin (1875–1928), rose from the bandit world. He became so powerful that he was able to expand into north China, with his Fengtian armies. He was assassinated by Japanese agents in 1928. His son Zhang Xueliang (1901–2001) might have been a great leader, except for the fact that he was held under house arrest by Chiang Kai-shek from 1937 until after Chiang's death in 1975.[20] The people of Manchuria lived with this misery of violence because they had no choice. To defend themselves many of the men became competent fighters. Occasionally there was an outburst of enthusiasm for autonomy – as during the civil war, when two of Zhang Xueliang's brothers tried to make a deal with the communists that would allow for greater autonomy for the region – but this came to nothing.

Canada has experienced no large-scale civil wars. The possibility of Quebec separating itself from the rest of the country has been avoided for the time being by negotiation. On the other hand, Canada was heavily involved in the two world wars. In the First World War, of the 600,000 Canadians who enlisted, two-thirds went overseas, and more than one in ten died. These figures are staggering, considering that Canada's population was small, under eight million in 1914. War losses left an indelible mark on the history of Canada, and on every community, each of which has a cenotaph commemorating the war dead. In the Second World War, the number of casualties was much lower, approximately 23,000.

One critical factor in the peaceful settlement of the Prairies was the early establishment of law and order. At the beginning of Prairie settlement in the 1870s, the North West Mounted Police, later the Royal Canadian Mounted Police (the Mounties), was established to deal with petty violence along the border with the United States. The many Prairie towns with the word "Fort" in their name were originally police posts and/or Hudson's Bay Company trading posts. The RCMP headquarters and training centre were set up in Regina; all of its members were (and still are) taught to ride fine black or bay horses. The characteristic scarlet uniform is still worn for ceremonial occasions. It is one of the best-known symbols of Canada.[21]

Modern Manchuria was in large part the creation of two nations driven by visions of greed and expansion. Russia and Japan contributed massive government and financial support over the first half-century of its development. In the 1950s, the new communist state, driven by its own

FIGURE 1.6 Canadian Mountie, postcard, 1970s.
*Source:* Collection of N. Smith

vision of industrial development along the Soviet model, made Manchuria its industrial pet. The model oil fields in Daqing were given enormous attention by propagandists, the whole country admonished to learn from Daqing (*In industry, study Daqing – Gongye xue Daqing*). At the start of the Reform Era (in the early 1980s), Manchuria seemed to slip backward, and its industries were no longer competitive. Today, the region is re-emerging, with considerable and enthusiastic help from trade with its North Asian neighbours – Russia (no longer the USSR), Korea, and Japan.

The Prairies also slipped into a form of economic decline several decades ago. Population has declined as the agricultural sector has faced greater international competition and as increased mechanization has meant that fewer farmers are needed to meet production requirements. A much-repeated saying in the 1980s in Saskatchewan was "Will the last person to leave please turn out the light?"

Economic decline now seems to be a thing of the past, in both regions. Manchuria is booming industrially. The resources of the northern regions of Canada (oil, liquid natural gas, gold) were seen by Ottawa until very recently as the future of the country's international trade; the effect of the current slump in oil prices is still unclear. In both countries, the winter has been defeated; people have learned how to live successfully in the cold. The threat of global warming has changed the status of all northern lands in terms of their economic possibility and – in Canada – has raised a new concern over sovereignty, this time over the Arctic. The two regions remain northern in one critical way: their natural environments are more fragile than those of regions farther to the south. Disasters are more difficult to control in remote regions. Northern regions take much longer to recover from environmental damage than their southern counterparts. The 1987 Black Dragon Fire, in a remote region of northern Heilongjiang, consumed ten thousand square kilometres of forest. The mountain pine beetle has destroyed over sixteen million hectares of timber in western Canada, mainly in British Columbia. The environmental damage caused in northern Alberta by the tar sands development north of Fort McMurray is extensive and may not be reversible. A chemical spill in Jilin in 2005 polluted a great stretch of the Songhua River and caused permanent damage to the river ecology. In October 2013, a super-smog enveloped Harbin, closing schools and businesses. Before the two great northern regions can fully yield their riches, there will have to be great improvements in conservation and in the protection of these vast but fragile worlds.

## Notes

1 The region was traditionally referred to in Chinese as Dongsansheng (Three Eastern Provinces), later as Dongbei (the Northeast). "Manchuria" is the most commonly used English term. The best-known English reference to Manchuria has

nothing to do with the region. The 1962 film *The Manchurian Candidate* is an improbable but gripping Cold War epic. Angela Lansbury, as a Soviet agent in deep cover, tries to take over the US government using her brainwashed son (played by Lawrence Harvey), who had spent a brief time in Manchuria as a captive during the Korean War. His former officer (played by Frank Sinatra) saves the day.

2 "Heilongjiang" is also the name of the province through which the river flows.

3 The tiger is now close to extinction, surviving mainly in zoos and Siberian Tiger Park (Harbin), where tourists can observe the tigers kill the small animals that the tourists have purchased.

4 Harold Adams Innis, *The Fur Trade in Canada: An Introduction to Canadian Economic History* (New Haven, CT: Yale University Press, 1930), is one of the seminal works of Canadian history. The Hudson's Bay Company has just been returned to its original name (from the anemic "The Bay"). It is the oldest commercial corporation in North America, already in its fourth century (though now under US ownership).

5 Rana Mitter, *The Manchurian Myth: Nationalism, Resistance and Collaboration in Modern China* (Berkeley: University of California Press, 2000), gives a careful analysis of what Manchuria meant as part of China.

6 Thomas Gottschang and Diana Lary, *Swallows and Settlers: The Great Migration from North China to Manchuria* (Ann Arbor: Center for Chinese Studies, University of Michigan, 2000), gives detailed analyses of the migration.

7 There is no English term for the concept of close ties based on place of birth, though words for the concept are found in many European languages (*Landsmann, compadre*) and in Chinese.

8 Xu Guoqi, *Strangers on the Western Front: Chinese Workers in the Great War* (Cambridge, MA: Harvard University Press, 2011).

9 Wang Ning, in Chapter 9 of this volume and in "The Great Northern Wilderness: Political Exiles in the People's Republic of China" (PhD diss., University of British Columbia, 2005).

10 Sir Clifford Sifton, "The Immigrants Canada Wants," *Maclean's*, April 1, 1922, 32–34.

11 The closest one can come to experiencing a Manchurian blizzard without actually being in one is to watch Kurosawa's magnificent film *Derzu Uzala* (1979), shot on location just north of Manchuria.

12 Carl Nathan, *Plague Prevention and Politics in Manchuria, 1910–1931* (Cambridge, MA: Harvard University Press, 1967).

13 Rosemary Quested, *Matey Imperialists? The Tsarist Russian in Manchuria, 1895–1927* (Hong Kong: Centre of Asian Studies, 1982), gives a lively description of Russian life in Manchuria.

14 Over a million Canadians go to Cuba every winter, over a million and a half to Mexico.

15 This line appears only in the English version, not in the French one, though the original version of the anthem was in French.

16 For a vivid description of the role of alcohol in Manchuria, see Norman Smith, *Intoxicating Manchuria: Alcohol, Opium and Culture in China's Northeast* (Vancouver: UBC Press, 2012).

17 Victor Zatsepine, "Russia, Railways and Urban Development in Manchuria," in *Harbin to Hanoi: The Colonial Built Environment in Asia, 1840 to 1940*, ed. Laura Victoir and Victor Zatsepine (Hong Kong: Hong Kong University Press, 2012), 17–36.

18 See Zhang Zhenglong, *Xue bai xue hong* [Snow white, blood red] (Changchun: Chunfeng wenyi chubanshe, 1993), for a detailed description of the siege. The author is a retired PLA officer. The book was banned in China shortly after its publication.

19 Alexander Hosie, *Manchuria, Its People, Resources and Recent History* (London: Methuen 1910), 69. Hosie was the consul at Niuzhuang (Liaoning) for many years.

20 The Zhangs, father and son, are now local heroes in Manchuria. Their three palaces in Shenyang (traditional, rococo, and modernist) have been restored and are now open to the public.

21 One of my favourite books as a child was *Susannah of the Mounties* by Muriel Denison, the story of an English girl sent to Regina to live with her uncle, a Mountie officer. A bowdlerized version of the novel was made into a film starring Shirley Temple.

# 2

# Rival Empires on the Hunt for Sable and People in Seventeenth-Century Manchuria

*David A. Bello*

In 1676, the Warka had become homesick during their involuntary removal southward from their Sahaliyan River basin homelands. Like many of their Indigenous contemporaries in North America, a group of at least 4,830 Warka was living in the path of oncoming empires.[1] Like the ginseng, river pearls, and sable pelts that enriched their boreal forest habitat in the vicinity of the mid-Sahaliyan, these Indigenous people were being hunted and gathered by both Romanov Russia and Qing China.

In 1676, Russia's Romanov dynasty (1613–1917) was reviving as its embassy under Nicholas Spafary arrived in Beijing. Cossack raiders, led by Vasilii Poiarkov, had initially spearheaded the empire's eastward Siberian expansion to the Amur River basin in 1643. After a major defeat by Chinese forces in 1658, Russians had steadily, and largely without authorization, infiltrated back into the basin during the 1660s. They had built an official Cossack stronghold at Yaksa (Ru., Albazin) on the Amur by the early 1670s. Spafary's mission was the latest in a number of futile and ill-conceived attempts since 1654 to obtain an imperial audience in China to formally assert Russian authority over the basin and its resources, including the Warka.[2]

In 1676, the Qing Dynasty (1644–1911) was pressed to defend its own authority against both Russian and Mongol challenges all along its northern frontier. In the fifteenth year of his reign, the Manchu Kangxi emperor (r. 1661–1722) was continuing to consolidate the 1644 conquest that had united the dynasty's homeland with China proper.[3] The re-emergence of the Romanov challenge compelled his state to incorporate the Heilong

River basin more fully than previous forms of indirect suzerainty had. The complexities of this incorporation are spelled out by the Qing removal of the Warka from a river basin that already had three distinct names in Manchu, Russian, and Chinese at the time (Sahaliyan in Manchu; Amur in Russian, Heilong in Chinese; "SAH" hereafter).

By the 1670s, the somewhat casual notions of regional domination initially entertained by both dynasties were transformed through rivalry over indigenous human and natural resources. The struggle to stabilize imperial frontiers and subject identities out of the basin's existing environmental interrelations between humans and the surrounding ecology resulted in a series of unanticipated changes for all involved. Imperial control of people required manipulation of their ties to basin flora, fauna, and climate – not just simply their links to other people. Many of these ties, however, could not be easily severed or reordered by imperial fiat. Moreover, even when changes made to these ties were ultimately successful, alterations could have unexpected consequences.

Ties between people and game, sable in particular, were critical for imperial incorporation of the SAH basin. The complex, mutually conditioning, and diverse interdependencies between human culture and basin ecology cannot be reduced to a single relationship. This chapter will focus on people and game as one example of how human-animal ties contributed to the formation of the Qing borderland Manchu military identity during the Kangxi period. At this time, the administrative core of China's Manchurian territories was established in response to Russian incursion.[4] Administrative centres were set up in Fengtian, in present-day Liaoning province in southern Manchuria (1646); Jilin in eastern Manchuria (1653); and Heilongjiang in northern Manchuria (1683).

In the latter half of the seventeenth century, the Warka and other basin inhabitants found themselves in the midst of a Eurasian imperial expansion that resulted in a more tangible borderland. Like contemporary steppe peoples, the Warka found themselves steadily hedged in – in this case by the Qing to the east and the Romanovs to the west. Forest and steppe responses to the Eurasian empires, however, varied significantly. One group of steppe Mongols, the Zunghars, responded with its own form of expansionism, which began in 1676 under the leadership of Galdan (r. 1676–97). The Zunghars briefly united a large Inner Asia expanse, including parts of Mongolia, Tibet, Xinjiang, and Qinghai, all of which lay between the growing Russian and Chinese empires.

In contrast, expansionism was not an option for the comparatively fragmented forest peoples of eastern Eurasia. Peoples like the Warka were unprepared to experience the seventeenth and eighteenth centuries – what Peter C. Perdue has called "the decisive turning point in steppe-settled interactions" – as anything other than imperial subjects.[5] During this time, the Qing Dynasty was consolidating these peoples as "Manchus," a categorization Russia would never have accepted. In effect, both indigenous territory and people were up for imperial grabs. The ensuing competition, however, was conditioned by an indigenous resistance, which shifted steppe-settled relations under pastoral and agrarian regimes into forested foraging terrain.[6]

Manchu identity and its eponymous homeland were recent Qing constructs. Indeed, the dynasty and its signature ethnicity had been formally declared only in 1635–36.[7] Both were initially consolidated through Qing attacks in 1644, when the conflict for the basin's human and other resources had been Manchu-Han (ethnic Chinese). Within a decade, conquest dynamics had reversed as the Qing was forced to go on the defensive in the basin against the Romanovs. This defence required the relocation, voluntary or not, of minimally subjugated Indigenous peoples away from northern Manchurian forests to much less bountiful hunting grounds further south around burgeoning administrative enclaves. The 1676 Warka move was possibly the largest of these disorienting relocations.

The process of imperial identity formation under both regimes was conditioned by yet another factor, the northeastern ecosystem, which also constituted a primary incentive for initial imperial incursion into the basin. Fur-bearing animals in general, and the sable (*Martes zibellina*) in particular, provided material incentives and informed cultural networks for the regional construction of empire and its attendant identities. In practice, subjugation, or the formation of imperialized identities, remained ritually and materially tied to the presentation of sable pelt tribute. The basin's foragers were especially well situated, by nature and nurture, to acquire such pelts. The implications for human identity on the pursuit of sable as game animals, however, have not been examined beyond very generalized statements – almost clichés – about the importance of hunting to Manchu identity, especially in its elite military mode.

The ritualized presentation of tribute itself is often examined within the context of inter-state relations, or as a mode of production, rather than as a relation of humans and nature.[8] This "anthropocentric" approach to tribute effectively excludes any study of how the ecology continuously shapes and informs these acts of human consumption. This is why, for example, standard works tend to reduce tribute relations to cultural dichotomies.

Such approaches unquestionably establish an important social context for tribute, but to the exclusion of its environmental context.

My analysis will track the role of sable pelt tribute beyond human relations of consumption by first considering the ecology in which the social practice of sable pelt tribute arose. I will then consider how this ecology informs an indigenous forager identity, like that of basin peoples such as the Warka. Both Qing and Romanov imperial orders attempted to "reinscribe" these foragers for exclusive exploitation within the wider environmental context of pelt tribute. While the Qing throne prevailed in the conflict with the Romanovs, it did not do so without environmental compromises to foraging. As the final sections of this chapter discuss, the transformation of the indigenous forager identity – necessary in order to secure the Manchurian borderland for the Qing – also required substantially severing this identity, nurtured by hunting, from its connections with forest game. This more centralized borderland isolated Indigenous peoples, literally reinscribed as "New Manchus" (Ch., *Xin Manzhou*; Ma., *Ice Manju*), from previous existential relations with game. The ongoing process of borderland consolidation rendered the dynasty's most valuable human military resources – that is, militarized foragers – an endangered species by the nineteenth century.

The SAH River is the ninth largest in the world. Approximately forty-four thousand kilometres in length, it carves out the largest basin in northeast Asia to form a watershed of over two million square kilometres, extending to parts of China, Mongolia, Russia, and North Korea. More than three thousand kilometres of the river's length form the present border between China and Russia. The watershed itself is sectioned north to south by three parallel mountain systems – the Greater Xing'an (Ma., Hinggan) the Lesser Xing'an, and the Sikhote-Alin ranges, all of which are generally low and heavily covered with mixed broadleaf and coniferous forests. Mountains make up approximately two-thirds of the basin, and mark its northern boundary along the Stanovoi range and its southwestern boundary along the traditional Qing homeland of the Changbai range (Ch., *Changbaishan*; Ma., *Golmin Šanggiyan alin*). The basin's remainder is primarily composed of alluvial plains and includes wetlands, steppe, and alpine tundra.[9]

Current and standard World Wildlife Fund categorizations recognize fifteen SAH basin ecoregions, each with distinct but interdependent species and dynamics. The basin is particularly rich in ecoregions because its locale exposes it to the combined climatic effects of monsoons, oceanic currents, and mountains. Even the basin's numerous tributaries

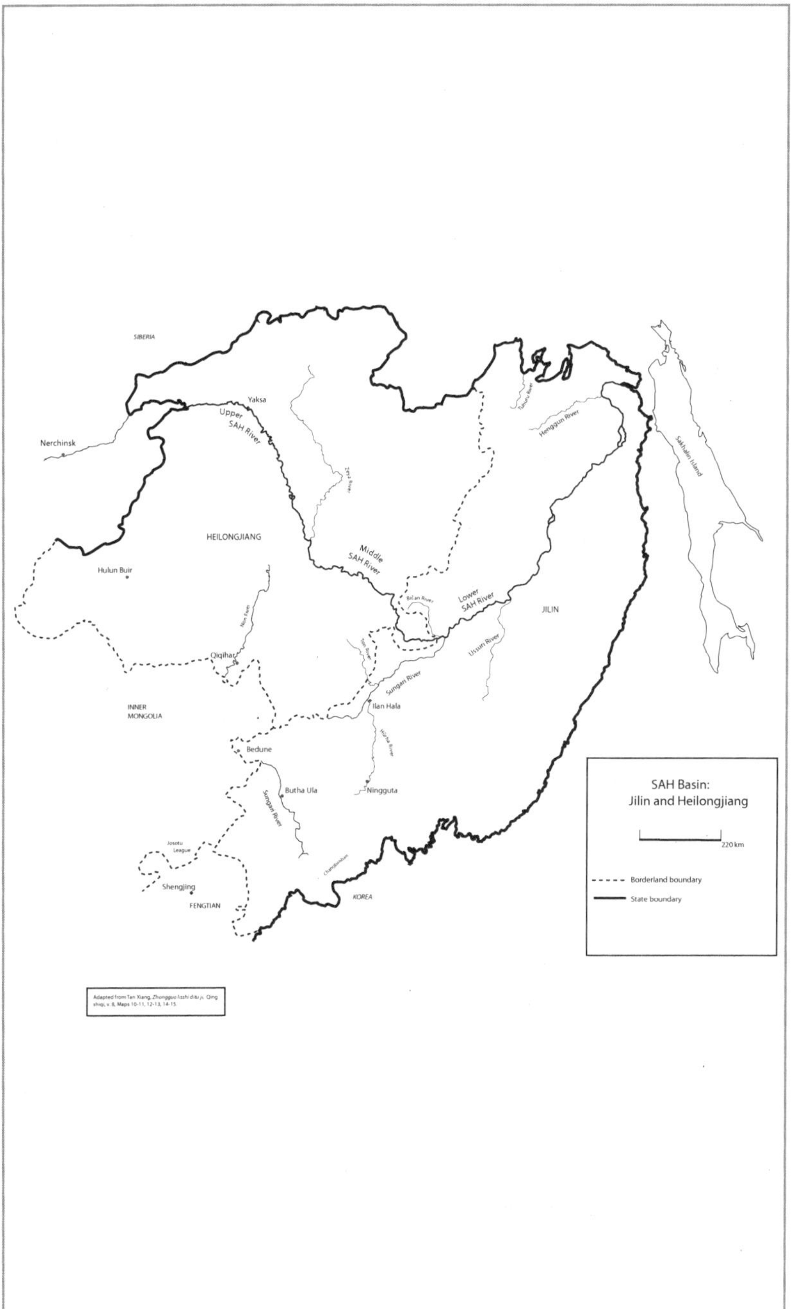


Map 4 Sahaliyan-Amur-Heilong Basin as Manchurian borderland, map.
*Source:* Collection of N. Smith

sufficiently differ (in hydrological terms) to form distinct aquatic habitats. The resulting species diversity is immense, particularly for a region so far north. Biologists have identified at least 6,000 plant species, and 130 fish, 600 bird, and 200 mammal species in the SAH basin.[10]

Boreal or taiga forest (which in the basin forms part of the largest tract of unbroken forest in the world) and the Manchurian mixed forest are the two most important SAH ecoregion cradles of this biodiversity. Among these forests' most attractive products are ginseng (*Panax ginseng*) and river pearls, two of Manchuria's three traditional "treasures." Changbai forests are mainly mixed stands of Korean pine (*Pinus koraiensis*) and a range of indigenous deciduous trees that foster rare species such as ginseng.[11] Boreal and Manchurian mixed forests also contain much larger stands of larch and pine conifers, which provide a food-rich habitat for rare mammals such as the Amur tiger (*Panthera tigris altaica*), the Amur leopard (*Panthera pardus orientalis*), and, most significantly for the seventeenth-century Eurasian empires, the sable.

The sable is a carnivorous mammal indigenous to boreal and mixed forest ecoregions. A recent study of sable in northern Heilongjiang province at the heart of the SAH basin found that the animals showed a marked preference for stands that contained both larch and birch, thought to facilitate resting and foraging activities. Overall, sable also almost exclusively favoured old-growth stands (which this study estimated to be an average of 265 years old, up to 518 years in the case of individual trees), which provided substantial cover, especially in winter, against predators. They tended to avoid unforested areas or places where early-growth pole-saplings predominated. This sensitivity of sable to variations in tree cover and type make the animal an important indicator of the effects of deforestation.[12]

In other words, sable are intimately and intricately dependent on distinctive tree species that combine to define boreal and mixed forests in particular, which in turn helps to constitute a characteristic ecoregions. Over time, humans have interacted with this biodiversity in ways that have augmented or reduced it, but always gradually integrate people into it. Human relations with sable exemplify this process in terms of immediate subsistence, elite consumption, and political development, all of which inform the environmental history of the Manchurian borderland. Here, human interaction with distinctive regional biodiversity created a space of "naturalized" hierarchical domination, which ranked China proper with all Qing borderlands while excluding rival Romanov practices of territorial formation.

## Sable Pelts and the Forager Identity

Sable pelt tribute can be traced back to around the second century BCE as one of the primary interactions between humans and the natural environment. By the seventeenth century, these interactions had produced a type of "sable ethnonym." For example, "Solon sable" referred to pelts from the primary group of tributary peoples in Heilongjiang. These people would become a main source of Qing elite troops, partly through the process of pelt tribute that monopolized the best sable. Qing records generally name basin peoples, including the Solon-Ewenki, Hejen (Ma., *Heje*), Fiyaka, Kiler-Ewenki, Orochen (Ma., *Orončo*), and Dagur (Ma., *Dagūr*), as "tribute sable tribes."[13]

The political significance of sable pelts was rooted in the necessities of basic subsistence in the boreal environment of the northeast. Although residents used the pelts for a variety of purposes, fur clothing made from sable pelts was a prerequisite for human habitation of the SAH region, especially for outdoor activities like farming and military operations. Officials requested fur coats (Ma., *jibca*) as vital for forty-seven Cossacks taken prisoner at Yaksa in 1684, as well as for five hundred Qing troops working the land in Heilongjiang in 1688.[14] Of course, furs, among which sable was the most prized, were also necessary garb for the trapping of fur-bearing animals, an activity ideally pursued during cold weather when pelts were at their maximum thickness. There were several methods for obtaining pelts, all of which necessitated extended periods of outdoor exposure. Hunters usually ran the animals down with dogs or smoked them out of their lairs, which were underground or in tree trunks.[15]

Sable habits and habitat demanded the acquisition of a complex range of foraging skills that culturally distinguished the inhabitants of the SAH basin from their agricultural and pastoral neighbours. These skills were developed through humans' extended exposure to basin ecology and were expressed in relationships with animals such as dogs, whose domesticated "identity" became specialized as "hunting dogs" under Manchurian conditions. Manchu identity itself correspondingly diversified as *niyahašara urse* (literally, "people who set dogs on game"), specialized hunters of deer and wild boar who supplied the Qing imperial table. Such human-canine interdependency also shaped what had been the relatively omnivorous practices of the former – resulting in, for example, the cultural taboo against the consumption of dog meat by some contemporary Manchurian peoples.[16]

Foraging skills formed a major part of the basin's human culture and were embodied in the composite ethnic identity of "Manchu." In this way, culture was environmentally conditioned, if not determined. Foraging relations between humans and animals (which, as in the case with dogs, was not always that of predator/prey), were recognized by overlords and Indigenous peoples alike as a basic component of regional ethnic identity. When the initial conquest of many of the Indigenous peoples of southern and central Manchuria was virtually complete in 1642, the Qing emperor Hong Taiji (r. 1627–43) exulted in a sable-ethnonym pronouncement: "Along the seacoast from northeast to northwest, to the Reindeer Herder (*Shilu*) and Dog Keeper (*Shiquan*) tribes … the lands that produce black fox and black sable, which do not plow and sow but customarily hunt and fish for their livelihoods … everywhere have been subjected."[17]

Pelt tribute was intrinsic to this process of subjection in a region where foraging had predominated for thousands of years.[18] Of course, the formation of a generalized SAH basin ethnic identity was dependent on a much wider system of human interdependencies with the region's biodiversity over a much longer period of time predating the Qing. These interdependencies could be intermixed with both pastoralism and agriculture. Yet neither alone could nurture soldiers whose identity was born of the chase, not of ploughing or grazing.

## Qing Pelt Tribute

By the first half of the seventeenth century, the basin's ethnic character was superficially imperial Manchu, because full dynastic control did not extend all the way into the basin's northern reaches. Nevertheless, the nascent Manchu state's incorporation of the basin up to that time had been padded with sable, presented in at least forty-nine of sixty (82 percent) tribute missions between 1626 and 1643, and sometimes in quantities as large as that of a 1634 Solon-Ewenki presentation of 1,818 sable pelts.[19] This practice continued as the Qing extended its authority northward. Official entries for thirty-eight indigenous tribute missions from the SAH basin from 1644 to 1673 indicate that all included the presentation of sable pelts, listed as the sole tribute items in twenty-eight (74 percent) of these cases.[20]

Qing pelt tribute usually required that a tribal mission travel to an imperial administrative centre, generally the town of Ningguta in Jilin during the Kangxi reign. Tribute was normally assessed at one pelt per adult male annually. The total probably usually amounted to approximately the 5,089 pelts assessed in a 1681 tribute list. Tribute sable pelts were generally categorized into three grades, with fixed quotas for each. The Qing officers receiving these pelts on behalf of the emperor reciprocated mainly with textiles from China proper, including robes, hats, boots, saddle cloths, belts, sashes, and fans. A feast was also a regular part of the proceedings. Once the ritual was concluded, the tributary delegation and their companions were permitted to trade the rest of their pelts and other goods with resident private merchants.[21]

Qing pelt tribute ritual was intended explicitly to subjugate SAH basin peoples rather than gain profit from trade. The ideal end of this process was the formation of "New Manchus" for military service across the empire, but this goal was at times suspended or obstructed by what appears to be indigenous resistance.[22] In 1675, tributary missions were dispatched from Hejen and Fiyaka peoples resident on the lower reaches of the SAH River and nearby Sakhalin Island, at the extreme and unstable northeastern edges of dynastic authority in the basin. The Kangxi emperor had bluntly recognized the vacuum of Qing norms among these two groups three years earlier in 1672. As part of a contemporary mobilization of dynastic forces against resurgent Russian incursion into the basin, he charged Jilin's first military governor, Bahai, "to spread a civilizing influence and employ all means to enlighten them," since "although they are submissive, they are actually savage, and it would be best to guard against them." Bahai had received a similar message from the emperor at about the same time, to civilize the equally "savage" and "crafty" Warka and Hūrha.[23] The "savage" identities of all these peoples were to be reconfigured by ensuring a more controlled interaction with their surrounding environment, which would ultimately result in a Manchurian borderland.

Bahai duly employed pelt tribute to effect this change, but a misunderstanding (possibly deliberate) revealed the ambiguities of tribute as an instrument of ethnic transformation. A mid-1676 report complained that Fiyaka and Hejen representatives had unilaterally violated tribute etiquette by not presenting their sable pelts at Ningguta but instead at an outpost much further north.[24] Because of this ritual impropriety, these "ignorant people" (Ma., *mentuhun urse*) would avoid witnessing Ningguta's impressive concentration of military power, which customarily formed the backdrop of pelt tribute proceedings there. Such a display was considered by imperial authorities to be particularly necessary at this time in order to discourage the Hejen and Fiyaka from plundering the "new people" (Ma., *ice urse*), a reference to the Warka, who were then being relocated south to Ningguta to protect them from Cossack plundering.[25]

Such Cossack assaults, however, could reinforce Qing subject identity – a fact that a particular group of Cossacks, sent on a mission to obtain the Qing throne's submission to the Romanovs, would fatally discover in 1653. While stopped at a "Ducher" village in search of horses and guides, the Russians casually announced their ruler's intent to divide Heilongjiang roughly in half. This declaration – to make "you east of the Amur our tributary and leave those Solon west of the Non River (Ch., *Nenjiang*) as tributary to the [Qing] emperor" – effectively ended the Russian mission. As the local villagers related later during an imperial audience in Beijing, they had realized that, in comparison to the value of dynastic gifts bestowed for their usual pelt tribute, the heavier and uncompensated Russian exactions would impoverish them. So they killed the Cossacks and appropriated their pelts to add to the village's own annual tribute. After admonishing them to spare further Cossack emissaries, the imperial throne sent off the delegation with their usual gifts of clothing.[26]

These Qing subjects proved willing, under the express motive of the pelt tribute system, to resort to violence to maintain a Qing Manchuria against imperial rivals. Groups like the Fiyaka and the Hejen, who had not been exposed to this imperial system for long, were less likely to uphold the Qing regional order so militantly. Here, then, is one indication that pelts were more than simply articles of commerce or consumption. Pelts were sovereign wares, commodities whose semi-ritualized exchange could establish and maintain hierarchical power relations, and their requisite suzerain-subject identities, without additional recourse to any form of direct coercion. The Russians, who were comparative newcomers to the region, were more heavy-handed in manipulating relations between Indigenous peoples and sable pelts. Such boorishness helps explain the villagers' violent reaction to their Cossack visitors in 1653 and the ultimately successful Qing mobilization of the SAH basin against the Romanovs over twenty years later.

## Romanov Pelt Tribute

Pelts in general and sable in particular also structured Russian expansion in eastern Eurasia to such an extent that one scholar has characterized this historical process as "the exhaustion of hunting grounds, the discovery of new ones and their subjugation to Russian authority." Fur certainly was the explicit medium of subjugation for Erofei Khabarov's 1649–53 expedition, the second major SAH incursion following Poiarkov's. Khabarov was specifically ordered to compel the peoples of the basin forests to pay "tribute" (Ru., *yasak*) in the form of fur. His previous operations on the steppe had given him no such opportunity to exact furs because, as his orders acknowledged, "there are no sable, foxes, beavers or otters in the steppe." Of course, Russian expansion in Siberia was not determined entirely by fur. Steppe dwellers were not exempted from *yasak*, which was to be paid "in whatever precious goods the land may offer." New Russian steppe subjects could "not think that because there is a scarcity of animals for *yasak* that they will not come under the [Russian] Sovereign's mighty hand."[27] In this way, Russia's Eurasian expansion was conditioned (rather than determined) by the biodiversity of distinct ecoregions, as reflected in the qualification in Khabarov's orders regarding the issue of fur-bearing animals.

Differences in the biodiversity of steppe and forest are a primary reason that Eurasian expansion was not fuelled by a single factor, such as fur. The way that the Russian empire was established within the semi-boreal SAH basin, however, was substantially conditioned by the presence or absence of fur-bearers. Human consumption of fur-bearing animals was thus decreed as the distinct, although not the exclusive, boreal idiom of incorporation within Russian imperial space, which remained in the grip of a veritable "fur fever" from 1585 to 1680. During this period, which nearly encompasses the span of Sino-Russian conflict on the SAH, pelts, mainly sable, represented about 7 to 10 percent of the Romanov state's total revenues.[28]

The economic significance of sable for public and private income created problems for both the Russian and Chinese regional empires. The Qing's dynastic predecessor, the Ming Dynasty (1368–1644), had enjoyed traffic in tens of thousands of sable pelts, despite its nominal degree of control of the area, mainly restricted to the southern fringes of the SAH basin. Imperial greed led to abuses of the Indigenous peoples, which undermined the Ming's own regional identity construction project. Ideally, Ming authorities held that tributaries should be "sent back with more than they came with," rather than cheated.[29] The Qing Dynasty had similar problems, as Loretta E. Kim has noted for many types of forage during the nineteenth century (Chapter 3 of this volume).

A variant form of extortion plagued the Russian pelt tribute system, which was in practice based on an exchange of pelts for Indigenous hostages taken by Cossack raiding parties. Khabarov's predecessor, Vasilii Poiarkov, imposed this practice on indigenous groups of Dagur and Hejen that he came across while collecting "the Sovereign's *yasak*" and searching "for new non-*yasak*-paying people." His expedition was hampered by the contradiction between his method of collection and his men's dependence on the locals for food. Poiarkov's violent basin forays had wound down by 1646 because of food shortages, which the alienated agrarian local peoples such as the Dagur did nothing to remedy.[30]

Khabarov continued to try to construct a Romanov Amur based on the crude direct exchange of hostages for pelts, as high as fifty pelts per head. This rudimentary tribute system was presumed adequate for a sustainable regime, based on maximum sable extraction. As an order to Khabarov's expedition made clear, Indigenous peoples were to be made to hand over as much *yasak* tribute "as circumstances allow" by taking "their leaders as hostage so that in future these tribes will pay the Sovereign's *yasak* ... for all time to come." Although the order defined a range of acceptable *yasak* pelts, including sable, fox, ermine, beaver, and otter, the vast majority of the Russians' boreal *yasak* was collected in sable. This preference was reinforced in subsequent orders.[31]

The brutality of this pelt tribute system is casually revealed in Khabarov's August 1652 report of attempts to impose *yasak* on a Dagur group living on the Zeia (Ch., *Jingqili*; Ma., *Jingkiri*) River. After pursuing and killing some of the locals they encountered, the Cossacks were able to seize approximately 170 villagers, in addition to "small children." Khabarov told tribal leaders through a translator that if the Dagur "would give the *yasak* to our Sovereign and be obedient and submissive in all ways," they would not be killed and would be protected "from their mightier enemies." Dagur leaders, including one "prince Tolga," reluctantly agreed. This whole group of Qing tributaries swore an oath to the tsar to accept "eternal *yasak* servitude." But the new tributaries soon escaped from their Russian captors. Khabarov angrily accused Dagur leaders (including Tolga, who had remained behind) of "treason." The Cossacks then tortured the Dagur leaders "by burning them," but not fatally, in an unsuccessful attempt to have them recall their people. This costly Dagur resistance soon forced the Russians to withdraw "because there was no food" available.[32]

Khabarov had overestimated the capacity of pelts, hostages, and oaths to create Romanov subjects. In the case of the Dagur hostage taking, the Dagur captives quickly adapted to the imperial challenge by manipulating Russian vulnerability to both nature and custom. The Dagur made Khabarov (who stated at one point during the incident that both sides had been "living together as one family") believe they had accepted Romanov domination by exchanging pelts for hostages, swearing oaths of loyalty, supplying food to the Russians, and "frequently" interacting with the imperial representatives. These cultural expressions, combined with the ecological conditions of the region, provided many Dagur with the opportunity to escape their imperial captors. The fugitives exploited Russian dependence on Dagur food production by denying Russian access to the requisite human resources, an act culminating in the suicide of Tolga and his fellow leaders, following their torture at the hands of the Russians. Even Cossack fishing on the SAH River could not eliminate Russian dependency on local provisions.[33]

The Russians could not dominate by the acquisition of pelts alone, but they depended on an intricate network in which pelts were definitely central. Plunder and extortion were not effective ways to tap this network sustainably. Regional domination would emerge only for the imperial power that could reorder all the critical network components for its own benefit. Consequently, after the initial, improvisational stage of Qing-Romanov conflict from the 1640s to the 1660s, both powers sought to adapt their domination of pelts and peoples more systematically in order to achieve a viable imperial borderland in the SAH basin.

## Preliminary Competition for Pelts, Peoples, and Empire in the Basin

Both Qing and Romanov forces received direct signals through their respective pelt tribute systems that the Indigenous peoples of the basin were under pressure to subject themselves to one side or the other in the 1650s. This decade saw the formal establishment of the Qing military administrative region of Jilin in 1653 and the outbreak of five military clashes between Romanov and Qing forces between 1652 and 1660. The Indigenous peoples most directly involved were those living along the Sungari River. The Sungari formed (roughly) part of Jilin's southwestern border with Heilongjiang, between its confluences with the Hūrha River (Ch., *Mudanjiang*) to the south and with the SAH River to the north. Mid-SAH ethnic composition is not entirely clear, but it probably included Warka, Hūrha (or "Ducher"), Dagur, Fiyaka, and Hejen. None of these peoples was entirely unified or subjugated to either Qing China or Romanov Russia, but by the 1650s these people had become some of the SAH basin's most compelling resources.

In 1657, Ningguta's commander, Šarhūda, submitted a memorial to the throne. He urged an expedition northward along the Sungari, both to gain control over various groups of unsubjugated Hejen and Fiyaka before the Cossacks did and to protect existing Qing vassals from subversion. He was prompted to make such a submission by the effect of Cossack raids on indigenous sable tribute. His warning that "it would be no easy matter" to resubjugate peoples who had been subverted is an indication of the contemporary limits of the tribute system's capacity to establish reliable Qing subject identities. Indeed, Russia had already "newly recruited" Dagur "Cossack servitors" in 1652, and scattered records indicate a subsequent expansion in servitor ranks.[34] Šarhūda's fears were confirmed two years after his own major 1658 defeat of the Cossacks. Bahai's victory in 1660 required the violent "pacification" of fifteen Fiyaka villages, totaling over 120 households near the confluence of the Sungari and SAH Rivers in western Jilin.[35]

For a generation starting in the 1650s, Qing sable tribute underwent a corresponding decline in terms of both quality and quantity that reflected dynastic losses in territory. Sable tribute shortfalls began to appear among the pelt tribute of the Dog-Keepers and others and persisted into the 1680s. Beginning around 1653, various incidents contributed to tributary shortfalls, including Cossack operations that scattered locals and disrupted their hunting and tribute activities.[36]

Cossack predation became less systematic after Šarhūda's defeat of the main, authorized Romanov forces under Onufrii Stepanov; however, independent, often rebellious, Cossack bands continued to move aggressively into the basin. Qing authorities were not really aware of such distinctions until approximately 1669. By this time, pressure had mounted for the formal mobilization of dynastic regional forces against the resurgent Cossack threat. One official proponent of mobilization complained of Russian violation of Chinese territory, "where the people who present sable pelt tribute live."[37]

Deterioration in Qing pelt quality and quota shortfalls continued largely unabated through the 1670s, but both empires were too distracted by conflicts elsewhere to achieve a decisive advantage. Nevertheless, tribute declines indicate that Russian incursion had made critical territorial inroads along the mid-SAH. When dynastic officials questioned the Solon-Ewenki about such declines in the statutory 5,089 tribute pelt quota due for 1680, for example, they were informed that Cossacks had cut off access to the best hunting grounds.[38]

Russian and Manchu progress in creating imperial space in the basin was in this way explicitly measured in pelt gains and losses, ultimately indicative of the efficacy of each empire's effort to transform Indigenous peoples into subjects. One of Khabarov's Qing prisoners stated during his interrogation that indigenous peoples had come to Ningguta to petition for protection, saying that if China would not defend them, "we will be forced to pay the *yasak*" to the Russians.[39] Such fragility was among the primary reasons for the dynastic mobilization of Indigenous peoples as ethnically remodelled "New Manchus."

## Indigenous Relocation, Reconstruction, and Resistance

The removal of the Warka and other peoples from the northern regions of the basin during the 1670s was a fundamental strategy for this mobilization. Even Han Chinese were not excluded from this process. Wang Ning notes in Chapter 9 of this volume that several thousand already in Manchurian exile suffered further relocation in 1676 to the Chuanchang dockyards on the Sungari in order to serve as sailors and boat builders. Indigenous peoples experienced drastic changes to their environmental relations, especially with game. Indeed, given the past history of human-sable interactions, mobilization of Indigenous peoples by both Qing China and Romanov Russia required the adaptation of foraging culture to the new conditions of the emerging borderland. New Manchus were the ethnic embodiment of these new conditions in the Qing sphere.

Pre-existing pelt tributary ties were the basis of both the 1652 indigenous appeal for Qing protection and the dynasty's decree for the Warka's removal southeastward to the environs of Ningguta nearly twenty-five years later. By this time, the Russians were receiving similar appeals from their own Indigenous tributary peoples for protection from the "subjects of the Bogdoi [that is, Qing] Tsar."[40] Such records, dating from the 1670s until the Russians were expelled from the SAH basin in 1687, suggest that more direct imperial struggles for the basin's human resources had polarized.

Although there is some general evidence to suggest that such extreme measures may have been imposed as early as 1653–54, details are scant prior to 1676. It may be that organized mass relocation began nearer this later time, when the Cossack ability to collect *yasak* also appears to have peaked. It is likely that the Russians sometimes misinterpreted the wholesale abandonment of villages prior to their impending incursions as a deliberate Manchu policy rather than a more spontaneous reaction to known Russian depredations. There is at least one example of the Qing administration authorizing such a move after the fact in 1653, when authorities became aware of the sudden abandonment of a village through the normal mechanism of pelt tribute management.[41]

Such expressions of forager-style resistance were not limited to responses to the Cossacks. As the 1676 relocation demonstrates, they could also be responses to imperial Manchus. Many problems in reconstituting indigenous identity within the process of relocation were related to disruptions in people's relationship with game, whose habitat had become restricted in some critical respects to the more northerly reaches of the basin.

For example, on their arrival at Ningguta, Warka officers petitioned resident dynastic officials for funds to purchase a large number of cattle, whose meat would be used to produce meat broth, an important staple that the Warka "were accustomed to eating." The officers requested the purchases because Warka agricultural duties, necessary to support enhanced Qing military operations, left no time for traditional hunting activities.[42] Here is a hint of the disruption in Warka foraging, which reveals one of their adaptations as New Manchus to an altered ecological and cultural habitat.

Another example was what pursuing Manchu troops (the imperial embodiment of what the Qing intended the Warka to become) called "desertion." Instead of formally petitioning for domesticated bovine surrogates, as any "proper" subject would do, some newly arrived Warka had reacted more like hunter-gatherers – stealthily returning to their homes along the Sungari while ostensibly out hunting, fishing, or herding around Ningguta. At least forty-six males from twenty-six separate households made tactical use of their old way of life to escape the new set of obligations that effectively required the Warka to leave off foraging and become full-time soldier-farmers.[43]

At least some of the fugitives were eventually caught heading back to the Sungari, and the record of their interrogation reveals the cultural gap that had opened between the two groups of "Manchurian" peoples. The Warka had remained in the forests; the Manchus' southern branches had left the woodlands to impose the Qing Dynasty on their nearer Chinese neighbours a generation earlier in 1644. When Qing troopers asked their Warka captives if they had deserted because residents in Ningguta had deceived or threatened them, the fugitives denied it. Instead, the Warka explained, "[W]e were unable to obtain [meat] broth and stew in this place. The grass and trees of our native place are abundant, and we can get broth and stew to eat. So, we missed it and wanted to desert."[44]

It is not entirely accurate to view this incident as evidence that the Warka were literally, purely, and immutably products of their pristine boreal environment – namely, that they were "primary hunter-gatherers" in more technical and controversial anthropological terms.[45] "The Warka" were not really a centralized people, in the sense that the Russians or Chinese were, nor did they all live in exactly the same way. Their diversity was, nevertheless, conditioned by their ecology, which tended to promote foraging as the most "natural" (if not the only) choice.

One Warka officer's report made it quite clear that relatively urbanized areas like Ningguta were ethnically unsuitable in his people's primarily foraging terms. He pointed out that, in their normal daily existence,

> our people are partial to [hunting/eating] wild animals and fish, working only a little in the fields. Some people do not work the fields at all, but are pastoral. Now since we moved to Ningguta, however, we have come to depend on fields, [but] oxen and plows are insufficient so we cannot engage in cultivation.[46]

These Warka in their natural habitat were not uniformly hunter-gathers, pastoralists, or agriculturalists. Instead, they engaged in a range of environmentally conditioned practices that maximized their chances of survival, with foraging constituting a common core. Such nuanced ethnic diversity was even instrumental in the dynasty's decision to move the Warka. During their initial deliberations, Qing officials observed that if the Cossacks moved towards the various Hejen groups on the lower SAH River, none needed evacuation, "since these people do not farm and so are able to move anywhere out of the way."[47] Practical exigency convinced the dynasty to use an essentialized agrarian identity as the determining criterion for indigenous removal. Problems emerged when Warka who were not essentially farmers tried to adapt to regions where agriculture was effectively monocultural.

There were practical pressures to make groups like the Warka "legible" (to use James Scott's term for state oversimplification of human and ecological complexity to facilitate control).[48] These offer us, perhaps, a pre-industrial glimpse of some of the dynamics visible in later agricultural developments examined by Blaine Chiasson in Chapter 4 of this volume. The Qing state, in the process of making the Indigenous peoples of the basin legible, viewed Warka agrarian practices in oversimplified terms. The resulting desertions, the exact extent of which remains unclear, reportedly occurred "one after the other" and became a serious security problem in Ningguta.[49]

Some of the desertions may have been related to the precipitate implementation of mobilization practices that prioritized the exploitation of the Warka's military skills over the allocation of time and resources to ensure the Warka's acclimation to the new conditions in Ningguta. As the Warka were being formed into military units, there were formal complaints that New Manchus were made to do agro-military tasks that "dispatched them back and forth between outposts without rest," depriving them even of "the time to obtain shelters." Two years later, in 1678, New Manchu troops were still asking to be moved from "temporary" residences in nearby villages to houses within the town in order to enable them to "understand its ways" and "acclimate" themselves.[50]

Further problems were emerging in the Qing pelt tribute system as a direct result of relocation. In 1678, two years after the Warka's removal, Qing officials sought to prohibit "new people" in Ningguta from trading pelts with nearby Hejen groups because so "many bad pelts have been discovered among the sable tribute collected this year, that if there is no prohibition, no good pelts will be obtained."[51] In this instance, Qing response to Romanov incursion provided an opportunity for the Warka to form new trade relations that were inimical to their old tribute relations. These new conditions required increased dynastic regulation of inter-tribal trade relations to preserve the integrity of tribute relations, and to maintain the throne's share of the best fur.

Dynastic officials, however, found they could not opportunistically reorder relations between people and pelts by fiat. In fact, sable pelts were so much a part of quotidian existence in the basin that some Qing military officers felt it was impossible to prohibit pelt exchanges (as envisioned by the 1678 prohibition proposal) because sable was used as dowry and to pay debts – in effect as currency for major transactions.[52] People-pelt interdependency was not wholly subject to state manipulation. The state may also have overlooked the simple ecological fact that Ningguta's environs were not prime sable habitat.[53] In this manner, intersections of indigenous culture and biodiversity, not just conflicting human agencies, complicated Manchurian borderland formation.

The Russians appear to have taken advantage of increasing strains on Chinese imperial power in the 1670s. The Qing strategy was to relocate Indigenous peoples to a less peripheral zone, where their "Manchu" identities could be reinforced or, in the case of many Fiyaka, Hejen, and Kiler-Ewenki groups, where they could be deployed against Russia. These processes would transform forager identities into more highly regimented military-administrative identities almost exclusively dependent on cultivation. Even after the Qing had secured the basin from Romanov incursion in 1689, contradictions in addition to those already manifest in hunting and town residence continued to emerge.

## Transformation of the Forager Identity

Indigenous discontent with dynastic relocation policies continued into the early 1680s, illustrated dramatically by the appearance of leaders of more than four hundred Qing tributaries at Nerchinsk, the Romanov administrative equivalent of Ningguta, around 1682. They were petitioning to transfer their pelts to the tsar, explicitly referring to Qing relocation policies as their motive. They explained that the dynasty was "driving them out and planning to take them from their encampments to their Bogdoi empire, together with their wives and children."[54] Romanov construction of indigenous tributary identity was, of course, precisely what the Qing relocation policies had been intended to avoid.

Qing forces were simultaneously massing for a counterattack back into the basin, with their primary thrust aimed at Yaksa, the Romanov empire's main, if modest, outpost on the SAH River. Once reconnaissance and logistical preparations were completed, Qing units moved forward in 1683 to clear out Cossacks from the basin. They began the assault on Yaksa in 1684, and it succeeded in 1685. The Qing, however, then withdrew, which enabled the Cossacks to reoccupy Yaksa's stockade until Qing forces expelled them decisively in 1686–87. By this time, direct negotiations between the two imperial powers had been initiated. These negotiations culminated in the historic 1689 Treaty of Nerchinsk that would cede most of the basin to the Qing empire.[55]

During these operations, Qing forces had several encounters with the Cossacks' indigenous allies, the Kiler-Ewenki, and took several of them prisoner. One encounter revealed a population of forty-seven Cossacks and nineteen Kiler-Ewenki, one adult male, two adult females, twelve boys, and four girls, all living and fishing together on the Tuhuru River, over 250 kilometres west of the mouth of the SAH River in northern Jilin. The officer in charge of the operation confirmed that "this is the first time we have pacified these ninety-nine males of thirty-one Kiler-Ewenki households ... The Kiler-Ewenki live in the mountains and forests like wild beasts and birds, [but] once they learn of Milord's civilization, they will then all submit in droves." Another pacified group contained both Kiler-Ewenki and Orochen. Yet another larger group of

Indigenous people was among the Yaksa garrison prisoners when it first fell in 1685. This group included eleven breastfeeding infants, along with forty-five adult males, sixteen adult females, nineteen girls, and twenty boys. Three boys and thirteen girls among them had been "sired by Cossacks," from "women taken captive" by them.[56]

These women, possibly hunted down like sable or any other valuable boreal animal, were being used to produce new human resources to develop a Romanov Amur, embodied by the Kiler-Ewenki men fighting with the Cossacks against the Qing. The reports make it clear that the regional order envisioned by the Qing was limited in its capacity to accommodate a foraging lifestyle in which people live "like wild beasts and birds" or are "widely dispersed." A foraging existence presented serious problems of ethnic fragmentation, which, Qing officials knew, made the region vulnerable. Indeed, the dynasty had even exploited the ambiguities of Cossack identity by recruiting its Cossack prisoners into Qing military ranks and deploying them against Romanov forces.[57] Even a few hundred semi-subjected people, given the right conditions, merited serious dynastic attention, as indicated in the Kangxi emperor's instructions to Bahai.

Foraging, a way of life adapted to the basin's semi-boreal ecology, was one such condition. Yet considerable dynastic resources were devoted to altering forager identities, before, during, and after the final conflict with the Russians. The Kangxi emperor acknowledged implicitly the incompatibility of forager and imperial identity when he remitted the penalties for the 1685 defaults in Solon-Ewenki and Dagur pelt tribute by recognizing that "the Solon and Dagur act as the great army of Heilongjiang and staff its military postal relay stations. In consideration of their efforts, they will be spared censure and receive their statutory rewards." Recognition of this incompatibility was even more explicit in 1690 when it was decided that one thousand new Dagur recruits for expanded regional garrisons "would be unable to hunt or fish for themselves, so their presentation of sable tribute should be stopped. They will be issued money for rations in accordance with regulations on provincial capital garrisons."[58] In such instances, subsistence foragers were transformed by imperial fiat into sedentary stipended consumers of grain, just individuals in the garrisons of China proper. The tribute of foragers had become far less important to the state than the dynasty's continued access to their persons as members of its standing army.

The obligations of this army were challenging for newly recruited foragers. Some, like the impoverished Solon-Ewenki relocated to Ningguta in 1690, required special "instruction" since they had "never understood working fields or living in houses" but instead "live like wild beasts and birds without resting places." Other impoverished Solon-Ewenki and Dagur, even when they were seemingly fit for duties like manning the five new stations of Heilongjiang's military postal relay system being erected in 1688, still had problems adjusting to a more regimented way of life. These men "did not know how to reckon the time" in apportioning their rations. Some nearly starved until additional grain was sent to these "ignorant new people." Just locating forager recruits proved time-consuming. In the same year, attempts to recruit 332 Solon-Ewenki and 86 Dagūr, all "willing" and "impoverished," had to be delayed for three months until their return from hunting.[59]

Foraging, poverty, and military enlistment seem to have combined to provide the Qing with voluntary, but unusually raw, recruits. The condition of the new recruits was exacerbated by the likelihood that many would be stationed in relatively sedentarized places where "poor people, because they fish and hunt for a living, would not be able to sustain a livelihood."[60] The incompatibilities between forager recruits and the Qing imperial infrastructure suggest that the SAH basin was in a considerable state of transformation in both cultural and ecological terms during the latter half of the seventeenth century.

## Conclusion

From a comparative perspective, pelt tribute takes on a wider significance as integral to the establishment and maintenance of SAH basin borderlands during the seventeenth century in ways similar to those apparent during the North American fur trade in the *pays d'en haute*.[61] Furthermore, imperial attempts to regularize pelt tribute are visible, if contested, mechanisms for the reduction of diversity in the process of spatial formation. These spaces have been variously defined as a "middle ground," a frontier, or a borderland and often as products of "creative misunderstandings" between actors rather than as a product of one actor's deliberate design.[62]

These concepts in general, however, seem united in their essentially ontological, and largely unexamined, emphasis on human agency. Such emphasis tends to underestimate the necessity for human adaptation to prevailing and limiting environmental conditions as a prerequisite for premodern borderland formation. The ongoing interactions between humans and the SAH basin's ecology make fur a necessity for the construction of an imperial borderland. This is not to say that ecological conditions are not subject to human manipulation; rather, human manipulation of the environment is limited by these conditions. Put another way, relations between humans and the environment are recursive.[63] From this perspective, the anthropocentric concept of agency, even as it is tempered by creative misunderstanding, is replaced by the more holistic concept of adaptation that integrates and transforms, both intentionally and by chance, but always within very material limits.

Imperial borderland formation asserts a particular human agency over a wider organic diversity that includes, but is not restricted to, other humans. State pretensions to the contrary, this assertion is contingent and subject to unpredictable change, mainly because it cannot absolutely determine pre-existing cultural and ecological formations. While it attempts to maintain a semblance of uniformity over what is actually a wide diversity, imperial authority is itself transformed over time, often inadvertently. Imperial assertion of uniformity on unincorporated diversity and the consequent adaptations necessitated by this process constitute new historical spaces – such as Qing Manchuria, which emerged fitfully in the closing decades of the seventeenth century.

This Manchurian borderland was a product of a comparatively flexible Qing adaptation to the situation created by a nexus of Romanov incursion, indigenous agency, and ecological conditions. Prior to the Romanov incursion, Qing authority was mainly indirectly expressed in the form of pelt tribute, which was based on pre-Qing relations between Indigenous peoples and sable. Russians endangered this tribute system and Chinese authority by asserting different, more rigid, relations between peoples and pelts. As it adapted to this crisis, the Qing began to alienate select Indigenous peoples from foraging by mobilizing them into New Manchus. Alienation, as part of both New Manchu identity and Qing borderland formation, continued even after Romanov incursions had ceased.[64]

Alienation also had unanticipated results. One of the most profound outcomes was the denaturing of what in the eighteenth and nineteenth centuries the Qing state itself would come to define as the proper Manchu identity – that of the pristine boreal forager-warrior. The consequences of the Qing practice of severing peoples from pelts found belated representative expression in the prominent Qing Manchu official Nayanceng's 1804 observation that Manchu soldiers "daily practice with diligence by hunting animals so that their military strength reaches the utmost level of power and skill." Proximity to more sedentarized and commercialized Han spaces to the south would erode genuine Manchu identity, however, as evident among Manchus resident in China proper. This was why Nayanceng judged the Manchu "troops of Heilongjiang superior to those of Jilin, as are those of Jilin in comparison with those of Fengtian."[65] Nayanceng's formulation challenges current conflicting views over Sinicization by asserting that Manchu identity was not constructed exclusively through human interaction but primarily through its hunting relationships with game.

In sum, and ironically, by the nineteenth century dynastic elites had conceived something like a "natural" axiom: the more northern, the more Manchu. What the seventeenth century saw as "savage" groups of "beasts and birds" now appeared as boreal paragons. Across this space of two hundred years or so, Indigenous peoples' adaptation to empire had become the endangered ideal in their sedentary Manchu overlords' collective, but selective, memory. Throughout these changes, however, the resistance of game, which could fight or flee, ensured that Manchus who remained on the hunt remained the most Manchu.

## Notes

1 "Zhongguo diyi lishi dang'anguan, Beijing, Ningguta fudutong yamen dang'an" [in Manchu] (hereafter NFY), Kangxi 2–1676: 74–79. The Warka, probably, comprise one of the main Donghai Jurchen Indigenous groups resident around the mid-SAH; Juha Janhunen, *Manchuria: An Ethnic History* (Helsinki: Fino-Ugrian Society, 1996), 101. I have generally followed Janhunen for transliteration and identification of (often ambiguous) indigenous ethnonyms, but I have modified this approach in some instances to link modern and Qing ethnonyms.

2 For overviews of Russia's Siberian expansion, see George V. Lantzeff and Richard A. Pierce, *Eastward to Empire: Exploration and Conquest on the Russian Open Frontier to 1750* (Montreal: McGill-Queen's University Press, 1973); and James Forsyth, *A History of the Peoples of Siberia: Russia's North Asian Colony, 1581–1990* (Cambridge, UK: Cambridge University Press, 1992). For primary documents on Spathary's mission, see John F. Baddeley, *Russia, Mongolia, China* (New York: Macmillan and Company, 1960), 2: 237–329.

3 For studies of Sino-Russian relations in this period, see Zhang Xi and Zhang Weihua, *Qing qianqi Zhong-E guanxi* [Sino-Russian relations in the early Qing period] (Jinan: Shandong jiaoyu chubanshe, 1997); and Mark Mancall, *Russia and China: Their Diplomatic Relations to 1728* (Cambridge, MA: Harvard University Press, 1971).

4 Niu Pinghan, *Qingdai Zhengqu Yan'ge Zongbiao* [Summary tables of changes in Qing administrative regions] (Beijing: Zhongguo ditu chubanshe, 1990), 79, 95, 110.

5 Peter C. Perdue, *China Marches West: The Qing Conquest of Central Eurasia* (Cambridge, MA: Belknap Press, 2005), 8. Perdue has persuasively argued for parallels between the Eurasian imperial consolidations by Russians and Manchus as "forest peoples": see *China Marches West*, 44–50.

6 Important borderland administrative studies include Robert H.G. Lee, *The Manchurian Frontier in Ch'ing History* (Cambridge, MA: Harvard University Press, 1970); Zhou Xifeng, *Qingchao qianqi Heilongjiang minzu yanjiu* [A study of Heilongjiang peoples in the early Qing] (Beijing: Zhongguo shehui kexue chubanshe, 2007); Matsuura Shigeru, *Shinchō no Amūru seisaku to shōsū minzoku* [Qing policy towards the Amur district and minorities] (Kyōto: Kyōto daigaku gakujutsu shuppankai, 2006).

7 *Qingshilu* [Veritable records of the Qing Dynasty], 60 vols. (Beijing: Zhongguo bianjiang shidi yanjiu, 1986–87) (hereafter *QSL*), TM 2:330b-31a, 360b-361a. All Qing reign titles are abbreviated as "Tianming" here.

8 John E. Wills Jr., *Embassies and Illusions: Dutch and Portuguese Envoys to K'ang-hsi, 1666–1687* (Cambridge, MA: Harvard University Press, 1984); James L. Hevia, *Cherishing Men from Afar: Qing Guest Ritual and the Macartney Embassy of 1793* (Durham, NC: Duke University Press, 1995); John Haldon, *The State and the Tributary Mode of Production* (London: Verso, 1993); and Fu Baichen, *Zhonggguo lidai chaogong zhidu yanjiu* [Trans-dynastic study of China's tribute system] (Changchun, 2008). For an environmental approach, see David A. Bello, "The Cultured Nature of Imperial Foraging in Manchuria," *Late Imperial China* 31, no. 2 (D2010): 1–33.

9 "Introduction," Amur-Heilong River Basin, http://www.amur-heilong.net/http/intro_chap.html.

10 "Amur Landscapes and Ecoregions," Amur-Heilong River Basin, http://www.amur-heilong.net/http/02ecosys_chap.html; "Species Diversity and Harvest of Biological Resources," Amur-Heilong River Basin, http://www.amur-heilong.net/http/03_species_chap.html.

11 Xiu Yang and Ming Xu, "Biodiversity Conservation in Changbai Mountain Biosphere Reserve, Northeastern China: Status, Problem, and Strategy," *Biodiversity and Conservation* 12, no. 5 (2003): 888.

12 Steven W. Buskirk, Ma Yiqing, Xu Li, and Jiang Zhaowen, "Winter Habitat Ecology of Sables (*Martes zibellina*) in Relation to Forest Management in China," *Ecological Applications* 6, no. 1 (1996): 323–24.

13 Mao Ruizheng, *Dongyi kaolüe* [Study of the eastern barbarians] (c. 1621; repr. Beijing, 1984), 43–44; *Qingchu Neiguo Shiyuan Manwen dang'an yi bian* [Translated compilation of Manchu archives of the Early Qing Inner Historical Office], ed. Zhongguo diyi lishi dang'anguan, 3 vols. (Beijing: Guangming ribao chubanshe, 1989), 1: 338–39; Yang Bin, *Liubian jilüe* [An outline of the Willow Palisade] (c. 1690; repr. Taibei, n.d.), 1.11b-12a; Cao Tingjian, *Cao Tingjian ji* [Cao Tingjian's collected works], ed. Cong Peiyuan and Zhao Mingqi, 2 vols. (Beijing, 1985), 1: 33–34, 178–79. For ethnographies, see Janhunen, *An Ethnic History*, 50–52 (Dagur), 61, 125 (Kiler-Ewenki), 68–70 (Orochen), 101–2 (Solon-Ewenki), 115 (Hejen), and 126 (Fiyaka).

14 First Historical Archives of China, Beijing, Heilongjiang jiangjun yamen dang'an [Manchu Archives of the Heilongjiang Military Governor's Office] (hereafter HJY), KX I–1684: 190–91, KX I–1688: 482–86.

15 Yang, *Liubian jilüe*, 3:8a.

16 Liaoning Provincial Archives, Shenyang, Hei-tu dang [Manchu Archive of Lateral Administrative Communications], KX 13:18.18; Ka-li-na, *Xunlu Ewenkeren wenhua yanjiu* [A study of the culture of the Reindeer Evenki] (Shenyang: Liaoning renmin chubanshe, 2006), 80.

17 *QSL* CD 2:804b-06a, 2:828a-30a. The terms "Reindeer Herder" and "Dog Keeper" conflate a number of ethnic groups; see Janhunen, *An Ethnic History*, 59–74, 126.

18 Jiang Fan, *Manzu shengtai yu minsu wenhua* [Manchus, ecology and folk culture] (Beijing: Zhongguo shehui kexue chubanshe, 2006), 10–11.

19 *QSL*, TM 2:30a, TC 2:54a, 55b, 56a-b, 60a-b, 64b, 72b, 99a, 104b, 123b, 124a, 125b, 174b, 198b-99a, 215b, 215b, 221a, 239b-40a, 270b, 278b, 280a, 287a, 301a, 307b, 308b, 358b, 360b, CD 438a, 438b, 448b, 506a, 511a, 518a, 519a, 530a, 542b, 549b, 580b, 585a, 587a, 591a-b, 592b, 593b, 643b-44a, 658b-59a, 706a, 714a, 719a-b, 728b, 793a, 793a-b, 793b, 797a, 798a, 807b, 835b, 878a-b, 857a-b, 882b-83a, 3:38b.

20 *QSL*, SZ 3:46a, 67, 225a, 229b, 313b, 362a, 394a, 519b, 571b, 579a, 608ab, 625b, 669b, 730a, 752b, 813a, 815a, 838b, 862b-63a, 865b, 894a, 944b, 959a, 972a, 983b, 1026a, 1036a, 1088b; SZ 4:70a, KX 4:142a, 171a, 195b, 237b, 249b, 263a, 380b, 572b, 581a-82b.

21 NFY, KX 1–1676: 205–10, KX 5–1678: 163–207; First Historical Archives of China, Beijing, Huke Shishu, 212:KX 20.8.24; Wu Zhenchen, *Ningguta jilüe* [Notes on Ningguta] (Shanghai, 1995), 731:608a. See, also, Lee, *The Manchurian Frontier,* chapters 3 and 4.

22 Important works on New Manchus include Yang Xulian, "Jianlun Qingdai Kangxi shiqi de xin Manzhou yu Bu-te-ha baqi" [A brief discussion of New Manchus and the Eight Hunting Banners in the Qing Kangxi period], *Shehui kexue zhanxian* 4 (1980): 192–96; Liu Jingxian, Guo Chengkang, and Liu Jianxin, "Qing Taizu shiqi de xin Manzhou wenti" [The New Manchu problem in the Qing Taizu period], *Lishi dang'an* 4 (1981): 102–7, 116; Zhang Jie, "Qing chu zhaofu xin Manzhou shulüe" [A brief overview of New Manchu recruitment in the early Qing], *Qing shi yanjiu* 1 (1994): 23–30; Zhang Jie and Zhang Danhui, *Qingdai Dongbei bianjiang de Manzu* [The Manchus of the Qing northeastern frontier] (Shenyang, 2005), 59–93.

23 *QSL,* KX 4:494b, 495a.

24 NFY, KX I–1676: 205–10.

25 NFY, KX I–1676: 143–48, 205–10.

26 Liu Minsheng, Meng Xianzhang, and Bu Ping, eds., *Shiqi shiji Shae qinlüe Heilongjiang liuyu shi ziliao* [Historical materials on the seventeenth-century tsarist Russian invasion of the Amur river basin] (Harbin: Heilongjiang jiaoyu chubanshe, 1992), 111–12; Zhongguo diyi lishi dang-anguan, ed., *Qingdai Zhong-E guanxi dang'an shiliao xuanbian* [Selections from the Qing Sino-Russian relations archive], (Beijing: Zhonghua shuju chubanshe, 1981), 1:9.11, 1: 13.14–15; *cf.* Basil Dmytryshyn, E.A.P. Crownhart-Vaughan, and Thomas Vaughan, eds., *Russia's Conquest of Siberia: A Documentary Record, 1558–1700* (Portland: Oregon Historical Society, 1985), doc. #84, 304, which lists the names of the five Cossacks. The villagers were probably known to the Manchus as ethnically akin to Hūrha people. For discussions of the identity of Ducher and Hūrha, see Janhunen, *An Ethnic History,* 101–5; and Zhang and Zhang, *Qingdai Dongbei bianjiangde Manzu,* 11–12.

27 R.H. Fisher, *The Russian Fur Trade: 1550–1700* (Berkeley: University of California Press, 1943), 34; Crownhart-Vaughan and Vaughan, *Russia's Conquest of Siberia,* doc. #75, 239–40.

28 Fisher, *Russian Fur Trade,* 29, 120.

29 Yang Hu, *Mingdai Liaodong dusi* [The Ming regional military commission in Liaodong] (Zhengzhou: Zhengzhou guji chubanshe, 1988), 131–32; Yan Congjian, *Shu yu zhou zi lu* [Record of a comprehensive inquiry into strange lands] (Beijing: Gugong bowuyuan tushuguan, 1984), 109.

30 Crownhart-Vaughan and Vaughan, *Russia's Conquest of Siberia,* doc. #67, 209, 213–15; Lantzeff and Pierce, *Eastward to Empire,* 156–58.

31 *Shae qinlüe Heilongjiang,* 178; Crownhart-Vaughan and Vaughan, *Russia's Conquest of Siberia,* doc. #75, 238–39, and doc. #81, 277–78, #86, 315; Forsyth, *A History of the Peoples of Siberia,* 41.

32 Crownhart-Vaughan and Vaughan, *Russia's Conquest of Siberia,* doc. #81, 264–67.

33 Ibid., doc. #81, 267, 269.

34 An Shuangcheng, "Qingchu zai Lafa dukou zhizao zhanchuan gaishu" [An account of the construction of war boats at Lafa Ferry in the early Qing], *Lishi dang'an* 2 (1995): 82–83; Crownhart-Vaughan and Vaughan, *Russia's Conquest of Siberia,* doc. #81, 273, doc. #92, 344–45; G.V. Melikhov, "How the Feudal Rulers of the Ch'ing Empire Prepared Their Aggression against the Russian Settlements on the Amur in the 1680s," in *Chapters from the History of Russo-Chinese Relations, 17th–19th Centuries* (Moscow: Progress Publishers, 1982), 62. The best known "defection" to the Russians is the 1667 act of the indigenous leader Gantimur; see Lantzeff and Pierce, *Eastward to Empire,* 171–72; and Zhang and Zhang, *Qing qianqi Zhong-E guanxi,* 45, 54–55.

35 *QSL*, 3:1068a-b.

36 *Qingdai Zhong-E guanxi dang'an shiliao xuanbian,* 1:1.2, 2.3, 7:10.

37 Ibid., 1:21.21.

38 Huke Shishu, 212: KX 20.8.24. Only 15 percent (74) and 75 percent (755) of regulation quotas for 500 top- and 1,000 second-quality pelts were turned in, all of which were deemed qualitatively unsatisfactory.

39 Crownhart-Vaughan and Vaughan, *Russia's Conquest of Siberia,* doc. #81, 271.

40 Ibid., doc. #106, 386. The Russians' regularized system of tribute collection from some Dagur groups, in operation for several years by 1654, had been stopped by local resistance and Qing forces no later than 1655; ibid., doc. #84, 306, 311.

41 Ibid., doc. #84, 305; Yang, *Qingdai Dongbei shi,* 86–87; *Shae qinlüe,* 178, 180.; *Qingdai Zhong-E guanxi dang'an shiliao xuanbian,* 1:10.12, 14.16–17; He Qiutao, *Shuofang beisheng* [History of the defence of the north] (Chengdu: Sichuan minzu chubanshe, 2002), 16:28a.

42 NFY, KX 2–1676: 74–79, 122–23, 197–99.

43 NFY, KX 2–1676: 91–93, 98–102, 225–29.

44 NFY, KX 2–1676: 39–54.

45 For general discussions of the issues, see Daniel Stiles, "The Hunter-Gatherer 'Revisionist' Debate," *Anthropology Today* 8, no. 2 (1992): 13–17; Robert L. Kelly, *The Foraging Spectrum: Diversity in Hunter-Gatherer Lifeways* (Washington, DC: Eliot Werner Publications, 1995), 14–35; Alan Barnard, "Hunter-Gatherers in History, Archaeology and Anthropology: Introductory Essay," in *Hunter-Gatherers in History, Archaeology and Anthropology,* ed. Alan Barnard (Oxford: Berg, 2004), 1–8.

46 NFY, KX 2–1676: 276–77.

47 NFY, KX 1–1676: 143–48.

48 James C. Scott, *Seeing Like a State: How Certain Schemes to Improve the Human Condition Have Failed* (New Haven, CT: Yale University Press, 1998), 2–3.

49 NFY, KX 2–1676: 188.

50 NFY, KX 2–1676: 197–99; KX 4–1678: 174–76.

51 NFY, KX 4–1678: 251–53.

52 Ibid.

53 An account of Ningguta in the 1680s noted that "ginseng and sable pelts were originally produced from the deepest and furthest regions" of a nearby mountain but had become completely exhausted. See Wu, *Ningguta jilüe,* 731: 607b.

54 Crownhart-Vaughan and Vaughan, *Russia's Conquest of Siberia,* doc. #117, 450.

55 Lantzeff and Pierce, *Eastward to Empire,* 173–80; Sun and Zhang, *Qing qianqi Zhong-E guanxi,* 60–115. For a study of the Nerchinsk negotiations, see Joseph Sebes, S.J., *The Jesuits and the Sino-Russian Treaty of Nerchinsk* (Rome: Institutum Historicum, 1961).

56 HYJ, KX 1–1684: 41–53; KX 1–1685: 133–37, 140; KX 2–1686: 69–71.
57 Huke Shishu, 241:KX 24.10.23; HYJ, KX 1–1684: 41–53; Ortai et al., ed. *Baqi tongzhi,* (Changchun: Dongbei shifan daxue, 1985), 1: 38. For Cossack identity, see Christoph Witzenrath, *Cossacks and the Russian Empire: 1598–1725* (New York: Routledge, 2007), 34–36.
58 *QSL,* KX 5:294a; Zhongguo diyi lishi dang'anguan, ed., *Xibozu dang'an shiliao* [Archival historical materials on the Sibe ethnicity] (Shenyang: Liaoning minzu chubanshe, 1989), 1:28.
59 HJY, KX 1–1690: 19–20; KX 2–1688: 306, 397–98; KX 1–1688; 180–83.
60 *Xibozu dang'an shiliao,* 1: 36.
61 Richard White, *The Middle Ground: Indians, Empires, and Republics in the Great Lakes Region, 1650–1815* (Cambridge, UK: Cambridge University Press, 1991), chapter 3.
62 See, for example, Richard White, "Creative Misunderstandings and New Understandings," *William and Mary Quarterly* 63, no. 1 (2006): 9.
63 My sense of the term "recursive" is related to the meaning intended by Anthony Giddens, *The Constitution of Society* (Berkeley: University of California Press, 1984), 25–26.
64 See, for example, HYJ, KX 11–1691: 123–24.
65 *QSL,* JQ 29:700b-01b.

# 3
# Inclement Weather and Human Error: Regular Irregularities in the Manchurian Tribute System during the Qing Dynasty

*Loretta E. Kim*

The German-English geographer Ernst G. Ravenstein (1834–1913) described "Chinese mandarins" calling on nomadic populations living near the upper reaches of the Amur River in his mid-nineteenth-century study on the Russian presence in and around the shared eastern borderland between the Qing (1644–1911) and Romanov (1613–1917) empires. These officials "took up their residence in some native village, and having collected the tribute and disposed of their merchandize [*sic*] to the best advantage, returned to their ordinary stations."[1] He estimated that the officials demanded 4,497 sable pelts valued at 6,746 English pounds sterling of the day as tribute and 557 pounds in taxes for every 12,000 Amurian Indigenes. If they complied with these coercive requests, the tributaries received tobacco and silk from their visitors, which, according to Ravenstein's perception, they prized as valuable goods.

This narrative of the tribute process merits attention for several reasons. Ravenstein judged the amounts of pelts and cash that the mandarins collected as "extortionate." His assessment is most likely based on the fact that Amurian Indigenous economies were subsistence-oriented. The primary occupations of hunting and fishing for most of the population provided enough food and material for clothing and housing structures. If they expended additional effort to hunt more than they would need for their own consumption, they could trade with outsiders to acquire cash, which they would use to buy processed commodities or the raw products of other regions. In the early nineteenth century, during which Ravenstein made his observations, populations of fur-bearing animals, as well as those of other flora and fauna that the Qing state requisitioned as tribute, were shrinking noticeably, as noted in many an official missive to the imperial throne. Therefore, it was more likely that the tributaries had less surplus, if any at all, than in previous times, and may have even compromised their own subsistence in order to supply tribute.

Although Ravenstein deemed the practice to be corrupt and exploitative, the actual collection of tribute and tax went smoothly in this instance. The tributaries satisfied the officials' demands, apparently without even a hint of resistance. The officials must have felt secure enough to stay in the native villages, and exercised enough clout not only to collect what was due but also to profit from their dealings with the villagers, which most likely included private trade in goods that they desired for personal consumption.

That all went according to plan, at least from the Qing government's vantage point, is thought-provoking because administrators in Manchuria frequently relayed many of the complaints and excuses that leaders of tributary groups put forward to justify why their communities could not meet a yearly quota for tribute.[2] Contrary to the depiction of mandarins and natives ostensibly achieving mutual benefit peaceably and efficiently in Ravenstein's account, officials responsible for collecting tribute faced obstacles in carrying out their duties and expressed such difficulties in several cases.

The fact that quotas for natural products were imposed year after year ensured that the Qing tribute system in Manchuria would at some point encounter problems.[3] The bureaucratic framework for gathering, assessing, and exporting tribute items out of the region developed over several decades; the fits and starts in the institution's earliest years are predictable. However, even after tribute collection procedures were running more or less like clockwork, signs of stress gradually became evident. Officials cited similar reasons to explain and justify why problems occurred in the process of collecting tribute throughout what might be considered the peak of the system's maturity, during the eighteenth century and in the last few decades of the dynasty. The common-sense argument – that after a long period of continuous extraction neither Mother Nature nor the harvesters could be expected to produce the same numbers of tribute items that had been available in natural abundance before the tribute system was implemented and during its early years – is inadequate as a principal or sole explanation for the institution's decline and eventual termination in 1908, just three years before the Qing Dynasty itself ended.[4] The tribute system endured for over two centuries and satisfied the central government's requirements for quantity and quality in the approximate majority of collection cycles.

This chapter investigates a diverse range of problems that affected the core mission of Manchuria's tribute system, which was to extract natural resources by the optimal employment of native populations. By examining the irregularities cited and the Qing state's responses to them in three types of causative scenarios – human error, natural causes, and environmental abnormalities resulting from human actions – this study considers environmental conditions and system participants' performance, as well as the logic used by officials to mitigate potentially punitive consequences for themselves and for tribute collectors.

Geographical variations and time-specific changes in the Qing state's administration of Manchuria, as well as the distinct traits of individuals featured in the evidence, distinguished each unique situation. The presentation and assessment of examples from different areas of Manchuria during various time periods is the deliberate approach of this study. By employing this strategy, I aim to reveal more variables in the tribute process than what has been brought to light from a microanalysis of just one or two specific cases. This approach will facilitate a comparison of the definition and treatment of aberrations in several contexts. The primary comparative details are the reasons given for not satisfying a quota and the official responses to such justifications. These data are found in Qing government documents that circulated at the time among different administrative bodies at the local, regional, and central levels.[5]

The first section of the study describes the general parameters of tribute in Manchuria, which are comparable yet also distinct from those implemented in other parts of the Qing empire. Then, the three causes introduced above are discussed separately. The third and final part puts forward conclusions about how officials presented problems stemming from different causes variously, and considers other possible reasons for the tribute system not always generating the required yields.

## Tribute in Manchuria

The prevailing interpretation of tribute (*chaogong*, *tugong*) has been as a significant tool of political and economic domination during China's imperial age.[6] This perception was shared by those who created and maintained the system, and is widespread among those who have analyzed it. Whether the tributaries in question were representatives of sovereign governments or of populations living within the Chinese empire, their presentation of goods to the Chinese imperial court (usually followed by the bestowal of "gifts in return," or *huishang*) has been commonly understood in the academic and popular imagination as a procedure invested with prodigious symbolic importance and as a cornerstone of pre-modern international relations among East Asian states.[7]

Tribute in Manchuria was primarily a form of domestic tribute, which I define as systems and procedures of tribute that occurred within the geopolitical boundaries of the Chinese empire. The tribute system in Manchuria was less about power projection and cultural prestige than the more well-known forms of tribute, which involved foreigners visiting the Chinese imperial centre. Sometimes equated with taxation, domestic tribute can also be understood more broadly as the imperial state's employment of the tributary populations to harvest natural resources.[8] Since the Chinese emperor was their sovereign, domestic tributaries had no choice in the matter of submitting tribute. Codifying appropriate expressions of loyalty and deference to the Qing emperor, a priority in foreign or inter-state tribute, was far less important than carrying out the timely collection of select goods that met requirements for quality and quantity. Through this fundamentally economic expression, imperial Chinese governments reinforced their authority over subject populations in the borderlands and gained valuable commodities indigenous to those particular regions.

As in other regions of the empire, two major factors affected tribute system operations in Manchuria: timing and tributaries. The Kangxi court (r. 1661–1722) instituted the basic mechanisms of schedules, quotas, and strategic employment of Indigenous tribes to extract various resources and deliver them to regional collection stations, which would then forward them to the imperial capital of Beijing. Most quotas were calculated based on how many able-bodied adult males (*zhuangding*) belonged to a given tributary population, and had to be fulfilled on a semi-annual or annual basis, with some quotas set for multi-year cycles.

Manchurian tributaries can be classified into two broad administrative categories.[9] Some tributaries maintained loose associations to the Qing government, such as the Hejen and Fiyaka peoples, who lived in the extreme north and northeastern corners of Manchuria, including Sakhalin Island (*Kuye dao*). The tributaries retained considerable political and social autonomy as long as they met tribute quotas (usually twice each year in the spring and autumn) and paid ritual visits to Beijing to receive gifts, including wives for their leaders, from the emperor.[10] These groups gathered the requisite tribute from the geographic areas where they would normally find goods for their own consumption.

The second type of tributaries included official and full-time servitors of the Qing state. These individuals worked in designated sites to harvest various resources, including sable, honey, pearls, and fish. These sites included the twenty-two mountain foraging reserves (*caigong shanchang*) and sixty-four river mouths for gathering pearls (*caizhu hekou*).[11] In addition to such goods, which were subject to annual quotas, the court also requested that tributaries periodically submit special items – such as sparrow hawk in the seventh lunar month, and, on the occasion of the eightieth birthday of the Qianlong emperor (r. 1735–96), a live, well-fed crane.[12] These tributaries were members of the Butha (hunting) banners (*Buteha qi*), a distinct division of the Qing military apparatus, and earned their livelihoods by performing these economic duties as well as their martial ones.[13] The most important duty of the Butha banners was to hunt a variety of fur-bearing animals, especially sable. The Kangxi-era statutes (*huidian*) cited that Butha companies had to meet specific quotas for sable pelts starting in approximately 1653, the middle of the preceding Shunzhi reign (r. 1644–61), but they were encouraged to surpass these minimum requirements if they could. For example, according to a 1662 regulation, a Butha commandant would receive a cash reward for submitting more than the quota, but would be commensurately fined if his group turned in less. The same directive also stated that a Butha soldier could earn pieces of blue cotton cloth for presenting more than his individual quota but would be subject to three lashes of the whip if he failed to meet it.[14]

To control the variables of timing and personnel, the Qing state developed procedures for collection and appraisal of tribute goods. Besides stipulating rewards for quotas met or exceeded and imposing penalties for quotas not met, the Qing centre also attempted to maintain continual productivity through close supervision of its tributaries, as can be seen in the organization of work teams within the Butha banners. Thirteen or fourteen able-bodied males comprised one unit called a *juhiyan* in Manchu (Ch., *zhuxuan*).[15] One *juhiyan i da* (*juhiyan* chief; Ch., *zhuxuan da*) and one assistant *juhiyan i da* (deputy chief of the *juhiyan;* Ch., *fu zhuxuan da*) led each *juhiyan*. Sixty-five *juhiyan* belonged to the "upper three" banners, fifty-nine of which were responsible for gathering pearls, honey, and pine nuts; six *juhiyan* caught fish.[16] Among the forty-five *juhiyan* of the remaining "lower five" banners, thirty-five culled pearls and ten *juhiyan* fished.[17] Each *juhiyan* had to yield at least sixteen pearls per collection cycle, which meant that the fifty-nine *juhiyan* of the upper three banners had to collect 944 pearls. Some substitutions were permitted, such as in the winter and summer seasons, when every banner had to delegate nineteen troops to fish for sturgeon. These forces were exempt from the fur tribute for the period in which they were primarily responsible for fishing.

Routine procedures for harvesting certain items, such as sable pelts, were also implemented and overseen by regional and local officials.[18] Hunting for sable was normally done in the tenth lunar month of each year. A lieutenant, accompanied by a corporal and several adept archers, would go out to hunt sable for three months in a designated location, such as the area from the Nen River (*Nen jiang*) towards the Greater Xing'an mountain range (*Da Xing'anling*). The hunters faced the initial challenge of finding the sable. The location of the animals would depend on the weather; they often preferred living in cold forests and would climb trees, swim, and burrow to avoid extreme conditions, sometimes for five or six days at a time. Once they located the sable, hunters employed two major strategies to capture them: using dogs, or using nets as traps. The traps would be placed near the sable burrows, which the hunters could recognize by the animals' tracks, and then the sable would be smoked out. Horses were also essential to the hunting process because they enabled the hunting parties to go farther in order to collect pelts. Two or three people could share one horse, taking turns riding it and using it as a pack animal.

Submission and evaluation of tribute produce occurred at designated times and places. For the Butha banners and semi-autonomous tributaries living within reasonable travelling distance, the main sites of collection were league markets (Ch., *chulehan;* Ma., *culgan*) such as the one held annually in the northern corner of Qiqihar (Ch., *Qiqiha'er;* Ma., *Cicigar*) by the Heilongjiang regional administration. The Heilongjiang general (*jiangjun*) and the Qiqihar garrison lieutenant-general (*fudutong*) presided over the market from their elevated positions on a platform, and regiment colonels and other leaders of the Butha tribes would sit to the east and west of the platform.[19] Company captains and regional officials assessed the quality and measured the quantity of tribute products. When the generals appraised some of these items personally, their evaluations of the tribute items' quality would be sent up to the organs of the central government involved in borderland administration and compared with those prepared by their subordinate officials.[20] For more remote populations, the Qing administration made different accommodations for tributary groups according to their geographic location. For instance, Hejen scattered around northern Heilongjiang had to travel to

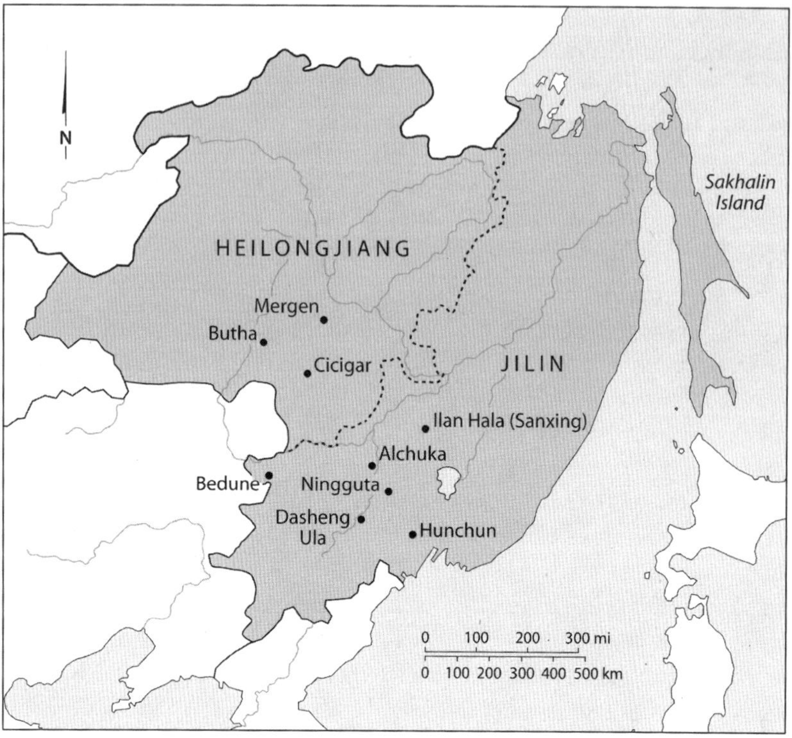


FIGURE 3.1 "Tribute Centers": Major tribute collection and processing stations in Jilin and Heilongjiang.

*Source:* Author's map, designed by William L. Nelson

the Ningguta garrison once each year, whereas Hejen and other populations on Sakhalin could wait for officials to visit areas closer to them for the exchange of tribute items and return gifts.[21]

The grading of tribute items was, like the collection process, based on two variables: quantity and quality. Evaluating quantity was fairly straightforward, whether the quota was set based on the number of individuals in a tributary community or as a production target for a work team. Judging quality was more complex, the simplest metric being whether an item was acceptable or unacceptable. Criteria for judging the quality of specific items varied, but items were generally categorized as high (first-class), medium (second-class), and low (third-class) quality. For example, the quality of sable pelts mattered greatly to both officials and tributaries. Even a collection that met the quota for quantity might fail to meet the prescribed quotas for first-, second-, and third-class-quality items if a significant number of the total was deemed sub-par.[22] If a particular individual's contribution was judged as unacceptable, he had to buy another person's (spare) pelt for a price of about three silver taels. Orochen within the Butha banners were considered the best hunters, delivering mainly first- and second-class pelts; their assistance was often sought by other troops who needed substitute pelts. All tributaries would mark their own pelts of passable quality with silk "seals" or signature symbols: red seals for Ewenki and Dagur, green seals for Equestrian Orochen (Ch., *moling'a Elunchun*; Ma., *moringga Oroncon*), and yellow seals for Foot Orochen (Ch., *yafahan Elunchun*; Ma., *yafahan Oroncon*).[23] Once pelts had been sorted, officers then distributed compensation for the tributaries, both Butha bannermen and non-banner persons. The Heilongjiang general arranged for the pelts of acceptable quality to be transferred to Rehe and other points to the south, and then to Beijing, where most of them would be distributed between the fur storehouse (*piku*) and the imperial wardrobe (*yiku*) of the Imperial Household Department (*Neiwufu*).[24] Pelts that did not pass quality assessment as proper tribute items could be sold in the Heilongjiang region's private markets. Sometimes each pelt that was considered sub-par in quality was marked by the removal of a claw.

When the process ran smoothly, as in Ravenstein's account, tributaries satisfied all minimum requirements for quantity and quality.[25] When one or both of these standards could not be met, officials explained what the underlying problems were in their communications to the regional administrative hierarchy and ultimately to the emperor. The problems fall into three major categories: errors committed by people harvesting or processing tribute goods; diminished animal populations due to irregular weather and other natural environmental conditions; and ecological changes caused by human action. The following three sections are arranged to reflect their prevalence in the early to mid-Qing, mid- to late Qing, and late Qing subperiods, respectively.

## Problems with Personnel

When the failure to meet tribute quotas was deemed to stem from problems with tribute system personnel, including the groups working to gather the items and the supervising officials, three broad causes were cited: incompetence or a "lack of effort," illegal or semi-legal sales of goods to official and private buyers, and the generally unspoken problem of official corruption. Ineptitude could be specifically detailed, or it could be cited as a generic explanation in reports that did not state particular causes for either a quantitative or qualitative deficiency, especially during the Kangxi, Yongzheng (r. 1722–35), and Qianlong reigns. In 1669, a Mukden (*Shengjing*) company captain (*zuoling*), Šindali, wrote that records of fresh ginseng and dried ginseng did not match because some of the ginseng had been stolen sometime between harvesting and processing, and also because the ginseng-pickers did not account for the attrition in size and weight of ginseng as it dried out. The throne responded that the Imperial Household Department should act accordingly to ensure that this kind of problem did not occur again.[26]

When the causes of particular shortcomings were either unknown or not identified, the Qing centre recommended the penalty of reducing the number of return gifts, usually by half.[27] Especially until the Jiaqing reign (r. 1796–1820), groups that submitted either the right amount of product, but at sub-par quality, or not enough product, but of adequate quality, were often penalized with reduced gift portions.[28] In one such example, the Qianlong-period official Yu Minzhong (1714–79) described a group of Solon troops as "not working hard enough" to catch enough sable for the pelt tribute. Ultimately, the errant Solon received half-compensation; even though the quality of the submitted pelts was poor, there were enough pelts to satisfy the minimum quantitative requirement.[29] Some courts occasionally exhibited greater stringency, disallowing the disbursement of any return gifts.[30] In 1814, judging another group of Solon troops that failed to submit pelts of sufficient quality, the Jiaqing emperor remarked that they were undeserving of any reward and that remunerating them with a reduced number of gifts would foster their bad habits and, in essence, deter them from becoming better hunters.[31]

In other cases, officials determined that misdemeanours were deliberate, and that "lack of effort" signified ulterior motives or problems beyond the transgressors' control. An investigation into the questionable bookkeeping practices of a Solon delegation going to Beijing in 1891, triggered by the "mis-writing" of Chinese numbers in the Manchu version of the accounts, revealed that the group was selling pelts on the private market for higher prices instead of turning them in as tribute. The final verdict was that the transgressors should receive only half of the customary return gifts and be admonished severely.[32] The penalty for this situation reflected the element of choice in the Solon tributaries' transgression. By comparison, in 1851, in response to discovering that Hejen and Fiyaka tributaries were dying pelts unevenly or purposely discolouring pelts with smoke and oil in order to satisfy the three-tier quota for first-, second-, and third-class goods, the Sanxing (Ma., *Ilan hala*) garrison lieutenant *yamen* (government office) sent out notices warning tributaries that doctored pelts would not be accepted in exchange for return gifts.[33] The reprimand also advised tributaries that they should find animals with thick and even-coloured fur because the pelts would be used for the emperor's wardrobe and should thus be carefully selected. By emphasizing the disadvantage of future perpetration instead of punishing individuals who had already submitted modified items, the announcement demonstrated some leniency, perhaps due to the state's understanding that animals with healthy pelts of the desired colour were becoming harder to find.

The Qing court and officials also acknowledged that some problems were not truly ascribable to a "lack of effort," even if lack of effort was initially cited as the cause of failure to meet a quota. Logistical factors impeded some groups, as seen in the predicament of a Dagur community that wanted to continue submitting tribute but could not afford to collect enough pelts because they did not have money to purchase horses and thereby cover the range of territory necessary to find enough fur-bearing animals.[34]

Awareness of such difficulties may have at least partially inspired the central government's efforts to reform the tribute system. In 1824, the Daoguang emperor (r. 1820–50) decreed that the Butha banners were no longer required to send representatives to deliver tribute goods to Beijing during the years when the court would not be in the capital during the tribute season. The imperial edict noted that the mission employed over seventy soldiers, who had to use their own provisions and spent six months to complete the round-trip itinerary. Such journeys disrupted the participants' agricultural work and hunting schedules, which would impact their production adversely and endanger their subsistence. The solution was that, as in the years that the tribute was due at Rehe, closer to Heilongjiang, the Butha banners could send three officials and ten troops to transfer the goods through postal stations.[35]

Other measures similarly granted relief in response to tributaries' recurrent failure to fill quotas, when it seemed this failure was unintentional but unavoidable. To mitigate the burden of satisfying a mutable quantitative requirement, in 1832 the court set an important precedent by enacting a set quota for furs rather than adjusting it on an annual basis according to the number of individuals obligated to give tribute. A few decades later, in 1878, the Guangxu court (r. 1875–1908) justified trimming down quotas as a response to long-term political instability and the poverty of the contributing communities, which was due to the declining amounts of food they could hunt and grow.[36] However, sometimes even exerting an acceptable amount of effort was not enough to merit leniency. In 1836, the Heilongjiang general Hafungga (?–1840) appealed to the central government to expend funds or reduce the quotas imposed on Butha troops who were not employed in an official capacity (Ma., *sula*) but who were nevertheless required to contribute pelt tribute. He cited as the basis for his request these troops' limited livelihood. The official response he received was that exceptional actions, especially those providing material aid, should not be undertaken without grave consideration.[37]

Another source of human error was the allure of the market. Whether they did so for survival or to satisfy a desire for lucrative returns, many tributaries engaged in the semi-official, or even illegal, trade of items that would otherwise be relinquished to the state as tribute. The gradual expansion of tribute collection stations into markets where a wide variety of commodities such as gunpowder, tea, and ground grains were available was both a help and a hindrance to tributaries.[38] Some tributaries shopped at these markets for items that they would then hand in to the authorities to meet minimum quotas. Others could sell surplus produce to officials, other tributaries, private merchants, and foreigners in order to earn income, in cash or in kind. Markets also distracted tributaries from honouring their obligations to the state. Whole delegations and individual tributaries contravened regulations about collection and item quality in order to earn more handsome profits than they anticipated would be given to them in return gifts. This perennial problem, exemplified above in the case of Solon troops selling in Beijing private markets, provoked different official reactions.[39] In contrast to the Guangxu court's one-off penalties of reducing the number of return gifts and censuring the offenders, the Jiaqing emperor's 1818 response to a similar situation of Butha banner troops who had reserved their best wares for Beijing private buyers was both corrective and preventative.[40] Until that time, tributaries were only punished for not meeting the numerical quota; they could receive some of the customary remuneration for substandard goods, as long as they submitted something. Based on his advisors' reasoning that this practice was not sufficiently rigid to enforce quality control, the Jiaqing emperor decided that subsequent submissions that failed to meet the quality quota would not be compensated at all.

Such a decisive judgment may have reflected the Qing centre's confidence that it could maintain its authority over the tributaries at the time; from the Daoguang reign onward, however, many submissions fulfilled the quantitative requirement but declined in relative and absolute quality.

From the tributaries' perspective, the government did not provide enough incentive to deliver premium goods. Since the return gifts for the quality quota were based on the entire group's collection and not individual contributions, there was no guarantee for any individual that other group members would all deliver enough items of the right quality to merit the gifts. Therefore, it is logical that many individuals deemed it would be more lucrative to sell things privately and ensure a good income, and supply the government with just the minimum numbers necessary to avoid any form of punishment. In the last two decades of the dynasty, such lack of accountability may explain why petitions for successive exemptions from annual quotas increased, and some tributary groups stopped appearing at the collection stations entirely.

Tributaries' preference to sell in private markets in and close by their home regions to Chinese and Russian buyers was in line with the reality that the imperial court could no longer afford to keep up its end of the exchange. Collection stations struggled to maintain an adequate supply of goods to distribute to tributaries.[41] As tributaries struggled to meet quotas, the Qing centre gradually reduced the quantity and types of goods provided as return gifts to perceived essentials such as tea, medicine, and ritual implements. It is equally possible that a decrease in quality of the tribute goods led to a lower quantity of return gifts, or that the imperial centre's goods began declining in quantity first, thus amplifying the unwillingness of tributaries to meet their quotas.

Another reason for tributaries to rely on buying and selling wares at market was that hunting and gathering had become more expensive and more logistically burdensome. Many people preferred to buy goods for tribute submission, either because they did not have the basic tools to carry out the necessary work (like the Dagur group introduced above), or because buying tribute goods was less taxing in terms of time and effort than hunting and gathering. Another hindrance to harvesting tribute goods was that military duties had become more time-consuming as the number of disturbances throughout Manchuria and in other areas of the empire increased in frequency. From the mid-eighteenth century onward, increasing numbers of Manchurian Indigenous troops were dispatched to campaigns outside of their home regions and were therefore unable to hunt. While individual and household quotas were reduced by increasing exemptions made for troops serving in remote posts and civilians assuming military duties without normal compensation, many tributary groups appealed for even greater leniency, claiming that there were not enough able-bodied males available to fulfill even the modified demands.

Trade in tribute goods was also difficult to discourage because it was beneficial for officials who acquired the valuable resources, through both legal and illicit means. Some officials in the Manchurian region wielded the privilege of buying pelts and other commodities at market prices approved by the central government. Other officials bought or otherwise appropriated items that should have remained in the tribute system. Some of these officials attempted to buy such goods at artificially low prices, or forced the delegations to hand them over as gifts. In other circumstances, tributaries and officials negotiated mutually beneficial transactions on an individual basis, or arranged multiple sales over periods of time. Officials could also derive unsanctioned profits from tribute, not only from the actual acquisition of goods but by abusing their position as intermediaries between tributaries and the regional and central collection authorities. For example, some Orochen company captains extorted extra fees for tribute-related matters. One such captain named Lunjišan, who collected a "scribal" fee in cash from each household when collecting the fur tribute and during fishing season, at one point ordered that each household was to provide five fishes to him for every one hundred caught.[42]

Officials had no incentive to reveal their own misdeeds or, likely, the misdeeds of their peers and superiors; thus, the extent and modes of corruption are difficult to assess. But one instance of whistle-blowing shows how seriously this type of human error was treated. Two Solon bannermen, a young company captain named Menguhutulingga and a vice-commandant named Cisan, protested the corruption they encountered at the Qiqihar collection station in 1795.[43] Menguhutulingga had prepared the requisite 5,457 pelts to satisfy the quota for his community, as well as 3,250 extra pelts, all of the highest quality. The station officials – Nayantai, a regiment colonel working closely with the Heilongjiang general Šuliang (?-1798) and his son, the company captain Šutongga (1777–1836) – coveted the pelts and declared that Menguhutulingga's delegation had not brought enough. In addition to demanding 1,000 more pelts over and above the existing surplus, and seizing other furs that the group had brought for private trade, the officials issued an statement that they would buy the "inferior quality" furs for eight copper cash each. One member of the delegation was beaten severely when he spoke out against this decision. In protest, Menguhutulingga and Cisan decided to present their case to the Qianlong emperor. They blocked the imperial carriage en route from Chengde and pleaded for justice. The emperor dismissed the two previous Heilongjiang generals, Dorja (1737–1805) and Mingliang (1736–1822), lowered Šuliang's pay but allowed him to retain his position, and ordered the garrison lieutenant-general, Anjing, to be imprisoned. Nayantai was executed by strangulation, and Šutongga received eighty lashes and was locked in a cangue for two months. Ironically, while their petition to the emperor revealed the misdeeds of the emperor's representatives, the two petitioners were punished as well. For the crime of impeding the emperor's passage, Menguhutulingga and Cisan were sent to Xinjiang, along with their commandant Šertu, to perform hard labour. This punishment combined the hardship of exile to a remote location in a harsh natural environment, where the condemned would have to adjust to unfamiliar weather and food, with forced servitude that would most likely end in death, much like the experiences of the exiles in Wang Ning's study in this volume (Chapter 9).

## Adverse Environmental Conditions

A second cause of failures in the Manchurian tribute system was extreme weather and natural phenomena, such as floods and fires, which reduced the population of animals to be hunted or fished. Qing central and regional officials discerned various factors and types of problems associated with the environment. These can be grouped into two main categories: (1) shortage of resources and (2) the effects of natural disasters in areas where sought-after flora and fauna were located.

Although the dwindling supply of natural resources became more pronounced from the mid-nineteenth century onward, local and intermediate-level officials routinely appealed for exemption throughout the dynasty, based on the generic excuse that tributary groups were attempting to hunt and gather in "barren mountains and forests" and could not produce the expected yields.[44] The officials submitting these requests for imperial mercy were invoking these problems as manifestations of exceptional natural circumstances because tribute requirements were set to ensure that items would be collected when they were most abundant and in prime condition, taking into account factors such as animal mating cycles. In 1691, the Dasheng Wula commandant Mandarhan reported that his troops could not find enough pine nuts to harvest. He received permission to buy fifteen *dan* (a unit of volume equivalent to about one hectolitre) from a Beijing market, which he would then remit to meet the quota.[45] In a comparable case, in 1778 the incumbent Dasheng Wula commandant stated that his team members could not find enough sturgeon of the required size (of eight to nine *chi*, or about 2.5 to 3.2 metres in length) and that they could find only smaller ones (of about four *chi,* or about 1.2 to 1.4 metres). To resolve this shortage, the Jilin general Fuk'anggan (1753–96), acting on the commandant's behalf, was instructed to buy some appropriately sized sturgeon in the imperial capital at the fish market outside of the Zhengyang Gate.[46] In a third example, when it was reported that Butha troops did "not [have] enough places to find fur-bearing animals" in 1908, central government officials did not demand further explanation or substantial evidence from the regional and local administrations.[47] They just decreed that the deficit could be rectified the following year. As it was, the order became moot because the pelt tribute obligation for Butha troops was abolished before the next quota was due. The common thread in all three scenarios, despite the fact that all three occurred in different temporal and logistical contexts, is that the imperial centre accepted the justification that a quota could not be met by customary means because of inadequate natural supply, and that alternative action, whether purchase or deferral, was taken.

Officials attributed some of these shortages to specific natural disasters, which usually warranted detailed investigations to confirm cause and effect. In 1896, the Jilin general, Yanmoo (?–1900) explained that the Sanxing fishing troops had caught only two large sturgeon (one measuring 7 *chi* 3 *cun,* or about 2.43 metres, and the other 6 *chi* 3 *cun,* or about 2.1 metres in length) because excessive autumn rain had flooded the river and disturbed the nets, allowing the large fish to escape. His superiors demanded more details about the exact deficiency, but acknowledged that this incident would be treated as an exceptional situation.[48] Twelve years later, when nature interfered with fishing operations in Jilin again, this time because the winter weather had been unseasonably warm and nets had been less stable in the relatively loose ice that had formed in the river, the imperial centre accepted this reason for the deficiency without question.[49]

In comparison to the cases in which human ineptitude or willful violation resulted in insufficient quality or quantity to satisfy quotas, in cases where environmental factors such as unseasonal weather were cited as the cause for the shortfall, these limiting factors were generally described more precisely in official correspondence. If the causes were proven to be beyond the control of any particular person or group, the involved parties were acquitted from liability. As they did with simple supply shortages, central and high-ranking regional officials focused on solving the problem at hand, usually by granting exemptions or deferrals, or resorting to alternative means of acquisition. In 1900, the Jilin general approved the purchase of pelts to compensate for a shortage caused by a wildfire that had ravaged a large patch of forest in the regional hunting grounds during the previous autumn.[50] The assumption that the fire had killed or driven away fur-bearing animals was implied in this decision; after a survey of the damage was complete, no further justification was necessary to rationalize buying rather than hunting for more pelts.

## Composite Causes

Many situations, particularly in the last two decades of the Qing Dynasty, reflected the impact of human activity on Manchuria's natural environment. The population in Manchuria had increased substantially from the mid-nineteenth century onward because the Russian imperial government pursued overt policies to enfold Manchuria into its domain. These policies included sending explorers and settler households to the region. The Qing responded by lifting prior restrictions on migration so that the new, mainly Han inhabitants would fortify the border territory and prevent the Russians from claiming and using the region's land and other resources.[51] This demographic shift also changed the economic dynamics of Manchuria. As discussed above, trade between Manchurian Indigenes and the increasingly greater number of Han and Russian merchants vying for their wares intensified the scarcity of goods that were more lucrative to sell than submit as tribute; this effect was more pronounced in circumstances where natural shortages became perpetual. The Heilongjiang general reported in 1899 that the number of pelts that could be harvested was already insufficient in 1898; since foreigners were buying pelts at high prices, they had become too expensive to purchase as a way to supplement what had been collected for the 1899 quota. The only course of action possible was to make up for the deficit the following year.[52]

The effects of human action and environmental change were also cited as factors in problems with foreign infrastructure projects. The Jilin general office's revenue division (*Jilin jiangjun yamen hu si*) reported in 1902 that it could not find enough sturgeon, lenok (*xilin yu*), or pine nuts. It blamed the Russian railroad-building project for the shortage of all three items. Since pine trees in the region were being cleared to make tracks, pine nuts could not be collected properly; fish were being caught in greater numbers for private consumption because their usual hiding spots in ditches and gullies along the path of the railroad were being filled in.[53] Two years later, another report sent to the imperial capital stated that wood and fish were becoming rarer because of the Russian railway project.[54] Besides railways, bridge building by Russians was also cited as a severe threat to the fish supply.[55] Russian defensive structures built in riverine areas during the lead-up to the Russo-Japanese War (1904–5) affected fish and other aquatic wildlife adversely as well.[56]

Rapid population growth in Manchuria, caused by the proliferation of Han farmer and foreigner populations, aggravated natural conditions and was another often-mentioned factor inhibiting the collection of tribute.[57] Increased population density led more people to hunt and fish in areas that had originally been reserved for tribute collection, and to compete for food resources by any means necessary. In 1896, the Jilin general asked that the sturgeon quota be modified because bandits had ambushed a fishing work team and demanded an eight-*chi* (2.67 metres) sturgeon that the team had caught. Overwhelmed by the number of bandits, the team surrendered the fish and was therefore short on the tribute quota by one sturgeon.[58] Amending the quota was deemed to be the most suitable remedy for the work team's failure to protect its catch; no other sturgeon of the right size was available, so it could not make up for the loss. This case does not seem to have been viewed as a matter of human error because the work team was not punished for its inability to fend off the bandits' attack. The ultimate decision to grant the exception suggests that evaluating officials took into account the natural shortage of large sturgeon as well. Likewise, in 1911, the Ula banner registry reported that the areas around the river mouths where lenok were usually trapped had become increasingly populated with people engaged in agricultural development. Despite the efforts of two officials to locate enough fish during the previous autumn, it was necessary, the report stated, to make up the shortfall by buying lenok available for sale in the markets.[59] The central government approved this proposal without calling for further investigation into whether there had been sufficient efforts made to locate the fish. This decision implies that the inadequate quantity of fish in the usual gathering areas, ostensibly an environmental problem, absolved the fishermen and their supervisors of any responsibility for the shortage.

The effects of population pressure gradually influenced not only official responses in discrete instances but also broader policy changes. The first preliminary moratorium on food product tribute from Jilin was enacted in 1903, based heavily on the argument that collecting food was increasingly difficult with the presence of Russians and bandits in the usual gathering areas.[60] The next year, the Jilin general emphasized that, due to deforestation, deer had been displaced and were harder to find. The court issued an edict suspending the collection of those items and others that were hard to acquire for the same reasons.[61]

## Conclusion

Like any other government institution, the Qing domestic tribute system in Manchuria did not always operate as expected. Tributaries, harvesting and collection procedures, and the supply of tribute goods were all influenced by numerous variables. When a certain quota could not be met, officials involved in the harvesting and collection of the goods in question had to identify the root cause of the anomaly, and often proposed what might be done to compensate (or simply account) for the resulting deficiency. An analysis of these problems and their solutions – using the categories of human error, abnormal natural conditions, and the adverse human impact on ecosystems – suggests that the central government's clear-cut expectations about quantity and quality were often subject to negotiation and modification in response to the conditions that affected resource supply and the tributaries' capacity to meet quotas. It is also clear from this analysis that failures to meet quotas in the mid-seventeenth through eighteenth centuries were characterized by more straightforward causes – such as errant or incompetent personnel and unseasonal weather. In the nineteenth century, human activity was more often the cause of disruption to the supply of natural resources earmarked for tribute; resources were either diverted to other beneficiaries or the environment was irrevocably altered by the installation of new military and civilian infrastructures.

The examples presented in each of the three categories reveal there were both similarities and differences in responses to exceptional situations. The adjudication of human error cases hinged on the assignment of fault. If certain individuals or groups were assigned blame, then managing officials administered suitable punishments and implemented measures to prevent re-offence. If there was no specific culprit, responses focused on resolving the particular inadequacies, whether through quota reduction, deferral, or supplementation through purchase. This approach also applied in cases where environmental elements generated shortages. For composite problems, judgments likewise stressed adjusting, writing off, or compensating for a problematic quota. In these cases, people were treated with greater leniency than in cases not attributable to natural causes.

Another major difference between the Qing government's treatment of human-induced problems and its treatment of environmental aberrations in the Manchurian tribute system lay in who took the lead in problem solving. The emperor and central authorities often played a greater, more direct role when the issue was delinquent tributaries. They often prescribed formulaic penalties and ordered long-term changes in operational procedures in order to prevent similar incidents. When it came to environmental issues, however, the imperial court often endorsed regional and local officials' proposals for handling problems rather than advocating particular strategies. Perhaps it trusted that the people who normally ran the system had a better sense of "ground conditions" and could therefore develop the most appropriate solutions, as seen in Ravenstein's narrative.

Given that Qing officials generally worried about incurring the throne's displeasure, which could mean being demoted or expelled from the bureaucracy, it is quite likely that they did everything within their power to solve minor problems or to hide them from their superiors. However, it is also plausible that they did not take the initiative to develop long-term solutions because they feared the consequences if such measures failed. If one makes conjectures based on what was *not* addressed in the documentary records, arguably three other factors also played a role in determining whether tribute quotas were met.

Technology was one such factor. Local and regional administrators in Manchuria did not suggest (or establish and maintain) new technologies for hunting and fishing, and they did not suggest (or establish and maintain) capital investments for the acquisition of such resources from other places. Unlike people exiled to Manchuria, who were forced to work in various capacities (as described by Wang Ning in Chapter 9 of this volume), the tribute-collecting groups were, to borrow business terminology, fit for purpose. They had the relevant expertise to make and use tools, as well as behavioural knowledge such as how to track animals through forests and mountains. However, it merits attention that the Qing centre expected them to employ more or less the same methods and devices from the mid-seventeenth century to the early twentieth century, such as the "fishing apparatus" shown in Figure 3.2. Although officials at the time noted that some individuals and groups pursued technological innovation of their own accord, such as buying more powerful firearms from the markets in towns bordering the Russian empire, they never suggested explicitly that imperial funds be used for wide-scale equipment purchases and training. Similarly, while those exiled to Manchuria who became slaves in Indigenous households undoubtedly served as auxiliary labour for hunting and gathering, when a group reported that it could not cover enough hunting ground or find enough goods naturally scattered over a

FIGURE 3.2 "Fishing."
*Source:* 1808 travel diary of Japanese explorer Mamiya Rinzō[62]

vast swathe of land, officials did not explicitly advocate increasing manpower for tribute operations.

A second problem that the Qing government did not remedy until the late nineteenth century was that the officials acting as intermediaries considered themselves to be (unsanctioned) shareholders in the enterprise of tribute collection. In the studies of Manchuria's Indigenous ethnic groups published from the 1950s to the present, the Qing-era hunters and gatherers of tribute items are praised much like the *rongjun* (honourable soldiers) described in Sun Xiaoping's exposition of agricultural reclamation from 1949 to 1975. These explanations, which plainly evaluate Qing historical conditions according to Marxist-Leninist ideological metrics, point out that these groups were productive *despite* rather than *because of* the officials appointed to supervise them. Officials who collected tribute items from the Orochen and Hezhen in remote locations were known in Manchu as *anda,* of which one meaning is "close friend." These individuals served three-year terms and often became (unwanted) "guests," which is another translation of the term. *Anda* are described in many post-1949 publications as "usurers" (*panbozhe*).[63] The term "usury" covers a broad range of sins – such as forcing Orochen into marriage or sworn brother relationships; creating financial arrangements that would cause several generations of a given Orochen family to be in debt; and even stealing Orochen horses, firearms, and other valuable property.[64]

Far from being advocates, or "friends," these officials made it even more difficult for the people they supervised to amass enough tribute items of the ideal quantity and quality. Their power had decreased by the final two decades of the nineteenth century because other parties, including private merchants, had successfully cultivated mutually dependent relations with tribute-harvesting peoples.

Certainly not all officials should be tarred with the same brush. As several of the examples presented in this chapter illustrate, many memorials to the imperial throne about tribute quota problems were relatively neutral. Reporting officials detailed the facts of a specific situation and sometimes proposed a course of action, such as deferring a quota or waiving it for one cycle. What was rarer (but what might have strengthened the tribute system as a whole had it been less rare) was the more courageous motion to revise some of the quotas altogether. In one Dagur folk history, an unnamed Dagur general took an aggressive stance to curb what he felt were unreasonable demands for seawater pearls as tribute goods.[65] In the folk narrative, the general killed a sea serpent that had been eating the divers responsible for collecting the pearls. He then brought the serpent to the imperial palace at Mukden and told the emperor that the pearls he desired were in the beast's stomach. In response, the horrified emperor eliminated the pearl quota. Though the tale is a folk narrative, and the sea serpent is a mythical creature, and the Qing court never suspended the pearl quota simply because an official remonstrated with the emperor, the story exemplifies how Dagur officials may have expressed grievances to higher authorities on behalf of their troops.

That tribute quotas did not change substantially over time in response to economic and environmental circumstances ultimately had severe political repercussions. The Qing and Romanov empires clashed in the mid-seventeenth century and designated their first legally valid border to clarify their claims over land and people. The 1689 Treaty of Nerchinsk was a decisive measure to restrain the fierce competition for these resources between the two states. But no diplomatic agreement could alter the fundamental reality that people living on both sides of the border were "Indigenous foragers," as introduced by David A. Bello in this volume (Chapter 2), and, as such, relied on each other to meet the demands of their respective governments as well as their personal needs. While the Qing centre achieved initial success in strengthening its control over Manchuria by imposing tribute quotas, it inadvertently also fostered closer relations between Manchurian Indigenes and their would-be foreign enemies. Indeed, many Russians became *anda* who sold or bartered commodities such as flour, cloth, and firearms, as well as pelts that would be submitted as tribute.[66] Some of these Russians maintained long-term relations with particular Manchurian households and engaged in the social rituals of eating and drinking liquor with them. Even though tribute collected from Manchuria was a yearly tax in form and function more than a symbolic gesture of obeisance, the outflow of the very goods that the Qing centre sought to monopolize via these Russian *anda* weakened the tangible power that it exerted over the Indigenous populations. To maintain its overall objective of being the dominant beneficiary of Manchuria's natural wealth, the Qing government accepted the recurring "irregularities" in the tribute system.

## Notes

1 Ernst G. Ravenstein, *The Russians on the Amur: Its Discovery, Conquest, and Colonization, with a Description of the Country, Its Inhabitants, Productions, and Commercial Capabilities; and Personal Accounts of Russian Travelers* (London: Trübner and Company, 1861), 75–76.

2 This chapter focuses on the Jilin and Heilongjiang administrative regions of Qing Manchuria.

3 For example, tribute collected once a year from the Heilongjiang general's territory included twenty wild boars, two hundred ring-necked pheasants, thirty lenok, thirty brown trout, forty sacks of wheat flour, two boxes of tinder, four hundred arrow shafts, and three thousand pieces of peach bark, as well as the sable pelt tribute discussed at length in Chapter 2 by David A. Bello. Other types of tribute items in Manchuria were seasonal, due in the sixth, tenth, and twelfth lunar months, or gathered directly from government-monitored sites (ginseng was one such type of tribute). In the area under the Jilin general's command, a total of thirty-one annual and seasonal tribute items were collected, including animal parts such as deer tails as well as fruits, vegetables, and timber. For detailed descriptions and numerical quotas of the Heilongjiang and Jilin tribute items, see He Xinhua, *Qingdai gongwu zhidu yanjiu* [A study of the Qing tribute goods system] (Beijing: Shehui kexue wenxian chubanshe, 2012), 111–18.

4 Scholars like Jonathan Schlesinger have already persuasively challenged the view that the tribute institution in Manchuria deteriorated mainly due to environmental degradation. See Schlesinger, "The Qing Invention of Nature: Environment and Identity in Northeast China and Mongolia, 1750–1850" (PhD diss., Harvard University, 2012).

5 See Ding Yizhuang, *Qingdai baqi zhufang* [Research on Qing Eight Banners garrisons] (Shenyang: Liaoning minzu chubanshe, 2003), 78–83, for a profile of the Jilin and Heilongjiang administrative structures.

6 Terms in parentheses are Chinese, unless preceded by the abbreviation "Ma." signifying that they are Manchu words, or "Ch." for Chinese transliterations of Manchu terms.

*Chaogong* connotes the physical act of participating in an audience with the emperor to present tribute, whereas *tugong* highlights the submission of resources from a certain region. Several other Chinese concepts have been roughly translated into the English language as "tribute," which reflects the imperfect synonymy.

7 The paradigm-setting work for these interpretations of Chinese imperial tribute was John King Fairbank and Ssu-yu Teng, "On the Ch'ing Tribute System," *Harvard Journal of Asiatic Studies* 6 (1941): 135–246, but countless subsequent studies have remoulded their argument about tribute as a mode of diplomacy and trade, including John E. Wills Jr., *Embassies and Illusions: Dutch and Portuguese Envoys to K'ang-hsi* (Cambridge, MA: Harvard University Asia Center, 1984), and Zhang Feng, "Rethinking the 'Tribute System': Broadening the Conceptual Horizon of Historical East Asian Politics," *Chinese Journal of International Politics* 2, no. 4 (2009): 545–74.

8 For a representative explanation of Manchurian domestic tribute as a type of tax, see Zhang Shuangzhi, *Qingdai chaojin zhidu yanjiu* [Research on the Qing tributary system] (Beijing: Xueyuan chubanshe, 2010), 210–12.

9 Tributaries of both categories were peoples speaking Tungusic languages and sharing traits such as the practice of shamanism as a religious and medical tradition and hunting wild game and fishing as major economic activities. See Juha Janhunen, *Manchuria: An Ethnic History* (Helsinki: Finno-Ugrian Society, 1996).

10 Journeys to Beijing were originally intended to be routine, annual affairs; however, various problems – including the expense (in money and the time required to make the trip) and the risk of contracting diseases such as smallpox – made the journeys more risky and onerous than meaningful, so they were discontinued after the Qianlong reign.

11 The customary use of "tribute" in English-language scholarship to describe the activities of these people is misleading, because the original Chinese term *caibu* is more accurately translated as "gathering and trapping," with more emphasis on economic production rather than political ritual. The three major resources to be extracted through *caibu* from the Kangxi reign onward were pearls, sable pelts, and ginseng. See *Da Qing huidian* [Statutes of the Great Qing Dynasty], Kangxi edition, 133:1a.

12 Xiqing, *Heilongjiang waiji* [Unofficial records of Heilongjiang] (1810; repr. Harbin: Heilongjiang renmin chubanshe, 1984), 5:1a.

13 See "Buteha tushuo" [Schematic description of the Butha (hunting) banners], in Tu Ji, *Heilongjiang yu tushuo* [Atlas of Heilongjiang] (repr. Taipei: Guangwen shuju, 1963), 77–85; and Yu Dinggong, *Buteha zhilüe* [Concise history of the Butha (hunting) banners] (1932; repr. Taipei: Taiwan xuesheng shuju, 1967). In the earliest years of this institution's history, the workers were also known generically as hunters (*busheng*).

14 *Da Qing huidian,* Kangxi edition, 133:2b.

15 A *juhiyan* refers to a group of workers that collected items for tribute, such as honey and pearls. Later in the Qing, a *juhiyan* consisted of thirty or more individuals.

16 The upper three banners (bordered yellow, plain yellow, and plain white) of the Eight Banners (*baqi*) were considered more prestigious, in part because of their origins as units that the emperor controlled directly.

17 *Qingchao wenxian tongkao* [Officially commissioned comprehensive examination of documents for the Qing Dynasty], comp. Xi Huang, Liu Yong, and Ji Yun (1784) 182: 24a-b. The lower five banners were the bordered white, bordered blue, plain blue, bordered red, and plain red units.

18 For a concise summary of this process, see Lü Guangtian, "Buteha dasheng bu de gong diao zhi" [The sable pelt tribute system of the Butha (hunting) tribes], in *Zhongguo shaoshu minzu lishi gushi* [Historical tales of China's ethnic minorities], ed. Li Songmao, Lü Guangtian, and Wei Wenxuan (Yinchuan: Ningxia renmin chubanshe, 1988; repr. Taipei: Taiwan shangwu yinshuguan, 1993), 87–89.

19 For a nineteenth-century description of this procedure, see *Heilongjiang waiji,* 5:8b. Also note that in Manchu, the letter "c" is generally pronounced as "ch."

20 Fang Shiji, *Longsha jilüe* [Concise records of Longsha] (n.p.: 1755), 1:18b.

21 Guan Jialu and Tong Yonggong, "Qingchao gong diao shang wulin zhidu de queli ji yanbian" [The establishment and evolution of the sable tribute and return gifts system during the Qing Dynasty], *Lishi dang'an* [*Historical Archives*] 6, no. 3 (1986): 94.

22 Specific quotas for the three classes of pelt varied, but as sample standards, the Heilongjiang general was responsible for transferring 500 first-class, 1,000 second-class, and the remainder of satisfactory items considered third-class to the central government. *Qingchao wenxian tongkao,* 38:3a.

23 See *Heilongjiang waiji,* 1:9a. The Equestrian Orochen (the Manchu term *moringga* customarily translated into Chinese as *ma shang*), were full members of the Butha banners. The Foot Orochen (the Manchu term *yafahan* translated into Chinese as *bu xia*) retained more political and social autonomy since they were subject to fewer requirements for military and tributary service and were led on a daily basis by their own leaders rather than banner officials. The Qing government did assign banner officials to supervise Foot Orochen clans and villages, but such personnel had to work cooperatively with clan heads and village chiefs.

24 For a brief profile of the Imperial Household Department, see Cao Zongru, "Zongguan neiwufu kaolüe" [An investigation of the Chancery of the Imperial Household Department], in *Guoli Beiping gugong bowuyuan shi yi zhounian jinian wenxian luncong* [Collection of documents and essays commemorating the eleventh anniversary of the Beiping National Palace Museum] (Beiping: Guoli Beiping gugong bowuyuan, 1936), 88. More extensive works on the same topic include Qi Meiqin, *Qingdai neiwufu* [The Qing Imperial Household Department] (Beijing: Zhongguo renmin daxue chubanshe, 1998), particularly pages 149–54, with details about who was expected to give, the distribution of quotas, and the end of the system due to its ill effects on the tributary populations; and Preston M. Torbert, *The Ch'ing Imperial Household Department: A Study of Its Organization and Principal Functions, 1662–1796* (Cambridge, MA: Harvard University Press, 1977).

25 Pinpointing the exact rate the quotas were met is difficult because not every cycle is represented in the same amount of detail in extant data. Based on samples throughout the dynasty, tribute collections yielded the minimum amount of adequate quality in about 80 percent of cases from the Kangxi to Jiaqing reigns, and then the percentage dropped incrementally until the end of the Qing, at varying rates for different items. The author of this chapter is continuing to collect and analyze quantitative data to refine these estimates.

26 *Shengjing canwu dang'an shiliao* [Historical archives on ginseng-related matters at Shengjing], ed. Liaoning Provincial Archives (Shenyang: Liaohai chubanshe, 2003), 2–3.

27 See Fuheng (?–1770), January 18, 1759, *Neige daku dang'an* [Archives of the Grand Secretariat] (Institute of History and Philology, Academia Sinica, Taipei), 055998–001; and Guwanboo, December 9, 1769, *Neige daku dang'an,* 084353–001.

28 Chengde, January 3, 1802, *Neige daku dang'an,* 062109–001.

29 December 12, 1774, *Qingdai gongzhong dang'an zouzhe ji junjichu dang zhe jian* [Memorials from the Qing palace archives and Grand Council] (Institute of History and Philology, Academia Sinica Taipei), 84353.

30 See Fuheng, January 24, 1757, *Neige daku dang'an,* 076129–001; Jin Jian (?–1794), October 28, 1789, *Neige daku dang'an* 068108–001; and Qinggui (1737–1816), October 15, 1812, *Neige daku dang'an,* 109593–001. In the first case, the imperial court ordered that the deficit in quality be redressed in the following year's cycle.

31 December 9, 1814, *Neige daku dang'an,* 111881–001.

32 December 18, 1891, *Qingdai Heilongjiang lishi dang'an xuanbian* [Selected historical archive documents of Qing Heilongjiang], ed. Manchu Division of the First Historical Archives of China and Institute of History, Heilongjiang Academy of Social Sciences, 2 vols. (Harbin: Heilongjiang renmin chubanshe, 1986), 2: 162–63.

33 May 5, 1851, *Sanxing fudutong yamen Manwen dang'an* [Sanxing garrison lieutenant-general *yamen* Manchu-language archives] (Liaoning Provincial Archives) 287, 223–25.

34 March 3, 1822, [*Da*] *Qing* [*lichao*] *shilu* [Veritable records of the Qing Dynasty] (Xuanzong) 29, 525. There were many other references to unspecified economic hardships that occurred on a regular basis in this source. This and other citations of this text follow the convention of identifying sections by the reigning emperor's temple name, such as Xuanzong for the Daoguang emperor.

35 *Qingdai bianzheng tongkao* [General investigation of Qing frontier administration] (Nanjing: Bianjiang zhengjiao zhidu yanjiu hui, 1934), 319.

36 Produce bureau (*Guozi lou*) to the Jilin general *yamen,* November 6, 1881, *Jilin gongpin* [Tribute items of Jilin], ed. Jilin Provincial Archives and Jilin Minority Nationalities Ancient Books Editorial Office (Tianjin: Tianjin guji chubanshe, 1992), 238–39.

37 June 4, 1836, *Neige daku dang'an,* 202995–001.

38 For a vivid description of a nineteenth-century market, see Thomas Witlam Atkinson, *Travels in the Regions of the Upper and Lower Amoor: And the Russian Acquisitions on the Confines of India and China with Adventures among the Mountain Kirghis; and the Manjours, Manyargs, Toungouz, Touzemtz, Goldi, and Gelyaks; the Hunting and Pastoral Tribes* (London: Hurst and Blackett, 1860), 349.

39 These behaviours appeared in official correspondence from the Kangxi reign onward. See, for example, July 5, 1667, *Heilongjiang jiangjun yamen Manwen dang'an* [Heilongjiang general *yamen* Manchu-language archives] (Beijing: First Historical Archives of China), 13–1722.

40 December 2, 1818, *Qing shilu* (Renzong), 349, 611.

41 Reports of shortage and requests for assistance to higher authorities were common. In one of many such documents, the Sanxing garrison lieutenant-general sent an urgent appeal to the Jilin general for more grain to pay Hejen and Fiyaka tributaries on April 23, 1866. *Sanxing fu dutong yamen Manwen dang'an* [Sanxing garrison lieutenant-general *yamen* Manchu-language archives], 318: 65.

42 Qiu Pu, *Elunchun shehui de fazhan* [The development of Orochen society] (Shanghai: Shanghai renmin chubanshe, 1978), 122.

43 For a summary of this case, see *Qing shilu* (Gaozong), 1487: 14a-16a.

44 Regional commanders verified such exemption requests and forwarded them to Beijing. In one such sequence, the Heilongjiang general Iktangga memorialized that the Orochen at the Xing'an garrison could not find enough animals to fulfill the pelt quota:

*Qingdai gongzhongdang zouzhe* [Memorials from the Qing palace archives], National Palace Museum of Taiwan, 130214. The imperial rescript on February 28, 1894, instructed the appropriate offices to confirm whether the situation was that serious. Two weeks later, Aisin Gioro Fukun, representing the Imperial Household Department, advised the throne to approve Iktangga's request for exemption (March 13, 1894, *Qingdai gongzhongdang zouzhe,* 130501).

45 Wang Jiping, ed., *Jilin sheng bian nianji shi* [Annalistic chronicles of Jilin province] (Changchun: Jilin renmin chubanshe, 1989), 1: 23.

46 Ibid.

47 Xu Shichang (1855–1939) and Cheng Dechang, January 20, 1908, *Qingdai gongzhong dang'an zouzhe ji junjichu dang zhe jian* [Memorials from the Qing palace archives and Grand Council] (Institute of History and Philology, Academia Sinica, Taipei), 408000218.

48 December 30, 1896, *Jilin gongpin,* 376.

49 Jilin provincial revenue division, January 20, 1908, *Jilin gongpin,* 314.

50 Jilin general *yamen,* February 26, 1900, *Jilin gongpin,* 381.

51 A recent and comprehensive publication that discusses the push-pull factors for Manchuria's population explosion is Kimitaka Matsuzato, "The Creation of the Priamur Governor-Generalship in 1884 and the Reconfiguration of Asiatic Russia," *Russian Review* 71, no. 3 (2012): 365–90.

52 Enje (?–1899) and Saboo (1793–1852), September 24, 1899, *Guangxu chao Heilongjiang jiangjun zougao* [Memorials of the Heilongjiang general during the Guangxu reign], ed. Xing Yulin, 2 vols. (Beijing: National Library Microfilm Copy Center, 1993), 2: 607.

53 July 7, 1902, *Jilin gongpin,* 89.

54 Jilin general to the produce bureau, May 22, 1904, *Jilin gongpin,* 241. The Jilin general submitted a palace memorial with the same content on June 1, 1904 (*Jilin gongpin,* 245).

55 July 17, 1902, *Jilin gongpin,* 89.

56 October 4, 1905, *Jilin gongpin,* 90.

57 The influx of non-Indigenous peoples accounted for the surge in Manchuria's population during this period. Groups that were already settled in the region before 1800 did not change significantly in size or composition, but because many tribute-collecting individuals and their families chose to evade the duty by hiding tribute figures from the mid to late nineteenth century outwardly suggests that the groups to which these people belonged experienced attrition.

58 December 24, 1896, *Jilin gongpin,* 85–87.

59 March 28, 1911, *Jilin gongpin,* 102–5.

60 This cessation of the food tribute requirement applied to only certain items. Jilin general to the produce bureau, July 9, 1903, *Jilin gongpin,* 239; Jilin general to the produce bureau, September 11, 1903, *Jilin gongpin,* 241.

61 May 31, 1904, *Jilin gongpin,* 249.

62 Japanese explorer Mamiya Rinzō (1775–1844) travelled around the Kuril Islands and in the Amur River basin in 1808–09. During his journey he kept a diary in which he made many observations of Indigenous peoples.

63 For a representative sample of the many similarly phrased references to the *anda* as being "usurers," see Lu Guangbin, Han Youfeng, and Du Yonghao, *Elunchunzu sishi nian, 1953–1993* [Forty years of the Orochen nationality, 1953–1993] (Beijing: Zhongyang minzu daxue chubanshe, 1994), 11.

64 Ibid., 12.

65 Kevin Stuart, Li Xuewei, and Shelear, eds., *China's Dagur Minority: Society, Shamanism and Folklore,* Sino-Platonic Papers no. 60 (Philadelphia: University of Pennsylvania Department of Asian and Middle Eastern Studies, 1994), 128.

66 Inner Mongolia Ethnic Minority Social History Survey Group, ed., *Heilongjiang sheng Huma xian Shibazhan minzu xiang Elunchunzu shehui lishi buchong diaocha baogao* [Report of the supplementary survey about the social history of the Orochen in the Shibazhan ethnic village of Huma County, Heilongjiang] (Huhhot: Nei Menggu shaoshu minzu lishi diaocha zu, 1963), 25.

# 4

# Producing Full-Fat Controversy: The Politicization of Dairy Production in Post-Colonial North Manchuria, 1924–30

*Blaine Chiasson*

*Thirty years ago – in the beginning of the nineties – North Manchuria was a desolate, sparsely populated, and little explored country.*

*Efforts of the Chinese administration to colonize the Amur region for the effort of creating a barrier against the steady and elementary eastward movement of the Russians ... were unsuccessful. The Chinese farmer, with his intensive methods, closely dependent on the city, was not able to sever connections with his base and remove therefrom to a distance of hundreds of miles into thinly settled wild country.*

*Russian genius, however, appeared in the Far East, it threw in a few years a steel bridge of rails through the heretofore desolate country and brought this country to life ... The Chinese Eastern Railway brought different parts of the country together – the cattle, lumber and agricultural districts, joined them into one unit, connected with the latter with the nearest markets of the Russian Far East, South Manchuria, and abroad, and brought North Manchuria closer to international life.*

*This great work was done by Russian hands, and thousands of Russians put no mean amount of tireless effort and energy into the creating of the present prosperous Manchuria.*

– *"Russian Pioneers in North Manchuria, 1924"*

*Marshal Chang referred to the aggressive foreign policy of the late Imperial Russian Government as exemplified in the building of the Chinese Eastern Railway and the colonizing projects that had been carried out in connection therewith. He inquired whether Dr. Schurman was not aware of the fact that the Russian Government*

> *had never paid anything for the land, a portion of which the Chinese now propose to recover. Marshal Chang advanced very much the same arguments used by General Chang and others in Harbin, and his remarks were characterized by considerable bitterness.*
>
> – *Excerpts from a report on the interview between Zhang Zuolin and the American ambassador Schurman in Mukden (Shenyang), September 10, 1923*

These two quotations illustrate a struggle that took place in northern Manchuria between its two colonizing populations, the Russian and the Chinese, in the early 1920s. The first quotation is from a commemorative history of the Chinese Eastern Railway (CER), commissioned by its still-Russian-controlled board of directors and printed in its Russian- and English- but not Chinese-language versions. The second is a discussion between Zhang Zuolin (1875–1928), Manchuria's leader, and the US ambassador on recent attempts by the Manchurian government to expropriate land from the CER's control. Together the quotations demonstrate how Manchuria's two colonizing elites could differ over the process of colonization and industrial and agricultural production. This struggle over land, environment and production, sovereignty, and what (or, more importantly, who) constituted Manchuria's "legitimate" and "modern" developer was not merely a discussion among the north Manchurian elite. On April 10, 1924, Chinese farmers occupied the recently built agricultural experimental farm located in Anda, a town on the CER's western line. Chinese farmers, claiming the land had been illegally expropriated one year before, invaded the dairy farm, tore down barns, ripped up fields, and destroyed crops. Armed with hoes and axes, they chased away the farm's Russian staff, and immediately began to erect homes, dividing the large fields, planting *gaoliang*[1] and soya beans. The Russians' expensive dairy cows became expensive draft animals.

The Anda farm was one of three experimental farms, and one of many different industries and services, created by the CER's board in the wake of the railway's 1920 reorganization. The CER, founded and subsidized by the tsarist government before 1917, needed to find a new raison d'être if it was to survive in a changed political environment. No longer could the CER rely on the military and economic strength of the Russian empire, nor could it rely on a complacent and disinterested Chinese population. The Chinese governments, national and local, insisted the CER perform what they, the Chinese, believed was its original role as a joint Sino-Russian business venture. In becoming a true partner in 1920, the Chinese insisted the CER have a new Chinese railway president and Chinese board members, and that it accommodate Chinese managerial staff. The CER general manager, the true head of the railway, was, along with half the board, Russian. Thus, the new CER was now headed by two nationalities, each with a colonial project in Manchuria, each vying for legitimacy through the administration and development of Manchuria's land. Anda's new commercial dairy farm had the specific mandate to transcend the CER's original Russian identity. It was created with the intention of commercializing the production of butter, milk, and sour cream, with the intention not only to sell to China's large foreign community, but also, through sales to Southeast Asia and the Pacific coast of North America, to internationalize the CER, raising it above its contested Russian and Chinese foundation. The creation of the Anda dairy farm, and the resulting disappearance of the Chinese leasehold farms, signified that Russian-controlled agricultural and economic space and a continued Russian presence was being legitimized in post-colonial north Manchuria. This is similar to how Japanese agricultural settlements both erased former Chinese identities and constructed new Japanese colonial legitimacy and colonial subjectivities during the Manchukuo period, as argued in Annika Culver's chapter (Chapter 6 in this volume).

The Russians represented a former colonial power, now orphaned from the mother country. In this discursive battle, they would claim to be inherently "modern," with fifty years of professional expertise in the development of the Russian Far East and Manchuria. Russian CER managers, as they themselves said, understood the global economy and how best to adapt Manchuria to it.[2] The Chinese, having tolerated thirty years of Russian incursions into Manchuria, were representatives of a rising national power. They rejected claims by the Russians and Japanese that Manchuria was an "empty" space, a region of international development opportunity best suited to the power that could optimize its development. The Chinese declared, in a manner that Russians and Japanese wrote was tinged with "old-fashioned" nationalism, that Manchuria was Chinese space.[3] The Chinese governing elite, the press, and the chambers of commerce argued that Russian-directed production was not the value-neutral business that Russian managers and the Russian press claimed. Rather, it was a neo-colonial project aimed to keep Russian managers, Russian workers, and Russian citizens at the helm of modern development in northern Manchuria and Chinese territory, and as the principal material beneficiaries of this production. In making this argument, the Chinese demonstrated limited appetite for Russian plans – and delicious sour cream.

This chapter, like others in this volume, examines how people claim and control space, how they make territory their own through territory management and the economic use and worth (both monetary and cultural) of the territory's resource production. I consider the economic and political meanings – in this case contested meanings – that are attached to the land and its use. Specifically, this chapter is about a seemingly minor issue: a dispute over the use of fertile land – between Russian agronomists committed to an industrial dairy farm and Chinese farmers, policemen, and administrators committed to upholding leases on small family farms. Although the Chinese administration quickly resolved the issue and restored control over the dairy to the CER, the dispute highlighted how Manchuria's Russian and Chinese populations invested agriculture and use of the land with specific national meanings – dairy for Russian, cereal production for Chinese.

The incident was symbolic of a deeper conflict between north Manchuria's two colonial peoples. Both the Chinese and the Russians saw Manchuria as a wild, undeveloped frontier to be exploited, modernized, and tamed. Each, however, had a different vision of what constituted the exploited, modernized, and tamed frontier. For the Russians, it was a continuation of the project begun with the construction of the CER railway, using Russian knowledge to develop Manchuria and bring its products to market. Prior to 1917, the Russians had dominated north Manchuria militarily, politically, and economically. Before 1917, Manchuria's Chinese elite could only watch what they saw as Russian encroachments on Chinese sovereignty. Any post-1917 discussion of economic development dominated by the Russians, aimed at the Russian market and diet, and paying Russian salaries, all the while justified by the invocation of "neutral" economic goals of development (ideas put forward by the Russians themselves), was suspect. For the Chinese in contested post-1917 Manchuria, as both groups sought to justify their position and make Manchuria "theirs," no control of territory, no economic development, and no publicly sponsored dairy farm could be viewed as "neutral." Everything was politicized: dairy farms, milk production, culturally constructed ideas of what constituted "proper" or "modern" agriculture, and how land was used and by whom.

As colonial peoples, both the Russians and the Chinese have a long history of justifying control by invoking the image of a superior civilization taming the savage. This Chinese colonial project in Manchuria stretched well into the twentieth century, as Sun Xiaoping's work on the People's Republic of China's post-1949 colonization demonstrates in this volume (Chapter 10). The trope of "civilizing the frontier" is well known in Chinese colonial historiography, but it is also an element of Russian colonialism. In a statement by Russia's foreign minister A.M. Gorchakov to foreign governments concerning Russia's recent acquisitions in Central Asia, the minister stressed that expansion was justified in Russia's bid to secure borders against "hostile tribes," and portrayed Russian rule as the only alternative to a "state of permanent disorder." Russia's expansion, however, was about more than national security; it was "the moral force of reason" in a land inhabited by "half-savage nomad populations" unaccustomed to "more peaceful habits." By claiming that Russia's "special mission" was "the civilization of these countries, which are her neighbours on the continent of Asia," Gorchakov justified Russian expansion not only by the traditional rubric of national defence but through an emerging discourse of justifiable (if somewhat forced) modernization of "half-savage nomad populations" in the name of civilization.[4]

Recent works have explored how the concept of Russian expertise in the fields of finance and of political and economic administration was the essential building block of Russian identity in the pre-1917 railway concession zone.[5] If the zone was built with Chinese labour on territory that was politically Chinese, Russian finance, administration, and development know-how was the basis of a legitimate Russian claim over the concession and its products. Sergei Witte (1849–1915), the architect of the Russian transnational railway system, of which the CER was a part, saw the railway as essential to the development of Russia's recent full acquisition of the Russian Far East. Its extension into Chinese territory was a project of Russian "informal" domination of the north Manchurian economy and society, through a process of development linking the region to the world market – a process that included significant Russian colonization along the railway concession. Russian understanding of the CER was as a non-political agent of development – non-political in the sense that Russia did not have full sovereignty over the concession zone and in ninety-nine years' time, the area would revert back to Chinese control. In the meantime, Russia was developing the region according to what it believed were positive goals of modernization and the establishment of links to the world economy.

The Chinese view of what constituted the basis for a legitimate claim over a frontier colony was remarkably similar to that of Russia, except for one important difference: Manchuria was, according to internationally recognized standards of political sovereignty, Chinese territory. It was an insecure Chinese territory, however, recently incorporated into the Chinese political system as three provinces, sparsely populated and associated in the popular imagination with its former separate status as the Qing homeland. The late Qing had quickly adopted the new global imperial benchmark for frontier and colonial legitimacy, direct control of a state's hinterland through political administration and colonization. By means of colonization, agricultural development (including the creation of agricultural experimental farms), the development of industry, and the creation of a Chinese provincial administration identical to that of China proper, the Qing had planned to modernize Manchuria as Chinese. The republican government in Beijing, and the regional government in Shenyang, inherited and expanded these projects in Manchuria.[6]

Chinese administrative, technical, and managerial expertise was expressed in the creation of two new institutions for post-1917 Manchuria. The first was the reorganized CER, under Chinese and Russian management with the patronage of the provincial and national Chinese governments. Chinese – primarily those trained abroad in railway administration and engineering – were placed in management positions. Boris Ostroumov, engineer and minister in the former Russian Ministry of Railroads, who was in the Russian Far East during the revolution, was appointed to replace the former tsarist manager, Dmitrii Horvath.[7] The railway dropped its identity as a tsarist-subsidized enterprise; its new task was to make money and develop Manchuria. To ease the transition, the Chinese allowed the Russian language to remain the working language of the line, and left most of the previous Russian workers in place.

As a professional bureaucrat, Ostroumov shared the interests of the Russian technocratic class that had built and maintained the line and, since 1898, had benefited from the CER. This class maintained its technocratic approach to modernization, associating itself with Russian modernity of the imperial period. To ensure that his modern vision of the CER was implemented, Ostroumov dismissed a number of CER employees and replaced them with "highly qualified" émigrés – engineers, technical specialists, lawyers, and other members of the "émigré intelligentsia."[8] Under Ostroumov, the CER management set up several experimental farms to conduct agricultural research and to mechanize agricultural production. The "agricultural laboratory" of these farms investigated various grains and farm products but focused especially on processing soya beans and milk products, and on dairy cattle breeding.[9] The CER manual published in 1924 suggested that Ostroumov's management style transcended the CER's contested origins as a Russian and Chinese company, demonstrating "an equal attitude toward all of its clients independently of nationality," and at the same time reminding the reader that the still-Russian-controlled CER was the vanguard of the civilizing Manchuria through "a maximum development of efforts toward the spreading of civilization." But the manual also prominently listed "attention to the welfare and wellbeing of its employees" – overwhelmingly Russian – as among the railway management's objectives.[10] In sum, the CER's new goals were to be modern, outward-looking, and focused on controlling production for domestic and international markets, another form of Witte's "economic colonization." In response to Russian claims that this was all done for the greater good, Chinese critics would counter that this "greater good" in Manchuria, like the CER's concept of modernity, was coded as Russian.

As part of the pragmatic policy of accommodation, to protect the railway from being broken up or dissolved, or becoming a source of competition between Jilin and Heilongjiang provinces, Zhang Zuolin created within the boundaries of the former CER concession the Special District of the Three Northeastern Provinces. The district preserved the CER as one company and kept its assets intact, at the same time asserting Chinese management over the area. Russian elements, such as language, religion, education, and the legal system were preserved for the Russian

FIGURE 4.1 Anda train station in the early 1900s.
*Source: Kitaiskaia Vostochnaia zheleznaia doroga* [The Chinese Eastern Railway], special gift edition (Harbin: n.p., 1909). Collection of Dr. O. Bakich

inhabitants, while a Chinese managerial class was appointed to supervise land administration, education, and municipal government in the district, all roles that had once been filled by the CER company.

Up to now, most research on the importance of modern managerial and technocratic values, education, dissemination, administration, and experience in the northeast Asian context has focused on the activities of the so-called Japanese technocracy in their south Manchurian, Korean, and Southeast Asian colonies.[11] Nevertheless, extending Prasenjit Duara's argument from *Sovereignty and Authenticity: Manchukuo and the East Asian Modern*, the creation of the "east Asian modern" was not only a Japanese project.[12] The same values of legitimation through colonization, settlement, and agricultural development were a hallmark of Russian and Chinese settlement in north Manchuria. This can be seen in the Russian-language CER reports and research institutes, and in the Russian press. It can also be seen in the Chinese-language press, in the reports from the newly created Chinese municipal governments, and responses to what the Chinese elite believed was a neo-colonial Russian policy of the CER to monopolize its development and production for the region's Russian population.

Agricultural production and its modern application through scientific farming techniques was one manifestation of the claims of technocratic administration of Manchuria made by the Russian, Chinese, and Japanese elites. Research into Japan's investment, financial and ideological, in the South Manchuria Railway agricultural stations is already underway. Less well understood is the place of modern agriculture and its application to the claims of Russian and Chinese rule over Manchuria. In China and Russia, changing concepts of agriculture and its professionalization would contribute to each side's argument over which Manchurian agricultural products were best suited to the world markets, and who should benefit from their production. In the specific case of Anda's experimental farm's Russian administrators, a choice was made to produce milk and milk products, revealing a bias towards the Russian/European diet. The farm's Chinese invaders, who planted soya beans, and their defenders among the Chinese CER managers, opted for a proven crop that was more familiar to Chinese farmers.

Around the globe, from the nineteenth century on, agriculture has come under the supervision of the modern, interventionist state. Agriculture became a state science – studied, regulated – a means to reproduce knowledge and a means to reproduce social order. Agriculture was part of the overall development of a more direct relationship between government and those governed, a concept Foucault called "govermentality," referring to how governments produce subjects, but also to the strategies governments employ to control, rationalize, and reproduce knowledge. The professionalization of agriculture, like the professionalization of other fields of knowledge (health, the family) created positive and negative forms, modern and pre-modern agricultures, as well as employing the language of science and progress in the reproduction of forms of knowledge beneficial to, and supportive of, the modern state. Farming, perhaps the oldest organized human activity, rooted in local social networks and specific knowledge, was standardized as agriculture became a science. Instead of seasonal fluctuations and local conditions, the new professional agriculture used technology to homogenize crops and, sometimes literally, flatten local landscapes. Modern agriculture (in its push to transform land and the economy through modern planting techniques), "modern" (as opposed to "primitive") farmers, and homogenized landscapes produced new orders, new subjectivities, and new ways for governments to legitimize their control of territory.[13]

Experimental agricultural farms were on the cusp of this transformation of agricultural knowledge, along with state control and implementation of this modern production from a globalized economy. These farms standardized and reproduced agricultural knowledge, creating new plant and animal monocultures, and served to connect the farm – the local – to the globalized agricultural market. Experimental agricultural farms were thus key to the legitimization and circulation of new forms of modern agronomic science (the designation given to agriculture as a science), at the same time turning agriculture into a tool of governance. Agronomic science and experimental farms were essential to "supplying needed information and stimulating the progress of agriculture wherever they [were] established."[14] These farms brought together diverse fields of knowledge (chemistry, zoology, geography) and were often paired with agricultural museums and schools as part of the pedagogical mandate to transform local conditions. The primary goals of experimental farms, whether in Russia or China, were the modernizing of agriculture and gaining a national competitive edge in a global agricultural market. Each of these modernizing states used a discourse of commercialized and standardized agriculture to control their frontiers, to present resource production as modern, and to link national production to international markets.

In late-nineteenth-century and early-twentieth-century Russia and China, the association between agriculture and modernity was strong. In Russia, following the perceived failure of the 1905 revolution, thousands of young people who had dabbled in revolution abandoned politics and went to agricultural school. They hoped to change the Russian countryside, the perceived Achilles' heel of Russian modernization, by modernizing both Russian agriculture and the Russian peasant farmer. Some twenty to thirty thousand rural professionals (agronomists, physicians, educators, instructors, managers of peasant cooperatives) entered public life between 1905 and 1914.[15] The number of agricultural journals and books on agricultural subjects increased greatly (by 1916, there were 13,744,000 published copies of such texts and journals).[16] The Russian government also saw a link between agricultural modernization and national development: after the 1907 Stolypin reforms, funding for the Agricultural Ministry increased from 20.9 million rubles to 58.7 million rubles by 1912.[17] Between 1907 and 1914, the annual increase in the number of employed Russian agronomists was between 25 percent and 95 percent.[18] Not only did they wish to earn a living as crop scientists, but these agricultural professionals also believed their education made them part of an international culture of progressive modernization. Entering *obshchestvennost* – the public sphere – through public associations and formal institutions of civil society, *zemstvos* (lower-level elected self-governing bodies in the Russian empire), and government employment, these people saw themselves as part of a transnational progressive culture, a link between Russia's modernity and its soon-to-be modernized farms.

In Siberia and the newly colonized Russian Far East, this agricultural and professional modernization process was linked to plans to transform these regions into the breadbaskets and dairies of the world, by using the new transcontinental railways to bring these products to the world market. New, modern forms of agricultural business were sponsored by the Russian government. In Siberia, the rural cooperative movement created dairy, creamery, and cheese cooperatives that concentrated on the production, marketing, and transportation of dairy products. Beginning in 1893, as the railway crossed Siberia, the cooperative movement transformed "primitive" local production into a modern dairy production for the international market, chiefly Holland and Denmark. Success could be measured by the number of cooperatives: 65 in 1908, and 2,015 by 1918.[19] Perhaps Siberian bureaucrats, who moved into CER administrative positions after 1919, were inspired by this model of dairy modernization.

China was not immune to global trends in the professionalization of agriculture. Chinese agricultural manuals for the maximization of farm output, and social control, had existed for two thousand years. Some argue that these manuals are the earliest examples of governmentality as applied to the production of standardized agricultural knowledge.[20] These manuals were based in the local, not the national, and were aimed at specific, locally produced forms of knowledge. It was the Qing that founded the first modern experimental farm near Beijing under the auspices of the Empress Dowager Cixi (1835–1908), who also planned for agricultural schools to be built in the newly colonized north Manchuria. Sun Yat-sen (1866–1925) lectured in the early 1890s on the connection between modern agriculture and the strong state, and spent much of 1896–97 in the reading room of the British Library studying mining, agriculture, political economy, engineering, and the techniques of modern cattle rearing.[21]

China's first modern republican agricultural college was founded in Nanjing in 1914 by American agricultural missionaries who hoped to improve the Chinese countryside through the reform of agriculture.[22] In northern China, Peking University established an agricultural department that focused on the improvement of native pig stocks.[23] In Manchuria, Zhang Zuolin had created agricultural experimental farms on government-owned land and financed by his Bureau of Industry. The goals of Zhang's farms were the modernization of agriculture (including cattle breeding), profitability, and food production for the military. By 1924, the farms were making a small profit for the Manchurian government.[24]

In each Chinese case, agricultural modernization was connected to national and economic modernization, done in the context of Chinese agriculture and the Chinese market. Reformers chose what animals to stress with an eye to the Chinese market; thus, swine and poultry dominated Chinese agricultural reform. Cattle were draft animals, pulling ploughs and carts, and were used only rarely for food. In Manchuria alone, eight hundred pounds of pork meat was imported monthly, the largest food import of the region.[25] Traditional Chinese agriculture had focused on intensive farming of relatively small plots, which historically discouraged large animal husbandry. Cattle, most often water buffalo, were used in the rice regions to carry loads, or to plough, but most labour on the Chinese farm was human. The main plant sources of protein were soy, and the main animal sources were pigs, which did double duty as household scavengers. The Chinese diet consisted of very little beef; Europeans seeking to buy old animals for slaughter were often refused, because in the southern Chinese tradition it was understood that an animal that had served a family faithfully should not be slaughtered and eaten.[26]

The Chinese cultural associations with cattle should also be noted. Cattle, especially large herds of cattle used for milk and food, were associated with the open plains of northern China and the nomads who lived there. For millennia, the sedentary, agricultural Chinese looked down on the nomads as uncivilized. One of the principal markers of their lowly status, other than illiteracy (which was common among Chinese farmers as well), was the nomadic lifestyle attached to their pursuit of cattle herds, and the consumption of those herds, in the form of meat and milk, especially fermented milk. This dependency on cattle for food was seen by the Chinese as unseemly; the supposedly rancid smell of those who consumed dairy products was often cited as proof of their culturally inferior status. During the nineteenth century, overpopulation pushed the Chinese into the grasslands of the northern plateaus that were conducive to cattle grazing. These pastures, seen by the Chinese as wasteland, were immediately ploughed under and made into "productive" farmland; this caused violent conflicts between farmers and the nomadic populations they had displaced, and served only to associate cattle farming with primitive peoples reluctant to embrace agriculture.[27] Dairy and dairy farming were also associated with the now-discredited Manchu dynasty, members of which enjoyed drinking milk and eating cheese. In contrast, the Han Chinese rarely, with the exception of those in some of the frontier areas such as Yunnan, consumed milk or milk products in their diet.

These were all traditional prejudices that the farmers who invaded the Anda experimental farm would have shared. There was also the form of Manchurian colonial agriculture, family farms, that would have made the Anda experimental farm even more of an imposition. In northern Manchuria, large-scale Chinese agriculture had existed since the 1907 decree allowing Chinese to own Manchurian property. The most common form of Chinese agriculture in Manchuria was the tenant farm. Wealthy government officials and businessmen purchased thousands of acres that were leased on long contracts to individuals and families, who reproduced the grain-based agriculture of densely settled northern China, on the Manchurian border. There were two reasons for imported, grain-based family farming techniques – one cultural and one economic. Manchuria's grasslands, as "empty" lands, were called the "Great Northern Wilderness" with "uncultivated land like America and Australia."[28] "Real" agriculture, productive agriculture, would require that these wild areas be put under the plough. Beyond the preference for grain farming over cattle farming was a long Han ecological history that expressed itself in a spatial preference for the enclosed family farm, with its association with family and community. As David A. Bello notes in this volume (Chapter 2), people's ecological ties to land are constructed according to national and cultural norms, and are sometimes subject to manipulation from above, from states that have as their goals the creation of certain "facts on the ground." *Huang* (waste) and *guang* (vast), words often used to describe Manchuria's "unimproved" lands, carried with them the negative meanings of emptiness, and a lack of domesticity and civilization, and were often used to describe border frontiers inhabited by nomads far from civilization.[29] By civilizing and domesticating these lands by means of the Han family farm, the frontier was made "Chinese."

There were two reasons the Chinese thought Anda odd. The first was that it went against the dense village-based farming patterns of south China. The second was that it ran against the logic of the Manchurian agricultural market; it produced a good with no big local market. In the early twentieth century, it was the international soya bean market that was making Manchuria rich. Soya bean oil was used by industry, and the byproduct was reprocessed into fertilizer. Other than millet for home consumption, most Manchurian farmers were growing soy. In 1909, soy consisted of 15 percent of tilled fields; in 1922, 26.7 percent; and by 1924, 31.3 percent. By the mid-1920s, north Manchuria was producing 40 percent of the world's soya beans, 3,295,000 million tons, of which the international market consumed 91 percent.[30] In 1924, agriculture was the mainstay of the Manchurian economy – 44.7 percent of the Manchurian

FIGURE 4.2 Cows in the CER stockyards, postcard, 1920s.
*Source:* Collection of Dr. O. Bakich

GDP, compared with the GDP of the combined Russian, Japanese, and Chinese railways, most associated with the Manchurian economy, 3.4 percent.[31]

The 1924 attacks on Station Anda took place in an atmosphere of growing Chinese suspicion concerning the control of Special District resources, especially its agricultural resources, and Russian long-term colonial plans for the region. In its original incarnation, the CER was intended to be the link between European and Asiatic Russia, and to carry the raw materials of the Russian Far East and Asia to production and to international market. Agricultural production by Russians in Manchuria was not considered important, and Witte specified that it was to be Russian control over manufacturing and trade that would solidify Russia's "informal" sphere of influence. In Witte's vision, annexation was not the goal. Rather, the goals were complete Russian economic control of the Manchurian economy as a source of raw materials, a market for finished Russian goods, and a Russian transit point into the international market through Vladivostok. Competition between nations, Witte believed, was primarily economic competition, and Russia could gain the advantage through trade, not its military "since the authority of the metropolis over its colonies now more than ever is exercised not by force of arms but by trade."[32]

Witte had already conceived of Manchurian industrial development as Russia's link to a globalized market. Instead of formal political control, Witte, creator of Russia's late imperial modernization project and the CER, argued for Russian control over Manchurian regional and international trade, finance, and the circulation of goods and people, modernizing Manchuria for what he and other Russians believed was a denationalized and depoliticized global market. Witte argued for the CER concession settled by Russian tradespeople and skilled workers, leaving agriculture to the Chinese. In 1923, an American diplomat described a Russian conference that took place in 1900 on the subject of land expropriation in the CER zone. He noted it was agreed that "excessive and unnecessary expropriation of lands might arouse dissatisfaction among the local population and involve too considerable expenses, the Railway receiving in turn lots unfit for agriculture and for the Russian population, which under no circumstances might compete here, economically, with the Chinese."[33] But something changed between 1900 and 1902. Perhaps it was the realization that, following the end of the Boxer Rebellion in August 1900, all Manchuria could, potentially, be annexed by Russia. Certainly, the Russians dug in deep, occupying Manchuria for three years and taking over its administration – aggressive actions that prompted the 1904–5 Russo-Japanese War. Perhaps it was after Witte's own visit to the CER concession in 1902 that he realized the area's rich possibilities.

For whatever reasons, land expropriations for the CER increased in size and frequency. Russian colonization, once seen as a danger to Russian commercial development in Manchuria, was now necessary "to enliven and colonize the area depending on the railway, which certainly was of a great importance for the development of its freight traffic; all these last named aims could only be carried out on condition that the area to be colonized belonged to the railway and the necessary elements of colonization were established there."[34] Land was bought at great expense – ten thousand rubles – and following Witte's resignation in 1903, the Ministry of Finance opened northern Manchuria to Russian agricultural settlement on generous terms. Manchurian land was leased on multiple ten-year contracts, and lease money was collected for the first ten years up front to encourage production. Only Russian subjects were permitted to apply for leases.[35] In the first surveys done of the concession, land not occupied by Chinese was "quickly alienated." From the expropriated 192,000 *desiatinas,*[36] 500 farms were laid out, with plans for 1,500 to 2,000 future farms.[37] Expropriations continued until 1911.[38]

What use was made of this property? To function as a railway, the CER required only 56,098 acres. The remaining 108,068 acres were used to build villages, towns, and one city – Harbin.[39] Land was leased out for development and colonization. Once the area was settled, it had to be governed and taxed so that its administration could be paid. This administration was carried out by the CER's Civil Department, "a state mechanism in miniature," according to Dmitrii Horvath, the CER's first CEO.[40] The Civil Department supervised the Russian-controlled municipal governments and taxed the concession's inhabitants. By the early 1920s, revenues from this department constituted the CER's profit and were the source of its investment into new enterprises such as Anda. These profits were also used to pay the salaries of the CER's primarily Russian staff. Chinese complaints that the CER's ownership and administration of this property was in violation of the original 1896 contract and the 1920 reorganization, which both stated that the CER was a business venture only, were met with a uniform Russian response: local administration was a business, not a political venture. According to the Russians, the Russians had put this successful administration in place and had the talent to carry it into the future. China's sovereignty was not affected, and it was guaranteed under both contracts; China would regain full control over the area after ninety-nine years. The creation of the Special District was China's answer to the problem of continued Russian control of the former concession, but full Chinese control would take place only once property deemed non-essential to the CER's functioning was no longer under the CER's control. When that happened, the use of that property and any profits derived from it would be under Chinese control.

Thus, from the point of view of Chinese administrators, whether they operated within the CER, the Special District office, or in Shenyang, the Russian CER board appeared to be investing in a dubious, Russian-centric project, one that benefited Russian agronomists, paid salaries to Russian farm workers, and produced dairy goods for a Russian diet. It had the appearance of being the newest incarnation of the Russian colonial project, which sought to justify itself through leaseholds and the rational exploitation of agricultural space and economy. The specific issue of displaced Chinese farmers in Anda united the Special District Chinese administrative class with its farmers. Long leases were the most common form of landholding for ordinary Manchurian farmers. Ownership was considered too expensive and uncertain. When one leased the land, the owner took on the financial burden of house building, digging a well, building pens, and so on. Leasehold, therefore, was a common practice, and signed lease contracts carried with them obligations on both sides. In the case of Anda, Chinese translators working for the CER had originally leased the property from the CER and then subleased it to more than sixty-five Chinese families. In 1922, the CER forcibly removed these families and offered them uncultivated land in return, in order to create the experimental farm as part of the railway's efforts to diversify its activities and make the company indispensable to regional economic development. The original leases were revoked.[41] Thus, Anda's experimental farm for the production of milk and milk products was built on the expropriation of what the local population considered to be rightfully leased land.

Station Anda's experimental farm included a dairy-processing plant that specialized in butter and sour cream. The rationale for creating a dairy-processing farm was that it would assist farmers who were having difficulty finding markets for raw milk. Dairy farming was a Russian activity (as opposed to a Chinese one), as was the consumption of milk, butter, buttermilk, and sour cream. Thus, the benefits of the experimental farm were directed at and enjoyed by Anda's Russian farmers and the Russian-controlled CER. Six months after it was founded, the farm was produced approximately eighty-five thousand bottles of milk, four thousand pounds of butter, and three thousand pounds of sour cream.[42] For the Russian members of the CER board, the Anda farm made *Russian* economic sense. By providing a processing point for an expanding Russian market in dairy products, Anda diversified the economy and created jobs. The farm also showcased an agricultural modernity that connected the CER to national and international markets, justifying and legitimizing continued Russian economic control, a model begun by Witte and carried on by his successors. From the Chinese perspective, Anda was an attempt by the Russian-controlled CER to expand job opportunities for Russians by displacing established Chinese farmers; an attempt to dominate the Manchurian landscape physically, economically, and – through claims of agricultural modernity – ideologically.

This was the background to the 1923 creation of the Chinese Land Department and the 1923–24 seizures of all lands deemed unnecessary to the functioning of the CER as a railway. It is also the background to the common popular Chinese demand that the land returned over to Chinese administration; the Anda occupation was one manifestation of this. The 1922 Washington Conference further highlighted conflicting Chinese and Russian understandings of land and its development. The conference had been called to resolve intertwined military, political, and economic problems in Asia, particularly as they touched on China, and to prevent a potential naval arms race. There were two foreign railway concessions whose ownership was in question at the conference: the former German Shandong concession and the Russian CER concession. The Japanese government, having occupied Shandong, argued that that concession should be placed under its control. The Russian CER board, having no national power to back it, argued that the CER should be placed under an international trusteeship, with the Russian board preserved intact. Republican Chinese foreign minister Wellington Koo, China's representative at the Washington Conference, argued that both concessions were Chinese property on Chinese territory. He was able to have both the Shandong and CER railroads placed under joint administrations, with the Chinese having the dominant position. Koo also won back important postal, tariff, and telecommunications concessions for China. Frustrated by Koo's evasive negotiations, the US and Japan, who were "the powers" at this conference, signed a treaty that promised first and foremost to guarantee Chinese sovereignty over the railroads.[43] The treaty established the precedent that the CER was half Chinese-owned, and it bound the Chinese to protect the CER in the absence of a defined Russian partner.

Each side had a different understanding of the Washington Conference. For the United States government and the Russian CER board members, the agreement preserved the status quo. The CER's land, including the Anda dairy, remained under the railway's control to use as the railway deemed best for business purposes. For the Chinese, the CER board members, the national government in Beijing, and the regional government led by Zhang Zuolin, the conference result confirmed China's sovereignty over its territory – and for China, sovereignty was the same as territorial control and administration. On this basis, in the summer of 1923, the Special District administration seized the offices of the CER Land Department and reorganized the leases under a Chinese administration. Both sides used the Washington Conference in their arguments. Chinese demonstrators and delegations of shopkeepers, in support of this action, petitioned for the return of this land to Chinese control, citing the Washington Conference and its guarantee of Chinese sovereignty. The Russian CER board members and foreign diplomatic corps invoked the Washington Conference's guarantee of the CER's assets; at the same time, Russian board members insisted that only they were capable of administering the CER and ensuring that it continued to make a profit.

The land debacle occupied the Russian and Chinese press and American diplomatic dispatches over the fall and winter of 1923–24. The Chinese position focused on how the CER had always been a colonial venture, despite two contracts that specified that it was to be a business venture only. The CER's administration of the concessions, now district farms, villages, towns, and a major city, constituted a political act against China's sovereignty. The towns and farms, Anda included, were noted as examples of this continued colonial project. The CER's investments, in lumber mills that produced fancy parquet flooring, the mines, and Anda's dairy farm, were cited as unprofitable job creation ventures for the Russian community. Interestingly, the CER's other experimental farms, which focused on soya bean production, were not mentioned.

The Russian CER board's response to the Chinese accusation was based in equal parts on the CER's having a rational business plan and the claim that CER's Chinese employees were not capable of implementing this plan due to their nationalist prejudices against Russian enterprise. The Russians argued that the CER was a business, out to make a profit and attract foreign investors. One had to invest in long-term businesses, like the Anda dairy farm, for the international market. Foreign communities in China were growing, providing a consumer base for locally produced goods intended for the foreign market. Russia, too, would need goods like milk and sour cream. The Chinese were not capable of administering a complex enterprise like the CER. They were blinded by nationalism; they wanted to strip the CER of its assets for their own corrupt goals. Only under Russian management could the CER be above politics and fulfill its destiny. Both sides repeatedly invoked the Washington Conference, emphasizing different aspects. This was the atmosphere leading up to the spring 1924 occupation.

There were three attacks, the first in early March 1924. The farm's head agronomist reported that Chinese "armed with spades came to the field, disposed labourers of the [CER] Land Department and destroyed the work already done. Thirty men began to break and to cover with earth flowering beds, and then came Chinese with two ploughs and began to plough the field again."[44] On March 21, after the experimental farm resumed work, the head agronomist and his team were again driven from the fields by Chinese farmers armed with spades. Further attempts to resume work were prevented by the Chinese, who built a cottage in the fields and threatened to burn the cattle barns.[45] Presumably, work at the farm resumed once more because the final occupation occurred in April, when the weather was milder. From the perspective of the Chinese farmers, their leases had been illegally terminated and their lucrative grain fields replaced by, in their opinion, useless cattle. Instead of using potentially useful and expensive Russian-made buildings, the Chinese farmers attempted to destroy these concrete symbols of the Russian dairy economy that had displaced them, asserting what they perceived to be their legal right to the land and to an agricultural system oriented to local Chinese – not Russian – settlers. Their actions found support with the local Chinese authorities, who instructed the police not to interfere. Russian observers noted that the Chinese farmers ignored the expensive newly purchased agricultural equipment, leaving it lying in the fields. The expensive dairy cows became stock animals, pulling traditional and – in the Russian opinion – primitive Chinese ploughs.[46]

Russian response was immediate and polarized. In a letter to the American ambassador, Ostroumov, the CER general manager, accused all Chinese of being against modern development and hindering the railway's attempts to modernize northern Manchuria. The region's Russian-language press, based in Harbin, criticized the Chinese for rude and uncivilized behaviour, employing words such as "mob" and "marauders" to describe what they called "wild Chinese action."[47] The Chinese response was to defuse. In an agreement signed in the summer of 1924, both sides preserved face but the Chinese essentially won their case. Land not necessary to the functioning of the CER was turned over to the Chinese Special District. Leases were reinstated and no one was expelled, except, arguably, Anda's Chinese farmers. The dairy remained CER property, the Chinese farmers were removed, and Anda became, as it is still known today, the "City of Cows."

The Anda experimental dairy farm occupation was a pivotal incident in a forty-year conflict between northern Manchuria's two founding colonial peoples. The clash at Anda was not just a case of cranky, dispossessed farmers reasserting their claims. It was symbolic of how a seemingly neutral, humble dairy was invested with different meanings by Chinese and Russian farmers, meanings that ultimately reflected their (in some ways similar yet in some ways tragically different) goals for Manchuria's post-colonial space, represented by Russian industrial agriculture and the Chinese family farm. Each side had its own vision of agricultural and economic modernity. The Russian vision was of a globalized market harnessed in Manchuria through Russian skill. This argument was bolstered by the precarious nature of Russian existence in post-1917 Manchuria, where the collapse of the imperial Russian state had intensified the need to prove continued Russian technological and economic control in Manchuria. These Russians also continued a pre-1917 legitimizing discourse that linked economic development with Russian skill, at the same time claiming that such actions held no political meaning since they took place in what the Russians defined as the neutral, depoliticized space of the market economy. Manchuria's Chinese administrators and farmers called into question the "fact" that Anda's CER-sponsored dairy farm was neutral economic territory. For Manchuria's Chinese population, this formulation of the CER's purpose was fundamentally contradictory: the CER had created a colonial enterprise that controlled the economy, the physical space, and even the utilization of that physical space under the guise of a business concession, repurposing for Russian, not Chinese, use.

Full Chinese political control of Manchuria after 1917 made it necessary for the region's Russian elite to ideologically stake its claim to Manchuria through its use of the land and resources. It did so by arguing that Russian human capital and talent were essential to Manchuria's cultural, technological, and economic life, asserting that continued Russian supervision was tied to economic reorganization, Manchuria's place in the international markets, and the "common good." This common good, from the Chinese perspective, was an attempt to retain control over CER resources in order to justify continued Russian supervision, through the development of an agricultural product that made little economic sense to the Chinese in Manchuria, and which seemed to be a work-creation project for the Russian minority. Although the relationship between the Russians and the Chinese in Manchuria was, for the most part, a good one, it fell apart when both groups, each wholly and rationally convinced of its own arguments, were unable to compromise in the fluid post-colonial space of north Manchuria.

## Notes

*Epigraph:* Chinese Eastern Railway Printing Office, *North Manchuria and the Chinese Eastern Railway* (Harbin: CER Printing Office, 1924), 455–56; US National Archives and Records Administration (NARA) Record Group (RG) M316, Roll 137, 861.77/3233, Memo on the Journey of the American Minister to Harbin and over the Chinese Eastern Railway, September 19, 1923.

1 *Gaoliang* is a form of sorghum common in Northern China.
2 Chinese Eastern Railway Printing Office, *North Manchuria and the Chinese Eastern Railway.*
3 "Stimulation of the People's Scorn," *Tribuna (Harbin),* February 20, 1924. NARA, RG 59, Roll 153, M316, File 861.77/3376.
4 S.C.M. Paine, *Imperial Rivals: China, Russia and Their Disputed Frontiers* (London: M.E. Sharpe, 1996), 116–17. See also David Schimmelpenninck van der Oye, *Toward the Rising Sun: Russian Ideologies of Empire on the Path to War with Japan* (DeKalb, IL: Northern Illinois University Press, 2001).
5 Blaine R. Chiasson, *Administering the Colonizer: Manchuria's Russians under Chinese Rule 1918–1929* (Vancouver: UBC Press, 2010); Chia Yin Hsu, "A Tale of Two Railroads: 'Yellow Labor,' Agrarian Colonization and the Making of Russianness at the Far Eastern Frontier," *Ab Imperio* 3 (2006): 217–53; Chia Yin Hsu, "Railroad Technocracy, Extraterritoriality, and Imperial *Lieux de Memoire* in Russian Émigrés' Manchuria: 1920s–1930s," *Ab Imperio* 4 (2011): 59–105; and Chia Yin Hsu, "Frontier Urban and Imperial Dreams: The Chinese Eastern Railway and the Creation of a Russian Global City, 1890s–1917," in *Russia in Motion: Cultures of Human Mobility since 1850,* ed. John Randolph (Avrutin: University of Illinois Press, 2012), 43–62.
6 See Robert Lee, *The Manchurian Frontier in Ch'ing History* (Cambridge, MA: Harvard University Press, 1970); Li Narangoa, "The Power of Imagination: Whose Northeast and Whose Manchuria?" *Inner Asia* 4, no. 1 (2002): 3–25; Gang Zhao, "Reinventing China: Imperial Qing Ideology and the Rise of Modern Chinese National Identity in the Early Twentieth Century," *Modern China* 32, no. 1 (2006): 3–30; Blaine Chiasson, "Late Qing Adaptive Frontier Reforms in Manchuria, 1900–1911," in *Entangled Histories: The Transcultural Past of Northeast China,* ed. Dan Ben-Canaan (Cham, Switzerland: Springer Publishing, 2014), 161–75.
7 Jamie Bisher, *White Terror: Cossack Warriors of the Trans-Siberian* (New York: Routledge, 2005), 377.
8 Hsu, "Railroad Technocracy," 59–105.

9 Chinese Eastern Railroad Printing Office, *North Manchuria and the Chinese Eastern Railway,* 280.
10 Ibid., 406.
11 For recent works, see Daqing Yang, *Technology of Empire: Telecommunications and Japanese Expansion in Asia, 1883–1945* (Cambridge, MA: Harvard University Press, 2010); Janice Mimura, *Planning for Empire: Reform Bureaucrats and the Japanese Wartime State* (Ithaca, NY: Cornell University Press, 2011); and Jun Uchida, *Brokers of Empire: Japanese Settler Colonialism in Korea, 1876–1945* (Cambridge, MA: Harvard University Press, 2011).
12 Prasenjit Duara, *Sovereignty and Authenticity: Manchukuo and the East Asian Modern* (Lanham, MD: Rowman and Littlefield Publishers, 2003).
13 James C. Scott, *Seeing Like a State: How Certain Schemes to Improve the Human Condition Have Failed* (New Haven, CT: Yale University Press, 1998), 253.
14 Department of Agriculture, *The Experimental Farms of the Dominion of Canada* (Ottawa: Department of Agriculture, 1887), 3.
15 Ilya V. Gerasimov, "Introduction," in *Modernism and Public Reform in Late Imperial Russia: Rural Professionals and Self-Organization, 1905–1930* (Hampshire, UK: Palgrave Macmillan, 2009), 1–11.
16 Ibid., 16.
17 Lazar Volin, *A Century of Russian Agriculture: From Alexander II to Khrushchev* (Cambridge, MA: Harvard University Press, 1970), 107.
18 Gerasimov, *Modernism and Public Reform,* 68.
19 Alexis N. Antsiferov, "Credit and Agricultural Cooperation," in *The Cooperative Movement in Russia during the War,* ed. Eugene M. Kayden (New Haven, CT: Yale University Press, 1929), 388.
20 Francesca Bray, "Science, Technique, Technology: Passages between Matter and Knowledge in Imperial Chinese Agriculture," *British Society for the History of Science* 41, no. 3 (2008): 319–44.
21 Audrey Wells, *The Political Thought of Sun Yat-sen: Development and Impact* (Hampshire, UK: Palgrave Macmillan, 2001), 15; and "A Plea to Li Hung-Chang," June 1894, in *Prescriptions for Saving China: Selected Writings of Sun Yat-sen,* ed. Ramon H. Myers (Stanford, CA: Stanford University Press, 1994), 7.
22 J. Lossing Buck, *Cornell International Agricultural Development Bureau Bulletin 25: Development of Agricultural Economics at the University of Nanking, Nanking, China, 1926–1946* (Ithaca, NY: Cornell University Press, 1973), 6.
23 Sigrid Schmalzer, "Breeding a Better China: Pigs, Practices and Place in a Chinese County, 1929–1937," *Geographical Review* 92, no. 1 (2002): 6.
24 Ronald Suleski, *Civil Government in Warlord China: Tradition, Modernization and Manchuria* (New York: Peter Lang, 2002), 109.
25 A.R. Agassiz, "Our Commercial Relations with Chinese Manchuria," *Geographical Journal* 4, no. 6 (1894): 539.
26 Ralph W. Phillips, "Livestock in the Lives of the Chinese," *Scientific Monthly* 60, no. 4 (1945): 276.
27 Ibid., 261.
28 Yi Zhou, "Dongsheng nongye zhi yanjiu" [A study of agriculture in the eastern provinces], *Dongsheng jingji yuekan* [*Eastern Province Economic Monthly*], February 15, 1930, 15, quoted in Fuliang Patrick Shan, "The Development of the North Manchuria Frontier, 1900–1931" (PhD diss., McMaster University, 2003), 48.

29 Dee Mack Williams, *Beyond Great Walls: Environment, Identity and Development on the Chinese Grasslands of Inner Mongolia* (Stanford, CA: Stanford University Press, 2002), 65.
30 Shan, "Development," 104.
31 Kang Chao, *The Economic Development of Manchuria: The Rise of a Frontier Economy* (Ann Arbor: University of Michigan Press, 1983), 16, 33.
32 I.U. Witte, "Dokladnaia zapiska Vitte Nikolaiu II" [Report on the letter of Witte to Nicholas II], *Istorik Marksist* [*Marxist Historian*], nos. 2–3 (February-March 1935): 133, as cited by Schimmelpenninck van der Oye, *Toward the Rising Sun,* 73.
33 Memo on the Land Question, Enclosure 8 in dispatch 382, Harbin, August 7, 1923. NARA, RG 36, Roll 157, M316, File 861.77/3181, August 23, 1923.
34 Memo on the Land Question. NARA, RG 36, Roll 157, M316, File 861.77/3198, August 23, 1923.
35 Chia Yin Hsu, "The Chinese Eastern Railroad and the Making of Russian Imperial Orders in the Far East" (PhD diss., New York University, 2006), 103.
36 One *desiatina* = 2.7 acres.
37 Chia, "Chinese Eastern Railroad," 115, 119.
38 Ibid., 115.
39 These numbers vary from officially stated numbers elsewhere, but are cited in Igor Mitofanov, *The Lands and Land Administration of the Chinese Eastern Railway Company and the Incident of August 1st, 1923*, Part I,"History and Legal Basis" (Harbin: n.p., n.d.), 4.
40 D.L. Khorvat, "Memoirs," unpublished manuscript (Stanford, CA: Hoover Institution, Stanford University, n.d.), Chapter 7.11.
41 "Chinese Farmers at Anda attack the Agricultural Experimental Station," *Sungari Daily News,* April 29, 1924. NARA, RG 59, Roll 153, M316, File 861.77 3459, May 5, 1924.
42 "Consular Report: Agricultural Experiment Station at Station Anda." NARA, RG 59, Roll 153, M316, File 861.77/3071, March 19, 1924.
43 Blaine Chiasson, "If the Railroad Is to Be Preserved, It Must Be Changed: The CER and Its National Ownership, at the Washington Conference, 1922" (paper presented at On the Front Line in East Asia: The Chinese Eastern Railway and Russo-Chinese-Japanese Relations, 1905–1935 Conference, George Washington University, Washington, DC, Institute for Russian, European and Eurasian Studies, January 25, 2013).
44 "Attacks at Station Anda," US Consul Hanson to Jacob Schurman enclosures in NARA, RG 59, Roll 153, File 861.77 3449, May 1, 1925.
45 Letter to the American Minister, Peking, April 24, 1924; "New Troubles at Station Anda," *Russkii golos* [*The Russian Voice*], April 23, 1924. NARA, RG 59, Roll 153, M316, File 861.77/3445, May 2, 1924.
46 Consular Report "Excerpts from the Harbin Russian Press. *The Harbin Daily News* and *Russkii golos* 14/15 April 1924." NARA, RG 59, Roll 153, M316, File 861.77/3445, May 2, 1924.
47 Boris Ostroumov, Letter to the American Minister, Peking, April 24, 1924. NARA, RG 59, M316, Roll 153, File 861.77/3445, May 2, 1924; "A Sad Page," *Russkii golos* [*The Russian Voice*], April 13, 1924. NARA, RG 59, Roll 153, M316, File 861.77/3439, April 20, 1924.

# 5
# "Hibernate No More!": Winter, Health, and the Great Outdoors

*Norman Smith*

In 1944, Huang Fujun (1890–?), Manchukuo's minister of agriculture, published *Manzhou nongye jishuzhe bi xie* (*The Spiritual and Technical Requirements of Manchurian Agriculture*), in which he criticized local people for their unhealthy tradition of lying "dormant" (*timian*) like animals through the winter.[1] Similarly, physician Wang Luo (1906–?), director of the Police Hygiene Division in Fengtian (present-day Shenyang) and subsequently minister of the People's Welfare Department, observed that while the people of Manchukuo experienced a climate similar to that of many Canadians, they suffered inordinate levels of scarlet fever, tuberculosis, and pneumonia, all of which he claimed Canadians rarely contracted.[2] Wang compared Canada and Manchukuo to heaven and hell, respectively, citing the different customs that supposedly advantaged Canadians with better health. Wang argued that people must adopt a "do not fear cold" (*wu wei hanleng*) attitude to overcome endemic disease and poor health.[3] Huang and Wang, two highly ranked Manchukuo officials, reflect a pivotal period in Manchuria's history, when winter lifestyles became a focal point of public debate, which revolved around the utility of incorporating "modern" ideologies and lifestyles. This chapter examines that era's shifting notions of what constituted healthful winter living, with a focus on the season, heating practices, and outdoor activities.

In this volume, Diana Lary has demonstrated that the latitude, topographical features, and wind currents from Siberia created an environment whose uniqueness was widely remarked by others (Chapter 1). In addition to regional names such as Manchuria, the Northeast, and the

Three Eastern Provinces, the territory was variously labelled "the Garden of China,"[4] "the Granary of Asia … the promised land of China,"[5] and, as described by the Manchukuo government, a "paradise land" (*letu*). The German geographer Ferdinand Freiherr von Richthofen (1833–1905), who coined the term "Silk Road," had lavish praise for the region.[6] Manchuria's riches were famously lauded by one of the greatest rulers of the Qing Dynasty (1644–1912), the Qianlong emperor (r. 1735–96), who waxed rhapsodic about his native land, romanticizing it and its vast array of animals, plants, and aquatic life; Mark Elliott has noted that the emperor's list of local fish and mollusk names was twenty lines long.[7] As David A. Bello and Loretta Kim have also made clear in this volume (Chapters 2 and 3), furs and other natural riches contributed to the region's enormous natural wealth.

The abundant natural resources and the relatively cold, long winters of Manchuria led to frequent comparisons of the region with North America, especially Canada – another northern region with a colonial heritage. This trend continued through the duration of Manchukuo as commentators attempted to enhance perceptions of the region. In 1938, Fujiyama Kazuo (1889–1975), associate director of Manchukuo's National Central Museum, wrote that Manchuria's environmental conditions were similar to those of the areas surrounding North America's Rocky Mountains and California, what he termed a "prairie" (*buleili*) style.[8] Fujiyama argued that Manchukuo was blessed with the very characteristics that made strong two of the world's greatest empires, England and France, as well as the strengthening characteristics of Australia and Canada, two of England's former colonies.[9] American geographer George Cressey (1896–1963) also argued that Manchuria resembled central North America – in its soil, climate, vegetation, and "newness."[10] The "pioneering" nature of that "newness" in local life had led British naturalist Arthur de Carle Sowerby (1885–1954) to compare Manchuria favourably with Alaska and British Columbia.[11] A.D. Woeikoff, a Russian director of the Chinese Eastern Railway–owned experimental farm in Jilin, even charted the migration of the Manchurian crabapple tree to Alberta, Manitoba, and Saskatchewan.[12]

As with Canada, perceptions of Manchuria often centred on the region's winters. The 1941 *Manchoukuo Yearbook* delineates the length of Manchuria's "non-frost" season: in the south, from 150 to 200 days a year; in the centre, around Taonan, 160 or more days; and in the north around Qiqihar, some 140 days.[13] Thus, at least half of each year was subject to winter conditions. Notably, these calculations are stated in terms of "non-frost" days – accentuating for readers that Manchuria did, in fact, experience weather that was not winter-like. But no matter how the editors of the yearbook sought to frame it, the length of the winter and the extent of Manchuria's cold temperatures did much to structure local life. As Thomas Gottschang and Diana Lary have observed, agricultural activities all but stopped during the winter, making labour demand "highly seasonal" as "winter brought virtual hibernation."[14] Ronald Suleski has also noted that in the "harsh climate ... the ground becomes as hard as cement and the rivers become highways of ice."[15] Such conditions made hunting, travel, and transportation "most favourable in the winter, when the ground is frozen and rivers are better highways than the land."[16] Yet despite the utility of frozen land and water for some, such winter conditions cemented Manchuria's reputation as "a harsh unromantic place."[17]

The extent to which such environmental conditions may structure human life has long been a topic of interest in Chinese culture. Assertions of environmental determinism in China date at least to the [*Writings of*] *Master Guan* (compiled from documents pertaining to the spring and autumn period by politician Guan Zhong [c. 720–645 BC]). The chapter "Shuidi" (Water and Earth), for instance, contains assertions such as: "Now the water of [the state of] Qi is forceful, swift and twisting. Therefore its people are greedy, uncouth, and warlike" and "The water of Chu is gentle, yielding, and pure. Therefore its people are lighthearted, resolute, and sure of themselves."[18] The linking of the environment with human behaviour reflects long-standing belief in *fengshui* and the merits of living in accordance with the seasons, as well as newer discourses on modern means to negotiate the tough climate. How to attain ideal human winter health was hotly debated in the first half of the twentieth century in resource-rich yet seemingly climatically challenged Manchuria.

In the early 1900s, Yale professor Ellsworth Huntington (1876–1947), influenced by social Darwinism, eugenics, and the shift in academic geography "from exploration to explanation," argued in books such as *Civilization and Climate* (1915) that the character of a race is shaped not only by inheritance but also by physical and social environments.[19] Although his work has since been rejected (in large part for its racist underpinnings), his arguments greatly influenced contemporaries around the world. Significantly, Huntington was effusive in his praise for Manchuria:

> In Mukden ... there is more activity and life than in any other Chinese city that I have visited. Activity and progressiveness are said to be still more evident in Harbin, and most of all in the far north, where the town of Aigun

> on the Amur River, opposite Blagoveschensk, is reported to be inhabited by Chinese who seem quite unlike their countrymen in their modern spirit of progress and in their bustling activity.[20]

Noting the "bustling activity" that he believed characterized local life, Huntington did not discuss the famed habitual winter "dormancy" that motivated others to critique the ways in which many Manchurian locals lived through winter. Indeed, Huntington credited the north Manchurian climate with fostering "an almost American progressiveness."[21]

In 1938, Fujiyama Kazuo explicitly cited Huntington's work in *Xin Manzhou fengtu ji* (*A Record of New Manchuria's Natural Conditions and Social Customs*).[22] Fujiyama argued that climatic determinism played a central role in Manchuria, in the development of what he perceived to be the Han Chinese "national character" (*minzu xing*).[23] He argued that the Chinese were unable to "conquer nature's conditions" (*kefu ziran de tiaojian*), leaving them with selfish characters defined by "egoism" (*liji zhuyi*).[24] Describing the Chinese as emotionless, cruel, and "beast-like" (*yeshou yang de*), Fujiyama warned Japanese migrants to be conscious of their environment and, unlike the Chinese, focus on conquering winter. The inherent racism of Huntington's work is explicit in Fujiyama's description of the Chinese; he cited the "simple" environment of the broad plains for "Chinese people's simple character" (*Han minzu de danchun xing*).[25] Fujiyama reflected the view of contemporary eugenics that one could improve the quality of human life, in his view through the proper use of Manchuria's "bright and clear" (*minglang*) weather, which he believed augured well for future development.[26] Winter may have long encouraged Chinese "hibernation," but that legacy, according to Fujiyama, had to be challenged in order to fully reap, and modernize, the region's riches.

Conquering the climate's influence in "long winter Manchuria" (*dong chang de Manzhou*)[27] necessitated the rejection of habitual hibernation – the "dormancy" cited by Huang Fujin – in favour of active engagement with the winter. Another writer, Dong Qiushui, praised the region for having China's greatest variety of weather, with more extensive winters than anywhere else in China, and detailed the positive influences that the season could bring to people's lives.[28] Dong criticized the long-term observance of the "Double-Nine Expulsions of the Cold" (*jiujiu xiaohan*) – such as wearing fur, sitting around a fire, drinking alcohol, and smoking tobacco.[29] In 1934, George Cressey also noted, regretfully, that "for four months the temperature makes most outdoor work impossible, and one can only sit by the fire and wait. How to overcome this winter idleness is an unsolved problem."[30] Increasingly through the 1930s and 1940s, "winter idleness" was denounced as a problem that people such as the Police Hygiene Division director of the capital city Xinjing (present-day Changchun), Bi Kegang, argued could be solved through a number of strategies to enliven the people, including winter outdoor activities, such as skating and skiing.[31] Ten specific recommendations were often made:

1. Participate in outdoor sports.
2. Ensure proper indoor air circulation.
3. Lower indoor temperatures [high indoor temperatures were thought to encourage illness].
4. Limit the burning of coal indoors [burning coal indoors harmed the lungs].
5. Prevent excessive coal heat from emanating from the stove.
6. Strive hard to prevent coal smoke.
7. Wait until a new house is fully dry before moving in.
8. Use clothing to protect against the cold.
9. Wear more clothing when the temperature is colder; reduce layers as necessary.
10. Eat more fruit and vegetables for nourishment.[32]

These ten recommendations, which appeared repeatedly in various media, were argued to increase one's chances of a healthy life in the winter. They also reflect essential elements of the modernist agenda in 1930s and early 1940s Manchuria – encouraging incipient consumerism and modernity through the purchase of clothing, food, and coal. The three most significant recommendations encourage increasing the safety and efficiency of heating practices, living the "low-temperature lifestyle" (*diwen shenghuo*), and participating in outdoor activities.

During the Manchukuo era, in the urban centres at least, there was increasing pressure to change long-vaunted heating practices. Dr. Miura Unichi (1896–1983), a professor of hygiene at Manchuria Medical College, observed that the *kang* (a low, wide stove-couch used for heating in North China) was the most common form of heating.[33] Dating as far back as seven thousand years, the *kang*, made of earth and brick, was used to fashion a cooking and resting space that maximized fuel efficiency. Hot air from the stove ran through the construction to heat the resting platform, which was covered with mats and bedding. The *kang*, at optimal warmth after two to three hours of heating, did not need to create very high

FIGURE 5.1 *Kang.*
*Source:* "Manchoukuo: A Pictorial Record," *Asahi Shimbun,* 1934, 21.

temperature (12–15°C) to make a living space comfortable; a well-constructed *kang* retained enough heat that it did not have to be re-lit during the night. Without doubt, the *kang* was a great comfort in cold buildings that had no central heating. Miura noted, too, that a variety of "Manchurian" *kang* existed (in addition to one Russian and two Korean styles), with different construction materials and sequences of pipes and vents.[34] He reasoned that the related *Chaoxian ondol* ([North] Korean underfloor heating) was superior to its Manchurian counterpart because of the large, thin stone slabs that constituted its uppermost level, which Miura preferred to the mix of earth and bricks that predominated in the latter. Miura wrote approvingly that Japanese pioneers sometimes employed *ondol*-type systems, with bricks, earth, and stone. Other observers also noted the benefits of local heating techniques, especially in rural areas, focusing on south-facing rooms (to maximize sunlight), the *kang,* and a few concessions to Japanese taste, contrasting with the Japanese-, European-, and North American–style housing that predominated in Manchukuo's largest urban centres. David Tucker has described how Manchukuo's planners dismissed Japanese-style housing as insufficient for Manchurian winters and instead promoted modified Chinese styles.[35] When it came to the *kang*, however, Miura preferred the Korean styles to the Chinese.

The *kang* was not without its critics. By the early 1940s, the *kang* was consistently blamed for health complications and economic difficulties. Tai Yuzhen, an editor at the popular journal *Simin* (*This People*), meticulously detailed diseases and costs associated with the *kang*, which he linked with potentially deadly and economically ruinous consequences. Tai argued that the *kang* had a "cold, damp nature" (*chao shi xing*) caused by moisture emitted from the earth and brick materials.[36] The damp was described as a serious problem, vividly expressed in a common saying: "can sleep cold, do not sleep damp" (*neng shui liang bu shui chao*).[37] To prevent moisture, Tai suggested applying multiple layers of oiled paper on top of the earth and bricks of the *kang* or adopting the Korean method of adding a top layer of stone.[38] Tai warned that if the *kang* were inadequately heated in the winter and the moisture remained, all sorts of health complications would arise, including the common cold, cold legs and waists, typhoid, leprosy, and skin ulcers.[39] Further, Tai argued that a famous (though unnamed) doctor had blamed the *kang* for making youth age prematurely, instilling in them a "greedy for sleep evil habit" (*tan shui de exi xing*).[40] Tai warned that the *kang* encouraged a hibernation that he believed hindered the ability to "conduct essential Manchurian local cold-resistance exercises" (*Manzhou difang bixu jinxing de naihan de duanlian*). Tai denounced the *kang* not only for being unhealthy and promoting unwise behaviour but also for its expense, due to its near-constant need for repair and the cost of heating materials, primarily *gaoliang* (sorghum) and grass, but also wood and coal when it was available. Tai was not alone in his criticism. In 1943, Ai Ren condemned the *kang* and Chinese people's winter hibernation as remnants of the Stone Age.[41]

Debate over appropriate heating practices extended beyond the *kang*. Miura Unichi, for example, urged Japanese to abandon their own customs and adapt to the mainland environment by ensuring "housing hygiene" (*zhuju weisheng*) that could "regulate the climate" (*tiaojie qihou*).[42] He explicitly warned Japanese that their clothing, eating, and, especially, housing customs were inadequate for local conditions.[43] Chinese housing did not meet Miura's approval either, however; he wrote extensively about what he considered to be Chinese housing's two main shortcomings: double windows, which blocked out healthful ultraviolet radiation, and inadequate ventilation, which resulted in noxious indoor air. Miura linked these two issues with respiratory illnesses and diseases such as tuberculosis, scarlet fever, diphtheria, and measles. Miura warned against the practice of sealing gaps in double windows with paper to prevent heat loss and penetration of cold outdoor air;[44] he cautioned that air needed to circulate constantly because of the dangers inherent in heating methods that involved burning. Even in coastal Dalian, where winters were quite mild, 72 percent of houses used braziers (*huopan*), which Miura criticized for producing worrying levels of carbon monoxide – a problem that affected much of the population.[45] Miura cites work by scientists such as Max Joseph von Pettenkofer (1818–1901), a Bavarian chemist and hygienist, who warned against the dangers of carbon monoxide.[46] Miura criticized Manchurians (that is, the Chinese, whom he believed were more worried about the cold than about poisonous fumes) for changing the interior air of their homes even less frequently than the Japanese. He cautioned that if indoor and outdoor temperatures differed by more than 30 degrees, usual heating practices necessitated changing the air in the room three or four times per night, which was even more than was customary for many Japanese. Miura recommended swing windows to increase air circulation while acknowledging that Manchuria's cold winters reduced the efficacy of windows for such a purpose.[47]

Miura also extolled Russian heating methods and recommended the traditional Russian stove (Ch., *peiqika;* colloquial Ru., печь / *pech* or *pechka*) as the best option for Manchuria.[48] This stove was typically the height of an internal wall separating two rooms, with a small opening towards the bottom and a little door through which firewood or coal could be loaded. The stove would heat the rooms by funnelling hot air through panels in the walls. Larger houses could have several stoves. In the homes of the wealthy, tiles often covered the walls, and these, when heated, were often leaned on by individuals who took advantage of their warmth. Miura also promoted the use of centrifugal fans (*fengche xing*), invented in 1832 by Russian military engineer Alexander Sablukov (1783–1857), by pointing to American debate in New York over preferences for the installation of ventilating fans to maintain steady air temperatures and humidity rather than the use of windows to get fresh air.[49] In his study of Manchukuo architecture, Tucker notes that Manchukuo architectural style "typically combined modern construction methods and a Russian stove or other coal-fired heating system with Western and Japanese-style rooms."[50]

In the 1940s, heating practices initially promoted in the 1930s became more common. In 1940, Deng Encheng noted that most urban homes were heated by a stove in conjunction with a fan to circulate the heat.[51] Almost a decade later, in 1948, Dong Qiushui praised northeastern architecture for its "tall and wide" buildings with central heating, a feature that he believed to be far superior to the *kang*.[52] In 1949, F.C. Jones observed

FIGURE 5.2 "Inside Manchurian soldier housing – warming the room equipment (*peiqika*)." *Source:* Fengtian, undated photograph. Collection of N. Smith.

that in the early 1930s Japanese settlers had "forced on them new styles of house-building and heating" to deal with "adverse climatic factors."[53] He wrote that popular, stand-alone heating stoves (*huolu*) were relatively ineffective – heat decreased as distance from the unit increased – and that ashes in the stove were a fire hazard. While he recognized that electric stoves or hotplates (*dianlu*) were less polluting than devices that relied on burning materials, he also acknowledged they were more expensive to operate. Jones recommended systems of pipes that ran warm water through walls to heat buildings. Dong and Jones highlight the multiple forms of heating that proliferated through the 1940s, yet neither explicitly links these changes with Manchukuo-era reformers.

Closely related to the issue of improving heating practices was the question of appropriate indoor temperatures and the emergence in the late 1930s of the "low temperature life" campaign or, as Deng Encheng phrased it, the "low cold life" (*dihan shenghuo*).[54] The Concordia Association (CA), one of Manchukuo's largest mass organizations and producer of the journal *Jiankang Manzhou* (*Healthy Manchuria*), was heavily involved in the promotion of this campaign.[55] In 1939, the Harbin CA debated policies regarding how best to deal with coal shortages, promoting five solutions that buttressed the low temperature life campaign:

1. Coal should not be available for heating until October 16 and then only until April 10, at the latest.
2. Coal should be used economically, by sealing windows and setting daytime temperatures to between 16 and 18°C and nighttime temperatures to between 10 and 13°C.
3. Make regular use of a thermometer in the home to measure indoor ambient temperatures.
4. Pay attention to media and radio reports.
5. Use exercise as a tool against the cold.[56]

There appears to have been general agreement among the campaign's promoters about the five points above and about the idea that high indoor temperatures were costly, wasteful, and unhealthy.

On November 26, 1942, at the "Low Temperature Life Forum" (*Diwen shenghuo zuotanhui*), a dozen doctors and officials promoted the low temperature life and linked it firmly with a coal efficiency movement. Miura Unichi pointed to American research on schools and children that demonstrated how overheated rooms were dangerous and led to increased levels of illness.[57] Miura warned that the average Japanese was "afraid of the cold" and preferred room temperatures of 25°C or higher. This, he reasoned, in combination with inappropriately thin clothing, made pneumonia and tonsillitis endemic among the Japanese.[58] Guo Songgen, who held a French degree in physiology, argued that temperatures above 20°C should be avoided as they caused sweating, which led to illness. Bi Kegang argued that different indoor areas should have activity-appropriate temperatures.[59] Citing research on human reactions to the cold by Americans F.C. Houghten and Constantin P. Yaglou, Miura argued that humidity should also be controlled, to between 30 percent and 70 percent, to lessen the potential for the spread of contagious ailments.[60]

While there began to emerge a consensus on the benefits of low temperature life, there was no agreement on the ideal temperatures that would ensure it. Citing the work of Miura Unichi and Huntington's *Civilization and Climate,* Deng Encheng argued that the most suitable temperature was 17–20°C when the body was active and 12–17°C when the body was at rest.[61] Similarly, Wang Luo argued that the most comfortable temperature was 18°C in the day and 14°C at night; however, he noted that, in support of greater coal efficiency, temperatures of 15.6°C and 11.2°C, respectively, would reduce costs and enhance health.[62] Zhang Jiyou, Director of Medical Information in the People's Department, suggested that, across the board, temperatures of less than 15°C should be mandated, and that the ideal would be 10°C, even lower if it could be tolerated. Zhang did, however, warn of the potential for frostbite on fingers and toes with lower temperatures.[63] Wen Nailang, Director of Public Epidemic Prevention, noted that the winter of 1942 was not especially cold and that his experience of low temperature life then had trained his body well. Such "perseverance" (*hengxin*) was praised and echoed in other media, exemplified by writer Fei Fei's bid to "do battle" (*zuo zhan*) with the "cold ghost" (*han gui*) winter through low temperature living, cold water rubbing, and outdoor sports.[64] He bragged about maintaining an indoor temperature of 6.5°C through the winter of 1940 to save fuel and money, in service to the country. He suggested that the most effective method to fight the cold was to begin in the autumn to rub cold water over one's body. Fei also asserted that the best form of transportation, in every season, including winter, was walking.[65] He recounted the experiences of Japanese general and third Governor-General of Taiwan Nogi Maresuke (1849–1912), whose parents had apparently trained him as a baby to strengthen his body's resistance to the cold. They rubbed cold water on his skin and never allowed him to complain about the cold. Fei credited this practice with being foundational to Nogi's skills as a general and leader of the Japanese forces in the Russo-Japanese War (1904–5).

Cold water rubbing and walking were not the only wintertime activities promoted by proponents of the low temperature life. Su Xipo observed that since people generally did not like to go outside when it was cold, it was necessary to actively promote "outdoor sports" (*huwai yundong*).[66] Blaming "fear of the cold" (*pa leng*) for making people lethargic, Su argued that fun activities such as making figures out of snow, skating, kicking balls, and hunting for rabbits could inspire people to go outdoors. Guo Wenhua agreed, contending that the poisons emitted from traditional heating methods made it necessary for people to abandon their "half hibernation life" (*ban zhe ju zhi shenghuo*) and spend more of their time outdoors.[67] Kawamura Yasuo (1908–?), a local speed skater who competed in the 1932 and 1936 Winter Olympics, also argued in "Manzhou de huabing" ("Manchurian skating"): "for good health, go outside" (*jiankang di'yi, yao dao waibian qu*). Kawamura suggested that skaters had a "conquering nature power" (*kefu ziran li*) and that to understand skating was to understand Manchuria's Japanese people; he bemoaned the fact that few "Manchurian" (that is, Chinese) people participated in skating, despite training in schools, noting that they preferred *mah-jong* to activities such as ping-pong or skating.[68] Kawamura remarked that despite constant promotion of outdoor winter sports, only skating was truly popular and even then only among certain people.[69] Wang Luo concurred, observing that few locations in Manchuria received adequate levels of snow for skiing, so that skating was most feasible. He urged those who could not skate to go outdoors and watch others skate. Without doubt, however, the most modern of all outdoor winter activities was film watching. In 1940, Zuo Mu noted that outdoor film watching was a community-building experience that was easier for workers to coordinate in the winter because the weather made transportation of the necessary equipment easier: frozen rivers became convenient highways. Zuo pointed out that in remote villages there were few amusements or cultural activities in the winter; people would sit outside in their thick coats and hats in temperatures of –20°C or –30°C to watch a film for two hours – and enjoy it. Zuo noted that while they made for an amusing sight all bundled up in their winter attire, audiences revelled in the entertainment nonetheless.[70]

Manchukuo officials and other commentators may have believed they were introducing a new lifestyle to the region, but they were overlooking lengthy traditions of indigenous winter activities and sports. Long-cherished Manchu winter outdoor activities included kicking balls, skipping rope, flying kites, and skating. According to Mike Speak, "one of the most interesting and spectacular developments in sport during the Qing Dynasty was the progress made in ice sports."[71] Skating activities that long predated the Qing were recorded as early as the fourteenth century, in the *Song shi* (*History of Song*, ca. 1343), and became more prevalent during the Yuan Dynasty (1279–1368). These winter activities came to be seen by Han Chinese as quintessential elements of northern minority cultures.[72] Nurhaci (1559–1626), founding father of the Qing Dynasty, ordered skate training for his troops and reviewed Eight Banners' performances on ice; ice inspections were an annual affair, although these gradually disappeared over the course of the dynasty.[73] The Qianlong emperor was famed for supervising winter activities and encouraged skating – the sport with "shoes with teeth"; the emperor and his mother watched skating in the Forbidden City, describing players as "dragonflies touching the surface of the water, or purple swallows flying over the waves."[74] Qing ice sport meets, with figure skating and races (from individual competitions to groups pulling sleds on ice), were held every year in what is now Beijing's Beihai Park. One of these meets has been immortalized in Zhang

Weibang and Yao Wenhan's magnificent Qing-era court painting featuring hundreds of skaters, *Bing xi tu* (Ice play painting).

The popularity of skating expanded greatly in the twentieth century. Formal skating rinks were first established in 1909, in Harbin's Daoli and Nangang districts. By this time, the Chinese had begun to view skating for leisure or sport as a "luxury item" (*gaoji shechi pin*), especially for foreigners, rather than as a peculiarity of northern peoples.[75] By the mid-1920s, Manchurian winter sports associations were also being established. In 1927, locals began to compete in skating and hockey competitions domestically and further afield; in 1930, a skating team from Manchuria Medical College competed in international tournaments in Europe. In January 1935, the all-Manchuria competition was established at Jilin (and was continued annually until 1939, in Xinjing).

In 1937, the Manchuria Skating Association elected Miura Unichi as its chair and announced plans to hire European, American, and (especially for hockey) Canadian trainers.[76] Local, regional, and international competitions were held in Manchukuo, and locals participated in international events.[77] One downside to outdoor skating in the region, however, was the ambient coal dust, which turned the ice black, reduced speed, and

FIGURE 5.3 Skating on the Songhua.
*Source:* Collection of N. Smith.

was derided by Olympic skater Kawamura Yasuo as an "evil star" (*moxing*) hexing speed skaters.[78]

Skating was not the only popular outdoor pastime. Zhang Yueqi argues that, contrary to popular opinion (and similar to skating), skiing – the "silver-coloured movement" (*yin se yun*)[79] – was not an "exotic" (*bolaipin*)[80] import but could be dated domestically to the Tang Dynasty (618–907), when it was called "ride the wood" (*cheng mu*) or "ride the wood and go" (*qi mu er xing*).[81] Even more suggestive of the local origins of the practice was a local name given to skis, "Tartar boards" (*Daban*).[82] The region's popular ski areas, including Tonghua, Acheng, and Mudanjiang, were praised as early as the mid-seventh century, in *Bei shi* (*Northern History*) (c. 649).[83] Indeed, some argued that Tonghua, which was already known as "the home of ginseng" and the "home of wood," should be named the "home of skiing" (*huaxue de xiang*).[84] Skiing has also been attributed to the Bei Shiwei (in the Xing'anling region around Qiqihar), who used it for transportation while hunting.[85] The Jurchen have been credited with transforming informal skiing practices into formal sporting events and the Suanzhe had contests in which they would ski from the top of one mountain up another and then down.[86] Not to be outdone, Manchu women participated in "group skiing" (*qun bu ping sha / jieban buxing*) and "walking on the snow" (*xue di zou*) on the sixteenth day of the first month of the lunar year.[87] Though a certain newness and modernity was often ascribed to it, skiing had long been a part of local culture in various capacities.

Skiing enthusiasts grew in number into the twentieth century; by the mid-1930s, the popular media reported regularly on domestic and international ski competitions. But two aspects of the sport kept large numbers of Chinese from taking up skiing – its association with foreigners, and the financial cost to the individual of participating in the sport. In 1941, Ya Qilin argued that recreational skiing was not known before the establishment of Manchukuo, and that even in the early 1940s there were not many participants – though the state did produce two famous Chinese skiers: Liu Baokun and Li Tieyan.[88] Liu Baokun, too, asserted that modern skiing was introduced to China by Japan. But he argued that even though the most popular stores selling ski equipment were Japanese, it was the German-made "Polar Bear brand" (*bai xiong pai*) that was most highly valued.[89] Liu recalled that in 1936 his school hired a Japanese teacher, Kumi Ryōzō, who convinced him as a young student to try skiing – for health benefits.[90] Kumi then persuaded the school principal, Satō (who apparently opposed Chinese students' learning to ski), to buy the necessary equipment on credit. By 1939, Jilin's Chinese students had begun to outperform their Japanese counterparts in long-distance competitions.[91] At the time, however, most of the population could not afford to ski because of the high cost of the equipment. A medium-quality set of skis cost more than forty yuan, and rentals in 1941 were half a yuan per hour – a fee which, for example, could have paid for 7.5 kilograms of cotton.[92] Japanese skiers had an advantage over the Chinese: they did not have to pay to ski at resorts owned by the South Manchuria Railway (SMR); its Excel Division (*Mantie liang cheng ke*) founded important ski resorts, including in Jiutai county, Jilin (1933); Acheng, near Harbin (1933); and Beishan, near Fengtian (1934).[93] Zhou Deqing, a Manchukuo ski instructor, recalled that SMR employees regularly arrived during the winters at Beishan on Saturdays, and paid for neither lodging nor rentals.[94] Chinese skiers were not so privileged.

## Conclusion

A 1942 article in the journal *Sakura* (*Cherry*) features a photograph of Russians in Harbin under a towering cross of ice, conducting religious activities outdoors in the winter. The photo is followed by text that boldly proclaims: "Civilization's corrupt and effeminate influences have failed to make any inroads here."[95] Harbin's winter is depicted as incorruptible, masculine, and timeless – a positive influence on the character of the region's people. In a similar vein, in a speech in 1869, Robert Grant Haliburton called Canadians "a healthy, hardy, virtuous, dominant race" whose "toughness, strength and vigorous energy had been instilled and maintained by the country's climate and geography."[96] Historian Gillian Poulter has documented how, by the 1860s, arguments emerged that Canada was "more than just a snowy waste and that, in fact, the weather was a positive attribute."[97] Cold water rubbing was promoted in Manchukuo, but even earlier, in 1866, William Rhodes explained in the *Quebec Daily Mercury* how Caucasian men could maintain their civilized habits "by washing the body daily in the snow." In this way, "an amount of comfort and cleanliness can be obtained which few people would suppose. The snow also makes the skin cold-proof."[98] In colonial Canada, resistance to the cold, and active engagement with it, was argued to produce healthy, hardy, hygienic bodies that could appropriately contribute to state building. In colonial Manchuria, too, new heating practices, low temperature life, and outdoor activities were promoted with the same promises.

FIGURE 5.4 "The Cross of Ice."
*Source: Sakura* [Cherry] 4, no. 9 (1942): 51

The twentieth century brought major changes in the ways that many people in Manchuria heated their homes and workplaces in the winter. The long-valued *kang* remained the dominant form of heating in rural areas, and many urbanites were doubtless loath to give up their *kang*. But in urban areas, especially, the *kang* was becoming a memory. To this day, many northern Chinese speak with great fondness of the cooking and sleeping platforms that dominated local architecture for so long. But large incoming populations of Chinese, Japanese, Koreans, and Russians introduced varied means of heating, and the new buildings erected to house these newcomers necessitated the adoption of heating equipment that was deemed more appropriate, both for economic and health reasons and for major urban development. These heating methods dramatically changed life in the region, especially in the 1930s and 1940s. By the end of the 1940s, tall buildings with central heating were praised by observers such as Dong Qiushui as the epitome of modernity and healthful living.[99] At the same time, increasing attention paid to carnivals, ice carving, and outdoor sports put winter in an ever better light, a time when there were "no distracting activities to lessen the severity of the season."[100] In 1991, editors of the volume *Zhongguo huaxue yundong ziliao xuanji* (*Selected Materials on Chinese Skiing Sports*) suggested that children born and raised in the Northeast are "richly endowed by nature" (*detianduhou*), and can further enhance their lives by building snow figures, having snowball fights, tobogganing, and engaging in other winter sports.[101] Such claims would have faced considerable skepticism a century before.

There remains debate over whether skating and skiing were indigenous practices or practices introduced by the Japanese, the Europeans, or the North Americans. While some studies describe skating in the distant imperial past, others point to its origins in early-twentieth-century Harbin, or further south in Tianjin. Skiing has been associated with northern Indigenous peoples, but popular ski resorts that catered to new migrants were founded by the Japanese, especially during the occupation era. Suk-Jung Han has shown that "exercise and sanitation were important fields in which the [Manchukuo] regime invested,"[102] to improve both the health of the population and its international image. As in other contemporary regimes, Manchukuo's rulers promoted eugenics-inspired policies in an effort to improve the health of local people – through exercise and hygienic housing – for a variety of reasons, often expressed in racist or condescending terms. Winter was increasingly depicted in positive ways to appeal to an expanding population and to realize varied state objectives. The growing popularity of winter sports such as skating and skiing depended on the emergence of a relevant infrastructure as well as a population that did not view outdoor activities in the winter as unseemly or unnatural but, rather, as healthful. This population also had to have the resources, time, and interest to pursue these activities. For some, this became possible in the 1930s and 1940s, when shifting notions of winter health, appropriate heating practices, and the necessity of outdoor activity left a permanent impression on local life. The state of Manchukuo may have collapsed in 1945, but shifting attitudes towards winter continued to serve essential functions in the post-1949 activities described later in this volume by Sun Xiaoping (Chapter 10).

## Notes

1 Huang, born in Fengtian, served in various financial positions in Fengtian and Chahar. From 1937 to 1939, he was governor of Andong Province and, from 1939 to 1942, governor of Heijiang Province. Subsequently, he was appointed minister of agriculture. He was imprisoned on the collapse of Manchukuo. Huang Fujun, *Manzhou nongye jishuzhe bi xie* [The spiritual and technical requirements of Manchurian agriculture] (Xinjing: Nong sui jinbu she, 1944), 70.

2 Wang Luo's original name was Wang Shigong. He was Taiwanese, a graduate of Xiamen's Xuying College and Manchuria Medical College, where he specialized in hematology. In Manchukuo, he served in various capacities, including as director of Fengtian's Police Hygiene Division and as an epidemic prevention section chief. In 1944, he was minister of the People's Welfare Department. This essay was originally published in *Jiankang Manzhou* [*Manchuria Health*] 2, no. 1 (1940): 10–12. It was republished in 1941 in *Kangde xinwen* [*Kangde News*] and in *Shengjing shibao* [*Shengjing Times*] and in 1942 in *Qilin* [*Unicorn*]. "Dongji baojian shenghuo fa" [Winter health care life method], *Shengjing shibao* [*Shengjing Times*], November 1, 1941, 5.

3 Wang Luo, "Dongji baojian shenghuo fa," 5

4 South Manchuria Railway Company, *Manchuria: Land of Opportunities* (New York: South Manchuria Railway, 1922), 13.

5 John R. Stewart, "Manchuria: The Land and Its Economy," *Economic Geography* 8, no. 2 (1932): 134–60, 140–41.

6 Fujiyama Kazuo noted that Richthofen named Manchuria "East Asia's First Temple of Earth." Cited in Fujiyama Kazuo, *Xin Manzhou fengtu ji* [A record of New Manchuria's natural conditions and social customs] (Xinjing: Man Ri wenhua xiehui, 1938), 5.

7 Mark Elliott, "The Limits of Tartary: Manchuria in Imperial and National Geographies," *Journal of Asian Studies* 59, no. 3 (2000): 603–46, 613–14.

8 Fujiyama, *Xin Manzhou fengtu ji*, 14.

9 Ibid., 15.

10 George Babcock Cressey, *China's Geographic Foundations: A Survey of the Land and Its People* (New York: McGraw-Hill Book Company, 1934), 232.

11 Arthur de Carle Sowerby, *The Naturalist in Manchuria. Volume II, The Mammals of Manchuria* (Tientsin: Tientsin Press, 1923), x.

12 A.D. Woeikoff, *What Can the Manchurian Flora as Well as the Flora of Neighbouring Countries Give to Gardens of Manchuria Itself and Other Countries with Cold Climates* (Harbin: A.D. Woeikoff, 1941), 7.

13 Manchoukuo Yearbook Company, ed., *Manchoukuo Yearbook: 1941* (Hsinking: Manchoukuo Yearbook Company, 1942), 48.

14 Thomas R. Gottschang and Diana Lary, *Swallows and Settlers: The Great Migration from North China to Manchuria* (Ann Arbor: Center for Chinese Studies, University of Michigan, 2000), 133.

15 Ronald Suleski, *Civil Government in Warlord China: Tradition, Modernization and Manchuria* (New York: Peter Lang, 2002), 2.

16 Cressey, *China's Geographic Foundations*, 221.

17 Gottschang and Lary, *Swallows and Settlers*, 133.

18 W. Allyn Rickett, trans., *Guanzi: Political, Economic, and Philosophical Essays from Early China: A Study and Translation. Volume II* (Princeton, NJ: Princeton University Press, 1985), 106.

19 Nico Stehr, "Climate Determines: An Anatomy of a Disbanded Line of Research," 1998, 5, http://www.hvonstorch.de/klima/pdf/Climate.determines.pdf.

A historiographical review of scholarly studies of eugenics can be found in Frank Dikötter, "Race Culture: Recent Perspectives on the History of Eugenics," *American Historical Review* 103, no. 2 (1998): 467–78.

20 Ellsworth Huntington, *The Character of Races, as Influenced by Physical Environment, Natural Selection and Historical Development* (New York: Charles Scribner's Sons, 1925), 200.

21 Ibid., 372.

22 Fujiyama, *Xin Manzhou fengtu ji,* 9.

23 Ibid., 15.

24 Ibid., 17.

25 Ibid., 16.

26 Fujiyama compared Athens's 83 percent clear days with Germany's 30 percent and England's 10 percent to assert the former city's brilliant history of philosophy and the latter's degeneracy (ibid., 17).

27 "Baojian" [Maintaining health], *Shengjing shibao* [*Shengjing Times*], October 11, 1941.

28 Dong Qiushui, *Dongbei fengtu xiaozhi* [A minor record of the natural conditions and social customs of the Northeast] (Hong Kong: Shidai piping she, 1948), 3.

29 Ibid.

30 Cressey, *China's Geographic Foundations,* 228.

31 Bi Kegang, "Dongqi huwai baojian" [Winter outdoors health care], *Jiankang Manzhou* [*Healthy Manchuria*], December 1940, 2–5.

32 "Baojian," 5.

33 Miura studied at Kyoto Imperial University and Johns Hopkins University, and became professor of hygiene at the Manchurian Medical College. The Miura Laboratory of Hygiene trained many scientists in public health; Miura's students became chiefs of public health organizations in Manchukuo after 1931. Miura Unichi, *Manzhou zhi zhuju weisheng* [Hygienic housing in Manchuria], trans. Lang Wenzong (Fengtian: Manzhou yike daxue weisheng jiaoshou shi, 1934), 34. This work was subsequently republished in three installments in *Dongfang yixue zazhi* [*Far Eastern Medical Journal*] in 1935.

34 Miura referred to local Chinese as "Manchurians" when the work was republished in *Dongfang yixue zazhi.*

35 David Tucker, "City Planning without Cities: Order and Chaos in Utopian Manchukuo," in *Crossed Histories: Manchuria in the Age of Empire*, ed. Mariko Asano Tamanoi, 53–81 (Honolulu: Association for Asian Studies and University of Hawai'i Press, 2005), 67.

36 Tai Yuzhen, "Huokang zai dongtian de bihai" [The evils of the *kang* in the winter], *Jiankang Manzhou* [*Healthy Manchuria*], June 1940, 11–13, 11.

37 Ibid., 12.

38 So, too, did Deng Encheng. See Tai, "Huokang," 11; and Deng Encheng, "Fang han yu jianzhu" [Defend against the cold and architecture], *Jian guo jiaoyu* [*Build the Country Education*] 6, no. 7 (1940): 75–79, 78; Miura, *Manzhou zhi zhuju weisheng,* 42.

39 Tai, "Huokang," 11.

40 Ibid., 12.

41 Ai Ren, "Fanghan xinshuo" [New talk on guarding against the cold], *Qilin* [*Unicorn*], January 1943, 50–52, 51.

42 Miura, *Manzhou zhi zhuju weisheng,* 2.

43 Ibid., 16.

44 Ibid., 17.

45 Ibid., 30.
46 Miura, *Manzhou zhi zhuju weisheng,* 18. David Tucker, too, has noted how Japanese urban planners studied Western industrial urbanization and housing policies in the 1920s. Tucker, "City Planning without Cities," 57.
47 Miura, *Manzhou zhi zhuju weisheng,* 21.
48 I am indebted to Olga Bakich for her explanation of the Sino-Russian pidgin term *peiqika.*
49 Miura, *Manzhou zhi zhuju weisheng,* 23–24.
50 Tucker, "City Planning without Cities," 68.
51 Deng, "Fang han yu jianzhu," 79.
52 Dong, *Dongbei fengtu xiaozhi,* 4.
53 F.C. Jones, *Manchuria since 1931* (London: Royal Institute of International Affairs, 1949), 87.
54 Deng, "Fang han yu jianzhu," 77.
55 "Diwen shenghuo zuotanhui" [Low temperature life forum], *Jiankang Manzhou* [*Healthy Manchuria],* March 1942, 2–7, 3.
56 "Shixing meitan xiaofei jieyue kuiding dongji shenghuo yaogang" [Outline for practising coal consumption economizing to regulate winter living], *Shengjing shibao* [*Shengjing Times*], October 24, 1939, 4.
57 See also Miura, *Manzhou zhi zhuju weisheng,* 29.
58 Ibid., 28.
59 "Diwen shenghuo zuotanhui," 4.
60 For example, T.T. Read and F.C. Houghten, *Cooling of Mine Air* (Washington, DC: US Department of the Interior, Bureau of Mines, 1923). Miura, *Manzhou zhi zhuju weisheng,* 29. These assertions were supported by experiments in Fengtian and Dalian that demonstrated that humidity below 30 percent was especially bad for spreading contagious diseases.
61 With constant humidity between 50 and 80 percent. Deng, "Fang han yu jianzhu," 77.
62 Wang, "Dongji baojian shenghuo fa," 130.
63 Ibid., 4.
64 Fei Fei, "Fanghan mihua" [Secret words for guarding against the cold], *Qilin* [*Unicorn*], December 1941, 39.
65 Ibid., 40.
66 Su Xipo, "Dongji de huwai yundong" [Winter outdoor sports], *Qilin* [*Unicorn*], November-December 1944, 66–67.
67 Guo Wenhua, "Mantan dongji weisheng" [Informal discussion of winter hygiene], *Jiankang Manzhou* [*Healthy Manchuria*], October 1940, 7–8.
68 Other notable skaters included Kitani Tokuo (1909–?) and Ishihara Shōzō (1910–93). Kawamura Yasuo, "Manzhou de huabing" [Manchurian skating], *Jian guo jiaoyu* [*Building the Country Education*] 5, no. 11 (November 1939): 50–54, 50, 55.
69 Ibid., 50.
70 Zuo Mu, "Dong yu xunying" [Winter and touring films], *Manzhou yinghua* [*Manchuria Movies*], January 25, 1941, 33.
71 Mike Speak, "The Emergence of Modern Sport," in *Sport and Physical Education in China,* ed. James Riordan and Robin Jones (London: E and FN Spon, 1999), 60.
72 I am grateful to Jonathan Schlesinger for this insight, at the workshop "The Manchurian Environment: Natural Resources, Climate and Disease," University of British Columbia, May 18, 2013.
73 "Qing dai huabing yu bingxie" [Qing-era skating and skates], *Xin Qingnian* [*New Youth*] 8, no. 11 (1940): 17.
74 Speak, "The Emergence of Modern Sport," 61.

75 From the second decade of the twentieth century, media began to promote the sport strongly. Norwegian Sonja Henie (1912–69), for example, was praised in a 1934 article in the *Shenjing shibao* [*Shengjing Times*], as were Canadians and Americans more generally. For example, see "Huabing baquan" [Skating supremacy], *Shengjing shibao* [*Shengjing Times*], March 22, 1934, 7.

76 "Manzhou kai zhong bing shang lianmeng huiyuan hui, jueyi Sanpu wei huizhang" [Manchuria establishes an ice sports alliance committee, decides on Muira as chairman], *Shengjing shibao* [*Shengjing Times*] 109, no. 97 (1937).

77 "Dong Ya bing shang jingji dahui bimu" [Lowering the curtain on East Asia's ice sports mass meeting], *Shengjing shibao* [*Shengjing Times*], February 10, 1943, 4.

78 Kawamura, "Manzhou de huabing," 52.

79 Ya Qilin, "Beishan huaxue shimiao" [Real depiction of skiing on North Mountain], *Qilin* [*Unicorn*], December 1941, 114–17, 115.

80 Zhang Yueqi, "Qianyan" [Foreword], in *Zhongguo huaxue yundong ziliao xuanji* [Selected materials on Chinese skiing sports], ed. Zhongguo huaxue xiehui (Changchun: Zhongguo huaxue xiehui, 1991), 1.

81 Ibid., 6.

82 Zhuang Feng, "Yi kanglian tiyu huodong pianduan" [Recalling snippets of the resistance physical exercise], in *Zhongguo huaxue yundong ziliao xuanji* [Selected materials on Chinese skiing sports], ed. Zhongguo huaxue xiehui (Changchun: Zhongguo huaxue xiehui, 1991), 25.

83 Zhongguo huaxue xiehui, ed., *Zhongguo huaxue yundong ziliao xuanji*, 1.

84 Shi noted that military ski training started in Tonghua around 1940, to combat the local Chinese resistance; the Japanese built five training camps around the city. See Shi Shi, "Tonghua shi huaxue yundong" [General narration of Tonghua City skiing activities], in *Zhongguo huaxue yundong ziliao xuanji*, ed. Zhongguo huaxue xiehui, 68.

85 Zhongguo huaxue xiehui, *Zhongguo huaxue yundong ziliao xuanji*, 1.

86 Ibid., 9, 15.

87 The latter in lengths of 100, 200, and 400 metres. Ibid., 17.

88 Ya, "Beishan huaxue shimiao," 116.

89 Jilin had three stores – including Fuji and Ueda. Xinjing, too, had three Japanese stores: Watanabe, Begonia, and Minakai. Liu Yongnian, "30 niandai huaxue mingjiang Liu Baokun tan wei Man de huaxue yundong" [Famous 1930s skier Liu Baokun discusses false Manchukuo's skiing activities], in *Zhongguo huaxue yundong ziliao xuanji*, ed. Zhongguo huaxue xiehui, 36.

90 Ibid., 35–36.

91 Ibid., 36.

92 Ya, "Beishan huaxue shimiao," 116.

93 Liu Yu, "Acheng: Huaxue zhi xiang" [Acheng: The home of skiing], in *Zhongguo huaxue yundong ziliao xuanji*, ed. Zhongguo huaxue xiehui (Changchun: Zhongguo huaxue xiehui, 1991), 77–80.

94 "Zhou Deqing tan 30 niandai Jilin shi de huaxue huodong" [Zhou Deqing discusses 1930s-era Jilin skiing exercises], in *Zhongguo huaxue yundong ziliao xuanji*, ed. Zhongguo huaxue xiehui, 31.

95 "The Cross of Ice," *Sakura* [*Cherry*] 4, no. 9 (1942): 51.

96 Gillian Poulter, *Becoming Native in a Foreign Land: Sport, Visual Culture and Identity in Montreal, 1840–85* (Vancouver: UBC Press, 2009), 3.

97 Ibid., 93.

98 Ibid., 107.

99 Dong, *Dongbei fengtu xiaozhi*, 3.

100 Gottschang and Lary, *Swallows and Settlers*, 133.

101 Zhongguo huaxue xiehui, *Zhongguo huaxue yundong ziliao xuanji*, 18.

102 Suk-Jung Han, "Those Who Imitated the Colonizers: The Legacy of the Disciplining State from Manchukuo to South Korea," in *Crossed Histories: Manchuria in the Age of Empire*, ed. Mariko Asano Tamanoi (Honolulu: Association for Asian Studies and University of Hawai'i Press, 2005), 171.

# 6

# Constructing a Rural Utopia: Propaganda Images of Japanese Settlers in Northern Manchuria, 1936–43

*Annika A. Culver*

From the late 1930s into the early 1940s, Japanese handlers of occupied Northeast China attempted to portray the new state of Manchukuo[1] as a utopian "paradise" for rural Japanese settlers and their families. In accordance with this initiative, numerous propaganda publications, including *Manshū gurafu* (*Manchuria Graph*, 1933–44), published by the South Manchuria Railway Company (SMR, or *Minami Manshū tetsudō kabushiki gaisha*), touted the region as fertile and productive, providing ample opportunities to raise numerous healthy children in the wide-open spaces and salubrious climate. Arable land in Japan's domestic archipelago was scarce; however, farmers with multiple children could cultivate larger plots of land freely on the spacious Manchurian plains and hillsides. This "virgin land" was supposedly empty, apparently free from the distressing claims of its original Chinese owners – a conceit belied by the fact that Japanese attempts at settlement in the early 1930s had met with failure for various reasons. Only a few years later, settlement schemes reappeared in earnest as war in China proper loomed on the horizon. These immigration efforts in northern Manchuria would help relieve pressures on domestic rural localities in Japan and provide an overseas base for the production of food and manpower for maintaining Japan's war effort. In addition, the settlers produced vital natural resources for the wartime Japanese empire while providing a "moral" bulwark against Soviet Communism immediately to the north across the Black Dragon River.

The covers of the *Manchuria Graph* pictorial feature bright, seasonal photographs, while articles, captions, and montages of photographs within depict the region's settlers happily cultivating or harvesting crops, with most of the younger women carrying babies on their backs. Such images promoted the ideal of peaceful, rural areas as a repository for Japanese values that could be translated elsewhere – that is, to any newly conquered multi-ethnic areas in Southeast Asia after 1940. In addition, from 1936 onwards, images depicting intrepid pioneers flourishing in their new environment and representing "Japanese" values demonstrated for both Japanese and Western audiences the success of the Manchukuo experiment.

These propaganda images, long neglected in English-language historical scholarship on East Asia, naturally elide the violence of Japan's wartime aggression. Yet they are fascinating in their portrayal of what I call a "right-wing proletarian" utopia desired by the Japanese framers of a fascist state – representing ideals they were unable to realize in domestic Japan. The settlements represent a kind of state-sponsored corporatist socialism (in other words, fascism), supposedly in the interests of the Japanese immigrant farmer. The agrarianism (*Nōhonshugi*) and communalism supposedly expressed in these Japanese farming settlements in Manchuria exemplified the beliefs of right-wing theorists such as Tachibana Kōzaburō (1893–1974). Yet the desire to create a utopia in the colonies (or within supposedly unexploited territories within one's own borders) is a common phenomenon among regimes on both ends of the political spectrum; it appeared in the colony of Ethiopia under Italian fascism and in the Soviet Union in Ukraine under communism around the same time through state-sponsored collectivization. A similar conceit reappears in the 1950s conversion of Northeast China's "Great Northern Wasteland" to the "Great Northern Granary" in the post-Liberation period, as detailed in Sun Xiaoping's chapter (Chapter 10 in this volume). A key difference in this case, however, was that the settler villages were to serve as a moral bulwark against the Soviet Union, and were even militarized, with the men patrolling the borders every few days. Men and women both played important roles – increasing their productivity and acting out purportedly traditional Japanese morals of frugality, endurance, and Shintō piety, with the shrine as the village nerve centre. Such propaganda images helped to legitimize the Japanese presence in Manchukuo, while the flourishing settler villages "proved" to individuals back in domestic Japan that the imperial government's settlement schemes were effective and promised a new way of life in tandem with contemporary moral values.

## Japanese Settlers and the "Continental Sublime"

In a common postcard from the Manchukuo period (possibly 1933),[2] issued by the Kantō Army as "military post" and featuring a watercolour print, a male Japanese settler surveys his land from atop a hill. A red sun rises in the top third of the image, with the caption *Manmō no kaihatsu wa suki to suki to de* (The development of Manchuria and Mongolia proceed from plough to hoe) in the top right-hand corner.[3] The sun appears in a white sky, like the colours of the Japanese flag, and the cloud hovering over the sun resembles the shape of Mount Fuji. Livestock graze peacefully in grasslands in the distance, while farmers cultivate the fields using tools or an oxen-driven plough. The sun's rosy rays softly grace the man's well-muscled physique while he hefts a spade and a hoe. Evoking the kind of harmonious atmosphere that Alan Tansman has described as reminiscent of the "fascist aesthetic,"[4] the postcard presents the receiver (possibly a relative back in Japan) with the image of a peacefully settled Manchuria full of possibility and blessed with the benevolent rays of a sacred sun – provided by imperial Japan. However, even without the military postmarks on the back of the image, this same aesthetic could also be representative of Socialist Realism, in a high modernism depicting utopian visions that could merge versions from both the political left and the political right.

The postcard described here, and other popular Japanese postcards from the Manchukuo period depicting empty, elegantly ploughed fields, could reflect a distancing – but also seductively intimate – awe that can be termed the "continental sublime," a concept developed by Kari Shepherdson-Scott in her art historical study of *Manchuria Graph,* in which she examines the practice of train travel in the region via the SMRC as advertised and depicted in the company's propaganda publication.[5] Though her study does not elaborate much on or give a clear definition of the "continental sublime," she references the term "to underscore a romanticized relationship with the Manchurian frontier"[6] in what I understand to be her version of a culturally specific "Kantian sublime." Such widely disseminated images may serve as idealized sketches or snapshots of a region deeply alien to the Japanese viewer back home, but they also depict, as Shepherdson-Scott argues, a "supposedly 'undiscovered country' [that] provides a space where *naichi* [domestic Japanese] readers could satisfy their appetite for discovery and excitement. This, in turn, created a strong emotional (and, arguably, national) connection to the distant landscape."[7] The irrational emotional bond of nationhood and feeling of personal connection instantaneously stimulated by postcard images and the photographs depicted in *Manchuria Graph* arose out of their nature as "epiphanic representations," described by Tansman as capable of "provok[ing] spontaneous consensus and aesthetic cohesion."[8] Whether an individual views such images from a train while entering a new homeland or from a hillock above a village, the Japanese settler (or Kantō Army soldier) observer looks down on the landscape from a privileged position, surveying this new territory now under his dominion by virtue of Japanese labour, in addition to the labours of other Chinese. Most photographs of the time show Japanese men perched atop state-of-the-art, American-made, 1930s McCormick-Deering or Caterpillar tractors; the people cultivating fields with mules symbolize the "less advanced" Chinese. Thus, in such images, the settlers also appear knowledgeable, well versed in the latest methods and techniques of cultivation, representing Japanese scientific prowess and agricultural know-how in the Manchurian space. By implication, they also suggest that the Japanese settlers are fashioning the environment according to their will rather than being controlled by it. Such a view was common in centuries of imperialist endeavours (or internal development schemes) in frontier areas throughout the world.

In the aforementioned postcard, the man overlooks the crenellated tile roofs of an enclosed Chinese village, or a landlord's compound, in a standard portrayal of what E. Ann Kaplan calls the "imperial gaze."[9] While Kaplan applies this concept to her investigation of how colonial subjects are portrayed in film from a feminist perspective, the idea of an imperial gaze is also relevant to the study of cultural and environmental history. The concept can theoretically explain how Japanese desires were depicted in visual materials; however, beyond these propaganda images, the reality of the "imperial gaze" and the practicalities of Japan's imperial relationship with the local population were often much more complicated. For example, as historian Rana Mitter argues, in order to maintain nominal control over land and resources, Chinese landlords cooperated in various ways with the Japanese administrators of the region that defied a simple resistance-and-collaboration paradigm.[10] Cooperation sometimes proceeded in apparently beneficial local alliances where money exchanged hands, trade opportunities expanded, or economic and cultural futures of areas and towns were planned, while tensions arising from unequal power relations continued to build and manifest in various forms, ranging from intransigence and "foot-dragging" to outright lawlessness and banditry. However, in cultural artefacts such as the postcard discussed above and the images of Japanese settler families in *Manchuria Graph*, the continental sublime and the imperial gaze capture fairly accurately Japan's visual representation of imperialism that I examine in the analysis that follows.

Ironically, the postcard image depicts a Japanese man during a time when he and his counterparts were scarce in northern Manchuria, and, indeed, non-Chinese settlers of any gender were few. Later, even these scant numbers would diminish as men were deployed to fight the Second Sino-Japan (1937–45), Southeast Asian (1940–45), and Pacific (1941–45) wars of imperial Japan.[11] The postcard thus advertises a peaceful utopia that was starkly different from the reality of domestic Japan at war, even prior to the outbreak of conflict in China. The postcard is one artefact amongst a plethora of Manchuria-related popular-culture products that were disseminated throughout the Japanese empire at this time. This trend developed after the Japanese Kantō Army's invasion of the region in September 1931, generally known in English-language scholarship as the "Manchurian Incident," which culminated in the Japanese government's declaration of the "independent" state of Manchukuo in 1932.[12]

In her now-canonical 1998 text featuring a cultural history study of imperial Japan's endeavours in Manchuria, Louise Young describes this post-1931 media saturation in a chapter titled "War Fever: Imperial Jingoism and the Mass Media."[13] The postcard and the images in *Manchuria Graph* were part of a frenzy of production that capitalized on interest in Japanese-occupied Northeast China following the Manchurian Incident, and then moved into enthusiastic support of imperial Japan's settlement schemes after 1933, a date marking the Kantō Army's alleged pacification of the region turned "independent" developmental state.[14] Young's seminal text has inspired several of my own studies, but much remains to be done in the realm of visual culture.[15] Since the 1990s, Japanese scholars have also published a flurry of books and articles on the region's history, yet historical research of Manchukuo's visual culture (besides the research of museum curators) has been neglected until relatively recently. In 2010, utilizing a rich database of images from his own personal collection, Kishi Toshihiko published a groundbreaking study on Manchukuo's visual media.[16] Two years later, Shepherdson-Scott completed a dissertation specifically on *Manchuria Graph* and its dissemination of romanticized, utopian images intended to meet the publicity needs of the SMRC.[17]

In my own 2013 study, I highlight the symbolic significance of multivalent images of Manchuria as expressed in the works of writers, artists, and photographers, which served collectively to promote a certain imperialistic vision of the region even when the individual works themselves expressed some equivocation about the Manchurian experiment.[18] For Japanese observers, Manchukuo served as a utopia not only for Japanese settlers, but also for urban Chinese laborers, who were promised steady employment, ample food, and adequate lodging. I read these images as part of an earlier trajectory of (Japanese) left-wing socialist advocacy of workers and farmers, an intellectual current of proletarianism that came under attack in domestic Japan by the late 1920s and intensified after the Manchurian Incident. However, in the Manchukuo context, these ideals are reinterpreted in what I call a "right-wing proletarian" guise, where the state functioned as a developmental utopia designed under fascist, corporatist principles and welcomed "converted" Japanese leftists.[19]

As a publication of the SMRC, *Manchuria Graph*'s "realist" images were actually the carefully constructed creations of the photographers of the Manchurian Photographic Artists Association (MPAA, *Manshū shashin sakka kyōkai*), under the direction of Fuchikami Hakuyō (1889–1960), who advertised utopian images of a region in the process of colonization as part of imperial Japan's settlement scheme. He and his cohort had earlier experience in the avant-garde photography world of the 1920s, while their later photographs maintained an aesthetic continuity with portrayal of earlier left-wing foci on labour and the working class. For example, Fuchikami patterned *Manchuria Graph* after the Soviet publication *SSSR and, here,* and here, images of workers in quasi-heroic depictions also abound.[20] Other MPAA members' photographs also appear in this publication funded by the SMRC. Published from 1933 to 1944, and edited by Fuchikami until 1941, the monthly periodical provided utopian visions of a region under the tutelary auspices of the Japanese, which included leadership by cultural figures, military men, and companies like the SMRC. The publication probably found its way into the homes of Japanese on the archipelago and throughout the empire. Copies were also sent to libraries in the United Kingdom and the United States as gifts from the Japanese consulate. (Incidentally, neither the UK nor the US recognized an independent Manchukuo; each country had sent a representative to take part in the League of Nations–appointed, British-led Lytton Commission that reviewed the Manchurian Incident, which culminated in a September 1932 report condemning both Chinese and Japanese actions.) Since *Manchuria Graph* was written mainly in Japanese, with few captions in English, its audience was largely Japanese; it probably had only a few Chinese readers beyond the small group of Chinese educated in Japan or at Manchukuo universities.

In Japanese, the word "propaganda," or *senden,* is the same as the word used for "advertising," but the word in Japanese does not suggest an association with fascism, Soviet communism, or the political messaging of authoritarian rulers as it so frequently does in the English language. Though the *Manchuria Graph* was published by the Japanese proponents of a Japanese-led Manchukuo regime (a fascist government serving imperial Japan, particularly after 1937), the publication should not be viewed as purely a piece of propaganda. It is an important primary source, a historical document that reveals the contemporary political concerns of the Japanese imperial government, as well as the commercial designs of the SMRC, the semi-public, semi-private company that owned the publication. In addition, the pictorial reveals how Japanese settlers served as an important moral bulwark against the allegedly spiritually detrimental communism of the Soviet Union bordering Manchukuo to the north, while also producing crops necessary for Japan's war effort and birthing children to replace the fallen. These important pursuits were chronicled in a "realist" fashion by the photographers of *Manchuria Graph,* who embarked on an important mission to glorify satellite regions of the Japanese empire.

International affairs historian Tomoko Akami argues that Japan's news propaganda had to appear "credible" to be effective, and that "the Japanese metropolitan state began a coordinated and systematic approach to news propaganda during the Manchurian crisis, and that foreign policy concerns, rather than apprehensions about general domestic thought control, were the main driving force for this move."[21] Published by the influential SMRC, *Manchuria Graph* certainly masqueraded as news for the Japanese at home and, seemingly, also for a Japanese diaspora and English readers abroad through consular gifts of the publication to university libraries. This motive must have had limited success, since, aside from sparse English text, captions were written primarily in Japanese. *Manchuria Graph* even included less-than-encouraging statistics about the initial settlement of the region, demonstrating that settlement endeavours were not without hardship and required great perseverance from all participants. All of these challenges, however, highlighted the intrepid Japanese settlers' strong moral values and seemed to hint that Japanese "spiritualism" and plucky, youthful strength could overcome any kind of adversity, not the least of which was the Manchurian natural environment.

## Setbacks to Settlement: Northern Manchuria and the Climate

Manchuria has a temperate climate broadly similar to that of the American upper Midwest (Minnesota and the Dakotas) or the mid-southern Prairie provinces of Canada (Alberta, Saskatchewan, and Manitoba) and, like its North American counterparts, is a land of extremes. However, Manchuria's extreme climate was also shaped by geographical elements quite different from those in the Canadian Prairies and the American Midwest. Areas to the south, like Dalian (Dairen while under Japanese control) and Jinzhou, enjoyed a climate somewhat tempered by their proximity to the sea, while inland areas to the north of Harbin suffered Siberian winters with winds blowing in from the northern steppes. Temperatures in the winter often dipped below –40°C, even on sunny days. Summer temperatures could be well over 30°C, with sandstorms in areas closest to the Mongolian border in the west. Intending to encourage settlement, publications like *Manchuria Graph* concentrated on the region's temperate seasons, glossing over the fact that winter in northern Manchuria could last for nearly half a year, from late October to the middle of April.

The climate of Manchuria greatly affected the nature of dwellings, clothing, and food, which all differed radically from those in Japan. Thatch-roofed houses in rural localities required double panes of glass, with thick brick walls and coal-based heating underneath a raised platform, or *kang*. In winter, people dressed in many layers of cotton-padded clothing, with multiple undergarments and heavy boots with fur inserts or linings. An archetypal, nostalgic image of the region in winter is that of a man in a fur hat with flaps and a padded, cotton jacket; today, most locals wear leather jackets or down coats. Food was on the hearty side, with heavy, wheat-based noodle dishes, soups with meat broth, pickled cabbage, and tofu. Families who could afford to do so, or whose menfolk were avid hunters, dined on chicken, duck, rabbit, game birds, and sheep – all alien foods to the Japanese newcomers. Fish, typically a staple to the Japanese immigrant, was not easily available and was generally limited to the bottom-dwelling carp or other less palatable fish, all of which tasted muddy due to their brackish, riverine habitat. Even Han Chinese were relatively new to this northern region, which had been "opened" for general settlement in 1902 by the Manchu Empress Dowager Cixi (1835–1908) to counteract Russian influence and proactively "Sinicize" the area to protect Chinese interests. Chinese had settled southern Manchuria since the Ming Dynasty (1368–1644); by the late Qing period (1644–1912), the area boasted a fairly dense population, considering that it was a borderland, but the northernmost areas still remained sparsely populated – for good reason. Even when the SMRC had laid tracks beyond the city of Harbin after acquiring the northern portion of the China Eastern Railway from the Soviet Union in 1935, the territory retained its negative reputation for banditry, remoteness, and a harsh climate.

Japanese rural settlement schemes for Manchuria began after 1932 in experimental villages, with settlers recruited from domestic Japan in rural areas supposedly suffering from overpopulation.[22] However, Young posits that factors like the drop in rice and silk prices, crop failures, and adverse weather prompted negative social conditions in rural localities, which then led to high debt in addition to tenancy disputes; these challenges translated into a desire to improve one's lot elsewhere.[23] Young also notes that areas traditionally involved in emigration sent out Japanese farmers in an attempt to settle northern Manchuria.[24] Yet, for a multitude of reasons, including climate, the area to the northeast of Harbin, in Pinkiang (Binjiang) Province, was not included in these initiatives for nearly a half-decade more. Even *Manchuria Graph,* a publication with clear incentives to present a voice of optimism, noted that as of 1936, 40 percent of the Japanese settlers in rural Manchuria had left their villages or had died.[25] Obviously, northeastern Manchuria was no utopia, and the harsh environment played a major role in quelling both enthusiasm and efforts for settlement.

In February 1937, under the first Konoe Cabinet (1937–39), more organized attempts to settle the challenging lands of northern Manchuria were begun. These consisted of sending bands of young men and youths (who were recruited by their schools in distressed rural areas of Japan like Nagano or Yamagata) to training camps in Manchuria – like the ones near Tieli or Sanjiang, located on rail lines between Suihua and Jiamusi to the northeast of Harbin. Saitō Toshie notes that media like *Anata mo giyūgun ni naremasu* (*Even You Can Become Part of the Volunteer Youth Corps*), a patriotic *manga* [illustrated cartoon book] produced by the artist Tagawa Suihō, glorified the national duty of these boys and portrayed the Manchurian trip as an adventure, without the obvious permanence that settlement implied.[26] Though the intention was to recruit youth aged fifteen to eighteen, the median age of recruits was fourteen to fifteen (some were as young as twelve), or boys who had finished only their second year of junior high school. The principal of each school in targeted areas of Japan arranged a quota of young boys, whose motives for joining up were usually the exhortations of their teachers; once they had been recruited, the boys would be readied for the journey following bureaucratic procedures in their home villages.[27] These young men, who were usually completely ignorant of the land that awaited them, would help pave the way for the Japanese farm families that would eventually colonize the region.

In the June 1938 special edition of *Manchuria Graph* dedicated to the immigrant youth corps (*seishōnen giyū imin*), the cover features a scene of boys on tractors cultivating the land, juxtaposed with an angular modernist map of Manchuria showing the locations of the seven northeastern training camps.[28] Japanese flags pinpoint the location of the camps, and arrows reveal the route the youth would take to get there, following the SMRC-run rail lines. In typical imperialist fashion, the flags and arrows highlight the protective Japanese military presence over the boys as they survey the land being cultivated from atop their mechanized tractors, in yet another evocation of the continental sublime. The huge, double-page spread that follows, with a Japanese caption that reads, "young warriors [*senshi*] developing the enlightened vast land," depicts two youth playing bugles at sunrise. In the bottom right-hand corner, the English caption headed "Youthful Japanese Immigrants – Pioneers in the Development of Manchoukuo," notes the following:

> In parallel with the plan for sending 1,000,000 Japanese households comprising 5,000,000 persons to Manchoukuo during the twenty-year period beginning 1937, the Japanese and Manchoukuo authorities have decided to settle 30,000 youths, ranging from 16 to 19 years of age, in the New State. With high hopes for the future[,] these young immigrants, who are destined to play an important role among the Japanese immigrants in Manchoukuo, are arriving in increasing numbers to make the "promised land" their new home.[29]

One of the more arresting images in the same edition shows a young man driving a massive Caterpillar tractor emblazoned with the Japanese slogan "young, robust power" (*wakaku takumashiki chikara*). A smaller English caption notes that "primary emphasis is laid on spiritual training." This phrase leads to the following explanation in even tinier text: "Although much attention is being paid to the technical and vocational training of youthful settlers, spiritual training is of primary importance and every effort is being exerted to develop them into patriotic citizens who will become the medium for harmonizing relations between Japan and Manchoukuo."[30] In a darker interpretation benefiting from historical hindsight, the boys have been placed (literally) in a liminal position between hostile Chinese, whose land was displaced, and Soviet troops patrolling the border. Their Japanese spirit seems their only defence (aside from their rifles, which are so large they often dwarf the boys' own physiques). The next page recounts the beginning of the boys' day after they rise at 5:30 a.m. – a flag-raising ceremony involving morning rites of the entire corps at sunrise, including praising the shining heavens and chanting "serve Japan" (*Yamato hatarake*).[31] These images serve as epiphanic representations designed to generate an emotional response of awe in the viewer and to advertise the Japanese-led government (and SMRC) schemes to develop the region.

The splendid efforts of the youth corps even made it into the literary culture of the late thirties through a *Bildungsroman* by the popular writer Tokunaga Sunao (1899–1958), who earned his fame with the 1929 best-selling proletarian novel *Taiyō no nai machi* (*Sunless Streets*). A decade later, he capitalized on Japanese interest in the rural settlers and published *Sengentai* (*The Forerunners*, or *The Trailblazers*), which detailed the efforts of the boys setting up militarized border villages close to the Soviet Union. In the novel, the youths register surprise at the large physical size of Chinese labourers, who seemingly subsisted on a meagre diet but worked much harder than the boys did in the hot sun and the bracing cold. In Chapter 8 of this volume, Ronald Suleski shares survivors' retrospective stories of hardship and later disillusionment, due in no small part to the harsh natural environment and strict disciplinary tenor of the camps.

If it was initially so difficult to settle the region, why did the situation change in the late 1930s, when ideologues and the general media touted Japanese rural settlement of northern Manchuria as the solution to a whole host of problems? Though the rural settlement schemes had been conceived before wartime, the eruption of conflict in China south of the Great Wall in July 1937 played a key role in accelerating the plans to "Japanize" the region – a motive not too different from that behind Cixi's twentieth-century move to counteract Russian influence. Because so many Japanese soldiers had been deployed, the task now rested with ordinary emigrant families to provide a bulwark – both physical and moral – against Soviet Communism immediately to the north. These Japanese settler villages potentially served to create a "buffer zone," preventing the infiltration of Soviet agents southward to "infect" Chinese populations. Indeed, during the Manchukuo period, the northern railway hub of Harbin served as a centre for espionage – due to its convenient location on the Sungari River and its proximity to Russia. The city harboured Soviet espionage cells, White Russians employed as agents of the Japanese, Korean independence activists, and anti-Japanese communist agents of the Chinese, so it was no surprise that the Japanese military police positioned their office across from Harbin's station.[32] In other words, the villages settled by Japanese rural families were to play a crucial strategic and political role, while settlers of this rather inhospitable region were recruited through attractive images of utopian bounty.[33] In sum, though the Japanese imperial government hoped to settle one million families (in total, five million people), only a fraction of the desired total actually settled there – 321,882 individuals.[34] The propaganda aims of *Manchuria Graph* may not have been met, but the pictorial leaves scholars with poignant and intriguing images of a failed social experiment, depicting what Nagoya Museum of Art photography curator Takeba Jō has called a "reflected utopia."[35]

## Vital Resources: Babies, Crops, and Wartime Productivity

This study analyzes three special editions of *Manchuria Graph,* which usually came out in December, along with a few images from other issues, as representative of some of the imperialist propaganda reflected in the periodical. The themes emphasized in the images include, first, the idea of great productivity through highly industrious labour performed by Japanese of both genders and all ages to create a utopia of Japanese values, and second, the idea of astounding fertility of both crops and cultivators, both of which could be reproduced on a large scale. On the whole, the images are a bit generic, which enhances their general appeal, as well as the portability and universality of the values they represent. Though generally emphasizing a broader rubric of *Kyōwa* (harmony or cooperation), the vision of "peace" portrayed in the photographs belies the true nature of military aggression displayed by the Japanese Kantō Army from 1931 to 1933 and the displacement of the region's largely Chinese inhabitants by rural settlers from Japan (who were themselves sometimes attacked and killed by Chinese locals) from 1932 until 1945.

Intriguingly, all of the special editions discussed here appeared in December, when the winter season was well established in northern Manchuria. Yet, in these issues, the negative aspects of winter were minimized. Covers do not depict wintry scenes, and the editions include no images of the preparations necessary for facing the unforgiving winter storms that would rush in from Siberia or the toll that inadequate clothing or lack of foresight could take on individuals who attempted to engage in hard labour outside during the coldest months of the year. However, the December 1940 edition on touring the rural development areas does feature a spread entitled "Winter is Even More Nice" ("Fuyu mo mata tanoshii"), which shows men, women, and children happily working outside in the sunny, icy daytime.[36] Bundled up against the bitter cold, men wear fur-lined *ushankas* (Russian-style caps) with flaps, boys are clad in Russian-style rabbit-fur hats, women don quilted and cotton-batted kimonos, and babies sport knit caps with pompons. Each photo shows a Japanese person engaged in productive labour: men with a toddler feeding chickens; women spinning or sewing with infants on their backs; a family drying fish; men overseeing mules transporting lumber; a youth chopping down a preternaturally large tree; a husband and wife bringing in firewood; and a father weaving *tatami* matting as his son looks on. Though the images appear posed, they suggest that the harsh winter proves no obstacle to the work necessary for running a village. In Chapter 5 of this volume, Norman Smith details efforts made by the Japanese to counteract the effects of a Siberian winter, and officials' attempts to improve public health during what was traditionally a time of inactivity. Nevertheless, and hardly surprising, enduring the bone-chilling northeastern Manchurian cold while engaged in productive labour was not as "enjoyable" as depicted here: the dry, frigid air made it difficult to breathe (especially for pregnant women) and the settler population experienced a rise in respiratory ailments during the winter months.

In *Manchuria Graph*, there are very few photographs that show the settlements in winter – possibly because urban photographers tended to visit rural settlements during milder weather (that is, from spring until the end of the harvest season). Those winter photographs that do appear are few and limited, often showing lumber harvesting, since that was an activity that usually took place in winter because transportation of the logs (on sledges across the ice) was easier at that time of year. Manchurian cities like Harbin, with their relatively good sanitation and central heating in Western-style buildings with electricity, boasted comfortable lodgings for tourists and enjoyable attractions such as ice sculptures, skating, and sled rides on the Sungari. As Kathryn Meyer demonstrates in Chapter 7 of this volume, however, complex ecologies of squalor lurked behind the thick walls of modern apartment complexes – housing quarters for heroin users, prostitutes, and petty criminals. It was also not unusual for cash-strapped addicts of all ethnicities to freeze to death after being thrown out of their lodgings. The less technologically advanced Japanese rural settlements lacked plumbing and private baths, and had only rudimentary sanitation with outdoor latrines and communal bathhouses in arrangements similar to those in farming towns in Japan.[37] Some parts of the villages had electricity, but this was very different from cities that boasted electricity and electric lights throughout the night. In the evening, settlements could be plunged into near complete darkness, illumined only by the stars above, in brittle air that froze any exposed skin in seconds.

Some of the initial experimental Japanese farm villages were railway guard settlements whose populations were composed of newly married men in their twenties with young brides. Interestingly, in northern Manchuria, outside the purview of the Kantō Army,[38] imperial Japan's self-defence needs for the SMRC in Manchukuo were outsourced to these ordinary people, who soon increased the size of their families despite various hardships and the proximity of the Soviet Union. A *Manchuria Graph* spread from August 1936, with a large heading in Japanese, states, "Japan's Lullabies Heard in Foreign Homelands." In English capital letters, the caption written on top of a photograph that takes up half the page reads: "Japanese lullabies over desolate Manchurian Plains." Here, two rural women, with plump-cheeked infants (one in the ubiquitous white knit pompon hat) wearing quilted padded kimono jackets and strapped to the women's backs, look modestly downward, away from the camera. A smaller text explains their moral duty and their important strategic role in extending the cultural reach of imperial Japan:

> In close cooperation with the near-by Manchoukuoan farmers, the Japanese settlers of the "railway guards" settlements shoulder a weighty and noble duty of being life-long guards of the railways. Consequently they are all married, well-built young men with a strong will to settle in Manchuria. As the safety of the railways enhances day by day, and as the cultivation of the settlers continues year after year, the number of the offsprings [*sic*] of the settlers increases, and today, just as in the peaceful villages of Japan, soothing melodies of Japanese lullabies can be heard over the virgin soil of Manchoukuo.[39]

A big block of text in Japanese on the right side of the spread describes the villages in more detail, highlighting how the homes are dark, unlike in domestic Japan, and the fact that, like the Chinese, the Japanese settlers eat millet (*gaoliang*) and wheat buns (*mantou*), foods unknown in Japan. They also raise livestock like Holstein cows, pigs, and chickens, and keep bees. The text notes that many of these unfamiliar creatures hail from an

SMRC experimental research station in Gongzhuling (Koshurei) and, that presumably, the SMRC had sent agricultural experts to teach the farmers how to raise them.

The cover of the December 1936 special edition, *Nihon imin no sōbō* (*Profiles of Japanese Immigrants*), shows a smiling young Japanese woman holding a sheaf of grain with a cute infant on her back. Another woman and child garbed in a similar fashion are engaged in harvesting work only a few metres away. The baby wears (the now very familiar) white knit pompon cap and padded cotton clothing, while his mother sports a multi-layered kimono and a padded *haori* jacket, with *monpe* work trousers (drawstring pants fashioned out of the bottom portion of the kimono). She appears to be wearing leather boots. The photo must have been taken in October, since there is no snow on the ground and harvests generally took place at the time of the first few frosts. Judging from the new baby and the bundle of grain, it is apparent to the viewer that both the woman and the land are fertile, producing abundant crops and plump offspring.

A few pages later, after an article entitled "Constructing a Utopian Homeland [*risōkyō*] for Japanese Immigrants on the Fertile North Manchurian Plain,"[40] another spread highlights common scenes from a settler village.[41] To the left, positioned beneath a bustling scene of thatch-roofed houses and elementary school students under the village gate, a midwife rides off on horseback to oversee a birth, and a new mother washes a newborn under another midwife's direction. The caption reads:

> All the first Japanese settlers were ex-soldiers, and they came to settle in Manchuria under arms. Since then they have been fighting with brutal bandits till at last they have established the peaceful life of today. Their families and brides have been called from home, and today the settlements are ringing with their triumphal songs and the encouraging cries of their new born [*sic*] babies.[42]

To the right, beneath a photo of a small country elementary school and a miniature one-storey brick hospital under construction, men are shown fishing and hunting, presumably to provide food for their growing families. The pictorial neatly depicts a gendered division of labour (birthing and child rearing versus teaching, hunting, and fishing). One page later, a large spread of an idyllic country scene, with women returning from the fields, is juxtaposed with an image of women washing clothing in a bucolic stream, where not even the mated pair of ducks (with half a dozen ducklings in tow) is bothered by their presence. An image of a backyard with a flock of geese appears to the side. The images communicate abundance, peacefulness, and the fertility of humans, livestock, and crops in a new "utopian homeland."

Echoing earlier themes of fertility in the settlements, the cover of the November 1938 edition shows two smiling women, both holding plump infants, with a little boy wearing a white cap in front of them.[43] Though

FIGURE 6.1 Cover of November 1938 issue of *Manshū gurafu* (*Manchuria Graph*). *Source:* SMRC

the issue came out in November, the scene was probably photographed in summer, since the children are wearing cool, white cotton clothing, and one infant's foot is bare. The cover (with the older woman looking directly at the camera and the other looking downward modestly) suggests that Japanese mothers are producing babies abundantly in Manchuria while their menfolk engage in local self-defence patrols or go off to war as soldiers in the intensifying conflict with China.

In a two-page spread entitled "A Glimpse of the Immigrant Farms," there is an even more conspicuous lack of men, as if the villages were exclusively populated with women holding infants (or carrying them on their backs), and with toddlers at their feet.[44] At this time, many of the men spent one out of every two days away from home, guarding the rail lines.[45] The next page (captioned in Japanese, "Children who grow up on the vast lands," and in English, "Children who grow on the New Land, the 'Good Earth'") shows a photo of two middle-school-aged boys in caps with another group of young girls holding their infant siblings and accompanying two boys in short pants walking down a path. The English caption seems to echo the title of American novelist Pearl S. Buck's (1892–1973) *The Good Earth*, which won the Nobel Prize for literature in 1938. Yet, unlike the "Good Earth" in Buck's story, this "good earth" is a benevolent land of bounty and abundance, blessing its settlers with food and offspring in exchange for their dedicated worship. In every village and Youth Corps training camp, a shrine welcomes the Japanese gods of the Shintō religion. In the main training camp of Tetsurei (*Tieli*), for example, there is a remarkable Shintō shrine gate, or *Torii*, made out of birch logs facing such a shrine.[46]

Towards the end of the issue, in the section "Listening to the Voices in the Immigrant Lands" (Iminchi no koe wo kiku), one can read the personal testimonies of the settlers, including that of a midwife, Ms. Yamamoto, who took care of 150 women who in a single year (1938) gave birth to (a remarkable) 250 infants in Senshin Village.[47] This birthrate appears almost unbelievable to the modern reader, but is not unlikely, considering that these men and women were in their late teens and early twenties – their most fertile reproductive years. It is not impossible for a woman to become pregnant three months after giving birth, making possible two births in a single year. A much smaller percentage of women may have had twins. However, these babies may not have all survived. In answer to the journalist's questions about (the state's policy) *Ume yo, fuyase yo* (Give birth and reproduce!), the midwife remarks that some women had given birth to no less than three children in four years. A grandmother, a former traditional fan dancer, also expressed to the journalist (in slang-laden language) that her greatest joy in Manchuria was to see the faces of her three grandsons every day. But the issue also hinted at the potential dangers in this fertile utopia (and the fact that the state was essentially using these farmers as border guards, or *kaitaku senshi* [rural development warriors]). The most poignant testimony was that of a widow caring for the young male heir of "a fallen warrior [*senshi-sha*] wielding a hoe" – a man who had been killed by bandits during a raging snowstorm. This "reflected utopia" was not without its perils, and those sacrificed in its service were accorded the same status as soldiers killed in action.

The December 1940 special edition on the settlers, or "Touring the Rural Development Areas" ("Kaitaku-chi wo meguru"), features more images of abundant harvests with men wearing military uniforms and caps, even as farmers. The spread inside, labelled "Battling with the Continental Soil" ("Tairiku no tsuchi to tatakau"), shows two women and boys next to a pile of massive daikon radishes, globular white beets, huge European cabbages, a large napa cabbage, and a bulbous pumpkin. Indeed, breaking up the land and coaxing it to produce crops in such a challenging environment paralleled mobilization for war. Nevertheless, the Japanese

FIGURE 6.2 Two-page spread in December 1940 issue of *Manshū gurafu (Manchuria Graph)*. *Source:* SMRC

spirit, even that of the women, could rise to the task of industriousness on behalf of the wartime state. On the facing page appears a more obviously militaristic photograph of a smiling wife holding a sheaf of wheat, while her husband in military garb sits atop a mound of harvested grain with a sickle in his hand. Their faces turn towards the rising sun, while the sickle seems to point towards a right-wing proletarian utopia (in opposition to the Soviet one). A similar scene, now including a child, would appear later in a 1943 edition, juxtaposed over enlarged rice stalks and overlaid with an edifying wartime poem.

The 1940 issue also highlights a section entitled "Women of the Development Areas" ("Kaitaku-chi no josei"), which includes a depiction of women readying the settlement for the harsh winter ahead by harvesting and preparing food for storage.[48] The women conspicuously embrace large sheaves of wheat, carry bundles of kindling for firewood on their backs (even while clutching their infants tied to their waists in front!), herd flocks of sheep, make dinner, and prepare daikon radishes for pickling. The women combine childcare with other necessary tasks, and were surprisingly productive despite their numerous children. This edition shows the increasingly militaristic nature of the settlements, as well as the high value placed on female labour (through work and birth) while men in the villages were being sent to the front.

Japan's war efforts further intrude into the Manchurian space by turning the region into a base of production of vital natural resources for wartime. The December 1943 special edition features "grain storehouse Manchuria" (*komekura Manshū*), its cover boasting a smiling young woman holding a sheaf of grain – a now archetypal image of harvest in the Manchurian space.[49] On one page of a two-page spread inside, we see a model Japanese family (husband, wife, and child) on the left, and on the right, the rare image of a Japanese man, who seems to wear military garb, juxtaposed against vastly enlarged stalks of rice.[50] On the facing page, a poem by Takahashi Akio appears in white over the rice stalks, describing how even the tiny grain of rice hopes for certain victory and even the tiny bean cries out for total victory, while warriors on the militarized fields expend the fruits of their sweat to increase production.[51] If the poem does not make the militarization of agricultural production in Manchukuo obvious enough, captions like "Agricultural Fields to Protect the Country" ("Hokoku nōjo") underline this initiative.[52] Following these photo montages is a two-page article titled "The Agricultural Production and Resources of Manchuria as a (Military) Supply Base" ("Heidan

Figure 6.3 Cover of December 1943 issue of *Manshū gurafu (Manchuria Graph)*. *Source:* SMRC

kichi Manshû no nôsan shigen") and several two-page maps and short descriptions of northern Manchurian counties supplying the troops and wartime Japan, indicating areas involved in natural resource production, livestock, forestry, mining, and crops, all linked by rail. This kind of imagery was not unique to wartime imperial Japan and transcends the wartime period. As Sun Xiaoping rightly points out in Chapter 10 of this volume, in the early 1950s, the communist People's Republic of China (PRC) used similar propaganda strategies to attract demobilized People's Liberation Army (PLA) veterans of the Korean War to the same area in Northeast China, hoping to convert a region then labelled *Beidahuang* (Great Northern Wasteland) into *Beidacang* (Great Northern Granary). Just as Manchukuo's Japanese overlords portrayed the land as "empty" by erasing its Chinese heritage, the postwar Chinese government also erased the legacy of Japanese rural settlement by initially presenting the region as a "wasteland."

The articles and photographs in *Manchuria Graph* are not the only elements of the publication that emphasize the surprising productivity of the land and its people. Advertisements for patent medications and vitamins to improve one's vitality, bodily health, and reproductive capabilities and organs had been appearing in the publication since 1940, but seem to have proliferated as the war intensified (Japan also added a new front in the Pacific against the United States after 1941). These various concoctions promised to increase labour potential and productivity by increasing vitality for both genders, while also improving reproductive capabilities. For example, the top of one page depicts a patent medicine named Metabolin, which promised to increase strength and virility in men; directly below it, a similar medication for women called Oivestin promised in its advertisement to regulate menses and normalize reproductive ability.[53] Presumably, domestic Japanese readers would learn about the surprising fertility of the settlers through the pictorial and wish to "reproduce" those feats back home – with a bit of help from the right over-the-counter medication. However, all of these ads suggest that, as the Japanese were in danger of losing the war and food was becoming scarce through rationing, their people's reproductive potential was adversely affected (thus necessitating the medication).

By 1943, the focus of the pictorial appears to have changed – from touting a rural utopia in order to exhort settlement and inflame Japanese pride to advertising the almost militaristic productive power of the rural settlements in north Manchuria, which would help the Japanese defeat their enemies and lead them to victory. Indeed, according to the images and the text, even a single grain of rice in Manchuria cultivated by Japanese settlers with a resolute will could help spur the empire on to certain victory! As the state's wartime goals intruded into the publication, its nature as an over-the-top propaganda device became increasingly difficult to ignore. Yet, of interest to the historian are these evolving themes indicative of the state's focus and its goals for these (willing) victims of Japan's imperial desires.

## Conclusion

Propaganda magazines featuring Manchukuo (such as *Manchuria Graph*) displayed scenes of a model rural settler population, where men were productive labourers and women performed the dual role of producing children and raising them as well as their crops. However, from these images, it becomes increasingly evident that as Japan's war efforts bogged down in China and intensified in Southeast Asia and the Pacific by the early 1940s, men began to disappear from view. Women became increasingly important to imperial Japan for their child-bearing and child-rearing capabilities, while still maintaining productivity in the rural workplace (in this context, northern Manchukuo settlement areas).

The photographic images featured in *Manchuria Graph* suggested that Japanese immigrant families served not only to produce vital natural resources in terms of food and children, but also spiritual resources that indirectly helped the war effort. Essentially, if "Japanese-ness" could be reproduced in the rural immigrants even in Manchuria's harsh environment, then Japanese colonization could succeed anywhere. In the introduction to a 1942 edited volume of multi-ethnic Manchurian literature (which included stories about the settlers), prominent Japanese establishment author Kawabata Yasunari (1899–1972) noted: "Because Pan-Asian ideals were first put into practice in Manchuria, if they are not attained here, then [it] might ... be assumed that they cannot be attained anywhere."[54] In other words, Japanese values were to be rooted here first, and then transplanted elsewhere. The Japanese "spirit" was thus of immense importance. The Manchukuo example is not unique: within studies of imperialism, similar examples abound of colonies serving as laboratories of social experimentation. Yet Manchukuo serves to set Japanese imperialism within a global context. In the pages of *Manchuria Graph,* the utopia presented grew against great odds, but revealed the absolute resolve of the intrepid Japanese settlers, even if only as a reflection of the desires of the new state's ideologues.

*Manchuria Graph* thus provides contemporary scholars of Sino-Japanese cultural history with important documentation of a failed social experiment. Many individuals actually believed it was possible to come to a "promised land" or a "new earth" to fulfill dreams otherwise unrealizable in a cramped and restricted authoritarian Japan. The vast Manchurian plains stretched ahead of them in a sublime vision of the future, where harmony and cooperation would assuage the social unrest in the decaying rural villages back home. However, once Japan's ongoing military conflicts intruded on settlers' lives, and the harsh Siberian winters began to take their toll, many settlers yearned to return to their former lives. As Suleski makes clear in this volume (Chapter 8), the settlers were ultimately abandoned to massacre by the invading Soviets (and even by their own men) in mid-August 1945 once the Kantō Army fled to grab the first trains heading to Japanese ships waiting at port in Dairen. The pivotal oral history research of Onitsuka Hiroshi has brought to life tales of tragedy and horror – including instances where Japanese mothers in the rural Sanjiang area killed their own children, and were then killed by their husbands in order to avoid death at the hands of Soviet troops – testimonies of which also appear in Chinese accounts.[55] The state had pushed the Japanese to produce more children and grain as wartime progressed, but then left settlers to their own devices in the wake of the empire's dissolution. Some Japanese survived to tell their stories in the recent Shimoina Project,[56] or began to look for family members back in Japan following Deng Xiaoping's post-1979 "Opening and Reforms." Called "left-behind orphans" (*zanryū koji*), Japanese children who had been too small to walk at the time of the empire's dissolution had been left with sympathetic Chinese families in the aftermath of Japan's defeat. Usually these children were not able to reunite with their parents or siblings in the chaos, civil strife, and revolution that followed.[57] In hindsight, the bright images on the pages of *Manchuria Graph* instead seem to preface the impending darkness, in which a government abandoned the people it had encouraged so enthusiastically to leave their homeland and chase a dream.

## Notes

*Acknowledgments:* This chapter evolved out of the presentation "Constructing a Rural Utopia: Propaganda Images of Japanese Settlers in Northern Manchuria, 1938–1943," given at "The Manchurian Environment: Natural Resources, Climate, and Disease" symposium and workshop held at the Institute of Asian Research at the University of British Columbia in Vancouver, Canada, May 16–18, 2013. Some portions of the chapter were adapted from the presentation "Japanese Women and Rural Settlement in Wartime Manchukuo: Gendered Reflections of Labor and Productivity in *Manshū gurafu* [*Manchuria Graph*], 1940–44," delivered at the Association for Japanese Literary Studies (AJLS) annual meeting at the Ohio State University (OSU) in Columbus, Ohio, October 12–14, 2012, and the subsequent proceeding published in August 2014, which contained an expanded version of that paper.

1 I will use "Manchukuo" to designate this state fashioned under Japanese auspices, since this is how English-language sources now most commonly refer to it. It appears to be a

pre-war Wade-Giles Romanization of the currently used *pinyin* Chinese Romanization *Manzhouguo*. (Pinyin was developed in 1958 during Mao Zedong's campaign to reduce illiteracy.) In Japanese, most writers labelled it either *Manshū* [Manchuria], which could refer to the region or state, or, more specifically, *Manshūkoku* [*Manzhouguo*, or "Land of the Manchus"]. Chinese sources at the time referred to it as *Manzhouguo*, while contemporary materials often add *wei*, or "fake," before it.

2 According to the dating of Japanese postcards developed by Urakawa Kazuya, "Kindai Nihon no Higashi Ajia, Nanyō shotō e no 'manazashi': ehagaki no rekishi kachi no 'ibunka' hyōshō" [The Japanese "gaze" on the peoples of East Asia and Micronesia: Archives' importance and the other race's representation in Japanese picture postcards] *Kokuritsu rekishi minzoku hakubutsukan kenkyū hōkoku* 140 (March 2008), 133. I thank Paul Barclay for making me aware of Urakawa's valuable study. An online version in English can be found at http://sites.lafayette.edu/eastasia/2014/09/04/how-to-ascertain-the-date-or-time-period-of-a-japanese-postcard/#more-639.

3 I thank Norman Smith for the gift of this postcard right before I published my first book on Manchukuo in 2013.

4 Alan Tansman, *The Aesthetics of Japanese Fascism* (Berkeley: University of California Press, 2009), 32.

5 Kari Shepherdson-Scott, "Utopia/Dystopia: Japan's Image of the Manchurian Ideal" (PhD diss., Duke University, 2012), http://dukespace.lib.duke.edu/dspace/handle/10161/5433.

6 Author's email correspondence with Shepherdson-Scott, September 6, 2013.

7 Shepherdson-Scott, "Utopia/Dystopia," 61.

8 Tansman, *The Aesthetics of Japanese Fascism*, 27.

9 E. Ann Kaplan, *Looking for the Other: Feminism, Film and the Imperial Gaze* (New York: Routledge, 1997). Kaplan's study falls under the category of women's studies and postcolonial theory in film, but can also be applied to the depiction of unequal power relations between colonizer and colonized in other media such as photography or art.

10 Rana Mitter, *The Manchurian Myth: Nationalism, Resistance, and Collaboration in Modern China* (Berkeley: University of California Press, 2000).

11 Even Manchukuo's mostly Chinese male subjects were eventually deployed by the early 1940s.

12 After 1905, the Kantō Army began as a small "police" force to guard the railway tracks administered by the SMRC. It grew increasingly independent of the company, and, by the late 1920s, began to involve itself in local politics. This invasion began with a plot allegedly masterminded by Kantō Army leader Colonel Ishiwara Kanji (1889–1949), when a bomb went off on a portion of track near Shenyang. The Japanese blamed Chinese communist conspirators; thus, this incident provided the pretext for a full-scale Kantō Army invasion of Manchuria to maintain imperial Japan's interests in the region. However, domestic politicians were not convinced, and Prime Minister Inukai Tsuyoshi (1855–1932) and his cabinet spent much time deliberating over the issue of recognizing Manchukuo. A disgruntled naval officer assassinated Tsuyoshi for his dithering.

13 Louise Young, *Total War: Japan's Total Empire: Manchuria and the Culture of Wartime Imperialism* (Berkeley: University of California Press, 1998), 55–114.

14 Annika A. Culver, *Glorify the Empire: Japanese Avant-Garde Propaganda in Manchukuo* (Vancouver: UBC Press, 2013), 126–30.

15 Ibid., and Annika A. Culver, "Japanese Mothers and Rural Settlement in Wartime Manchukuo: Gendered Reflections of Labor and Productivity in *Manshū gurafu* [*Manchuria Graph*], 1936–1943," in *Motherhood and War*, ed. Dana Cooper and Claire Phelan (New York: Palgrave Macmillan, 2014), 95–113; Annika A. Culver, "Japanese Women and Rural Settlement in Wartime Manchukuo: Gendered Reflections of Labor and Productivity in *Manshū gurafu* [*Manchuria Graph*], 1936–1943," *Proceedings of the Association for Japanese Literary Studies* 14 (August 2013): 195–211; and Annika A. Culver, "The Making of a Japanese Avant-Garde in Colonial Dairen, 1924–1937," *History Compass* 5, no. 2 (2007): 347–61.

16 Kishi Toshihiko, *Manshūkoku no bijyuaru medeia – posutaa. Ehagaki. Kitte* [Manchukuo's visual media – posters, postcards, and stamps] (Tokyo: Yoshikawa kōbun-kan, 2010).

17 Shepherdson-Scott, "Utopia/Dystopia."

18 See Culver, *Glorify the Empire,* 1–10.

19 Ibid., 9, 45, 50.

20 Ibid., 120. The publication was called *Soyuz Sovietskihk Sotsialisticheskikh Respublik* (CCCP in the Russian Cyrillic alphabet), which can be translated into English as USSR, or the Soviet Union.

21 Tomoko Akami, *Japan's New Propaganda and Reuters' News Empire in Northeast Asia, 1870–1934* (Dordrecht, The Netherlands: Republic of Letters Publishing BV, 2012), 206.

22 For a poignant contemporary photographic exploration of the surviving Japanese rural settlers and how the areas they settled in Northeast China look now, see Munekage Tadashi, *Kaitaku-min – kokusaku ni honrō sareta nōmin* [Immigrant settlers – Farmers at the mercy of government policy] (Tokyo: Kōbunken, 2012).

23 Young, *Japan's Total Empire,* 330.

24 Ibid., 331.

25 Yumani shobō reprint, *Manshū gurafu* [*Manchuria Graph*] 4, 12 (no. 29) (Tokyo: Yumani shobō, 2008), 226.

26 Iida-Shi Rekishi Kenkyūsho [Iida City Historical Research Institute], *Manshū imin – Iida Shimoina Ina kara no messeji* [Manchurian immigrants – A message from the Ina Valley in Iida] (Tokyo: Gendai shiryo shuppan), 56–57.

27 Ibid., 64–65.

28 Yumani shobō reprint, *Manshū gurafu* [*Manchuria Graph*] 6, 6 (no. 47) (Tokyo: Yumani shobō, 2008), 123.

29 Ibid., 124–25.

30 Ibid., 130.

31 Ibid., 131.

32 This location is now an elite junior high school in Harbin, and its nefarious history was recounted to me in an interview in January 2000 with Tang Xiang, a history teacher there.

33 Intriguingly, a sizeable portion of the internal immigrant population of Shenzhen, a city of nearly eleven million people in China's southernmost Guangdong Province, hails from Northeast China and, especially, the areas to the northeast of Harbin. As soon as the city began to develop as a Special Economic Zone (SEZ) after 1980, it attracted people from all over China due to its economic opportunities. Its mild tropical climate, humid warmth, and relatively clean air made it a huge draw for northern immigrants who had suffered six months of winter and coal dust every year. The popular film "Durian Durian" (2000), directed by Fruit Chan, follows the experiences of a young woman from Mudan-

jiang named Yan, who migrates to Hong Kong from Northeast China during three winter months to earn money as a sex worker. It poignantly details some of the hardships but also opportunities for newcomers, showing the woman's heartwarming friendship with Fan, a little girl living near her alleyway lodging.

34 Statistic quoted in Young, *Japan's Total Empire,* 321.

35 See Takeba Jō, *Ikyō no modanizumu: Fuchikami Hakuyōto Manshū shashin sakka kyōkai* [Modernism on foreign soil: Fuchikami Hakuyō and the Manchuria Photographic Artists Association] (Nagoya: Nagoya City Museum of Art, 1994).

36 Yumani shobō reprint, *Manshū gurafu* [*Manchuria Graph*] 8, 12 (no. 77) (Tokyo: Yumani shobō, 2008), 122–23.

37 One of the horrors for children might be falling into a communal latrine in the dead of night during winter. Usually, families made do with chamber pots, which would be emptied during warmer and brighter times of the day. In remoter parts of Heilongjiang, close to the former Japanese rural settlements, this was still the case in the villages I visited near Tieli City for fieldwork in 2000 and 2005.

38 Initially, the first forays into developing northern Manchuria after 1931 came from the Kantō Army, which opened up an immigration department to manage the rural development endeavours.

39 Yumani shobō reprint, *Manshū gurafu* [*Manchuria Graph*] 4, 8 (no. 25) (Tokyo: Yumani shobō, 2008), 148.

40 Yumani shobō reprint, *Manshū gurafu* [*Manchuria Graph*] 4, 12 (no. 29) (Tokyo: Yumani shobō, 2008), 226–27.

41 Ibid., 232–33.

42 Ibid., 233.

43 Yumani shobō reprint, *Manshū gurafu* [*Manchuria Graph*] 6, 11 (no. 52) (Tokyo: Yumani shobō, 2008), 243.

44 Ibid., 254–55.

45 See Yumani shobō reprint, *Manshū gurafu* [*Manchuria Graph*] 8, 12 (no. 77) (Tokyo: Yumani shobō, 2008), 258–61.

46 Ibid., 112.

47 Ibid., 258–61. The midwife's and widow's testimonies are on page 258.

48 Ibid., 120–21.

49 Yumani shobō reprint, *Manshū gurafu* [*Manchuria Graph*] 11, 12 (no. 113) (Tokyo: Yumani shobō, 2008), 59.

50 Ibid., 61–62.

51 Ibid., 62.

52 Ibid., 68.

53 Ibid., 85.

54 Kawabata Yasunari, Yamada Seizaburō, Kitamura Kenjirō, Gu Ding, Kishida Kunio, and Shimaki Kensaku, eds., *Manshūkoku kaku minzoku bungaku senshū 1* [Anthology of literary works by each ethnic group in Manchukuo, volume 1] (Tokyo: Sôgansha, 1942), 5.

55 Chinese accounts show how Japanese men even turned on their own families in a mass suicide (and massacre of those who had no choice). On August 12, 1945, after the Soviet invasion, in the village of Mashan in Dongan Province, the entire Japanese settlement group [*kaitaku-dan*] of 412 people, including women and children, were stabbed to death by their menfolk in the Mashan Incident. Chinese observers thought this was particularly cruel, even though the Soviets, whose army was composed of released ex-convicts,

were no less brutal. See the chronology of Manchurian settlement history by Nishida Masaru in Nishida Masaru, Sun Chiwu, and Zheng Min, eds., *Chūgoku nōmin ga akashisu "Manshū kaitaku" no jissō* [Chinese peasants witnessing the true face of "Manchurian development"] (Tokyo: Shogakukan, 2007), 371.

56 See *Iida-Shi Rekishi Kenkyūsho, Manshū imin – Iida Shimoina Ina kara no messeji.*

57 For more information on the issue of the "left behind orphans," see Rob Efird, "Japan's 'War Orphans': Identification and State Responsibility," in *Sino-Japanese Relations: Critical Concepts, Volume II,* ed. Carolyn Rose (New York: Routledge, 2011), 281–304, and "Japan's 'War Orphans': Identification and State Responsibility," *Journal of Japanese Studies* 34, no. 2 (2008): 363–88.

# 7
# The Garden of Grand Vision: Slums, Deviance, and Control in Manchukuo, 1940–41

*Kathryn Meyer*

One summer day in 2006, a team of Japanese architects spent twenty-four hours observing the activity along one city block in Harbin's Daowai section. They recorded the comings and goings of the people who lived in the area, concentrating on the busiest hours in the morning, at noon, and through the evening. The block interested the architects because it occupied the centre of a neighbourhood that had escaped the recent Chinese urban-renewal frenzy. Houses along the street predate the Second World War. The authors called them Fujiadian courtyard houses, "Fujiadian" being the old name for the neighbourhood. The layout of these brick buildings consists of two storeys of flats surrounding an open-air courtyard. An arched entryway connects the street with the courtyard common area. The architects praised this kind of structure because it brings communities together, both in the common courtyard and out onto the street. They contrast this with high-rise housing that is appearing in cities worldwide, where modern apartment life can leave people feeling isolated.[1]

The architects may have been surprised to learn that some six decades earlier, when China's northeastern provinces were occupied by the Japanese empire, three Japanese men – two policemen and one scholar – made a survey of a neighbourhood only a few blocks away from the location of their own study. The focus of the earlier survey was a courtyard house called the Garden of Grand Vision (*Daikan'en*). It gave shelter to a floating population, the poorest of Chinese people displaced by war, revolution, bad luck, or bad choices. The men who conducted the older investigation found in the building a dark atmosphere that nurtured crime. Every kind of vice flowed from the courtyard through the gate and into the street. Within this atmosphere of vice, people living in the building created a vibrant, informal economy that allowed them a chance at survival. Their activities did not remain confined to the slum. Products moved in and out of formal society, crossing the boundaries of legitimacy. This chapter will place the slum and the individuals who studied it within the dynamics of war and occupation. It will ask for whom the study was undertaken, and for what purpose, whether there a hidden message in the report, and, finally, what effect the study had on the men who undertook it.[2]

The Garden of Grand Vision was a vision of hell. Very much a courtyard house, built in 1927 as apartments and a theatre, the structure was converted by enterprising after-market engineers into a warren of flophouses and improvised services for a floating population of as many as two thousand desperate souls. Here men and women toiled in an underground economy. They made money as pilferers, gamblers, grave robbers, junk and rag pickers, and prostitutes. They recycled everything. Cigarette butts became new smokes; rat and cat pelts became hats and gloves. A meagre income bought them a bit of protection from the elements in Harbin, a city where winter temperatures can reach –30°C. In the Garden, a bunk under the jerry-rigged eaves of a flophouse could be found for as little as a dime a night. Outside the door of their lodgings, along the dark, fetid halls, lodgers could find fleeting comfort in the arms of prostitutes, brief excitement in a game of chance, or a cheap meal of soya bean milk, rice, and fish. Yet most of the lodgers saved their scant earnings to buy heroin from the unlicensed narcotics sellers who lived in the better rooms in the Garden.[3]

In 1940, as the Second World War escalated, the Garden of Grand Vision and its neighbourhood festered under the jurisdiction of a Japanese-run police bureau answerable to the puppet government of Manchukuo. Using the slogan "Harmony between Five Races," Manchukuo publicists touted the new state as a showcase of Asian modernism. They claimed to be operating under a Confucian moral order they called the Kingly Way. Yet unlicensed drug dealers operated openly in the Garden. Corpses of addicts lay on the street for hours, sometimes days, until an offal truck came to cart them away. This was not the picture that Manchukuo publicists presented to the world.[4]

In 1940, three Japanese investigators – Gotō Reiji, Satō Shin'ichirō, and Katō Toyotaka – began an inquiry into life in the Manchurian slums. They spent more than a day conducting on-site sociological research.

Over the course of a year, they looked into the causes of crime and narcotics abuse in the slums of Harbin. Two of the men spoke Chinese, one of the men spoke Russian. They interacted with their informants, and even considered several of them to be friends.[5] The community they discovered was not one of neighbourly interaction. The Garden was an environment of Darwinian struggle, mutual exploitation, and indifference in the face of tragedy. Even as the Second World War expanded, the inquiry into conditions at the Garden of Grand Vision revealed systematic social breakdown on a smaller scale.

The final 300-page report, *Autopsy of the Garden of Grand Vision*, was turned over to the Japanese military authorities running Manchukuo in 1941. A fascinating study of a slum, the report is at once sophisticated in its approach, empathetic in its descriptions of the urban poor, and blatantly racist in many of its statements. It contains cutting-edge understanding of the roots of crime in migration, social breakdown, and desperation alongside of the worst kind of voyeurism. It is a document that exposes the nature of Japanese occupation, both the paternalistic gaze of the agents of empire and the conditions to which the most vulnerable of the occupied were reduced.

## Setting

The invasion of Manchuria beginning in the fall of 1931 could arguably be interpreted as the first battle in the lead-up to the Second World War. The international community condemned the September 18 bomb blast (subsequently named the Mukden or Manchurian Incident) and the Japanese military occupation that followed. Yet for the Japanese military activists on the scene, men whose handiwork created the short-lived state, Manchukuo was a noble experiment in multi-racial Asian modernity. It offered a Confucian alternative to Western forms of social organization. It was also meant to reform the cabinet-based Japanese government system in Tokyo. Men like Colonel Ishiwara Kanji (1889–1949), who planned the military action, considered the cabinet-based parliamentary government to be corrupt, weak, and too Western to serve a nation of warriors.[6]

Manchukuo was not a static state. In 1933 and again in 1935, its borders expanded southward into China proper, laying the groundwork for more conflict to come. The Marco Polo Bridge incident of July 1937 brought Japan and China into an all-out shooting war; this was followed by the brutal Japanese invasion of eastern China.[7] In December 1941, the bombing of Pearl Harbor escalated tensions even further; mobilizing the population in an industrial centre like Harbin became crucial.

Late in the 1930s, even as Asia spiralled into war, there were signs that Manchukuo was not the modern, multi-ethnic beacon for Asia that the phrase "the Kingly Way" might imply. In Harbin, slums spilled into the Songhua River, where shipping brought goods to market. Addiction was rampant. Unlicensed heroin dealers sold their goods openly, and addicts died in the streets. The *Autopsy* dramatically presents these cracks and fissures in Manchukuo society. In doing so, the three investigators went beyond describing the traffickers, the gamblers, and the prostitutes. They also studied the dealers' clientele. Their curiosity drove them to question where the slum dwellers had come from, how they earned their money, and how they organized their lives. The resulting document provides a chronicle of the forgotten people of the war. These were not the heroic resistance fighters, nor were they the traitor Chinese who collaborated with the enemy. These were the individuals who fell through the cracks during invasion and civil war. The three investigators followed the people, the money, and the goods as their subjects survived – barely – by living by their wits or by sacrificing the capital of their bodies in the creation of a gritty, nefarious underground economy.

## The Garden and Its City

The Garden of Grand Vision was, in its architectural details, a typical Fujiadian courtyard house. It had a main gate leading to a courtyard, which is a hallmark of the building style. Yet beyond this, the Garden of Grand Vision was neither grand nor garden-like. Gotō Reiji, the lead investigator, said of the building, "A person passes through the gate into a fearsome dark world of evil and corruption."[8] Instead of bicycles parked by the door, or mailboxes in the passageway, prostitutes stood in the shadows calling out to those passing by. Inside the common area, gamblers set up impromptu games to attract potential players from the stream of customers passing through on their way to buy drugs. Gotō and his colleagues found informants and guides in the Garden – like the attractive prostitute Chun Fang, flophouse owner Lin Jitang, failed scholar Feng Dengdiao, and gamblers like Braying Mule Wang (*Wang da jiao lu lü*) and his shrewish concubine.

The Garden of Grand Vision was in the Chinese Fujiadian section of Harbin. Even today, tourists who visit the city marvel at the collection of building styles. During a leisurely stroll in the city centre, a person will come across Russian churches and homes, some built in a distinctive Siberian style, art nouveau buildings erected during the construction boom of Harbin's early years, and eclectic Chinese residences in the baroque style, all surrounded by the modern glass and steel of recent building projects. Gotō lived and worked in the Nangang section of town, a much cleaner and more gentrified section of the city. But not far from Kitaiskaia Street, with its international banks and high-end shops, was Fujin Street, along the river and not far beyond the iron railroad bridge: the location of the Garden of Grand Vision.[9]

Gotō described in detail for his readers a tram ride from his own neighbourhood into the Chinese slums: a descent from order into chaos. The main gate of the Garden of Grand Vision faced south, which, according to the building principles of *fengshui,* should have been auspicious. Just inside the front door, broad hallways ran along each side of the first floor; a grand staircase gave access to a second floor, also flanked by two long halls. Along these hallways were the doors to the flophouses – thirteen in all. In the doorways and on the stairs, prostitutes advertised their wares from early afternoon until late at night. At any given time, a visitor might have to step over a dead or dying body.[10]

A small alley separated the main building from a secondary rear section called the Village of Grand Vision. The Manchukuo government located Harbin Opium Dispensary Number Thirty-Four on the first floor of this building. Behind the entire complex on any evening, one could find a lively thieves' market, offering the would-be buyer everything from carpets to used clothes.

Gotō met his informants in the Spring Forest Lodge on the second floor of the main building. The franchise owner, Lin Jitang, was one of his informants. Lin paid the building landlord M$200 in monthly rent for his share of the second-floor hallway. He then let out nine rooms for M$0.50 a night, and bunk spaces in a lower or upper dorm for M$0.10 per night. Lin supplemented his income from lodging fees by fencing stolen goods. Otherwise, he spent his days smoking opium with his concubine.[11]

Lin left the running of the business to his cousin and a series of underlings, who, in the flophouse slang, were all called "boy" (in English). The boy's job was to eject any tenant who could no longer pay daily fees. Those guests who seemed to be close to death – and there were frequently many who seemed to be in this state – would be tossed naked into the freezing streets with little ceremony, their clothes stolen and resold. One of these boys, named Feng, was himself an opium addict. When Gotō first met Feng, he was a strong, rough fellow who excelled at his job. The residents called him "Feng Beats People" (*Feng Daren*). Within eight months, Feng's addiction had worn him down to such an extent that a new boy was given the job of pushing a naked Feng out into the street.[12]

The nine rooms in the Spring Forest Lodge were rented by pimps and heroin sellers. Each room was no more than a cubicle with a warped door and a rough plank for a bed. Yet in the Garden, this passed for luxury. The income from selling vice would cover the lodging fees.[13]

In Chapter 2 of this volume, David A. Bello describes a culture of Indigenous fur trapping that suited the climate and terrain of the Manchurian north. Arguably, the "culture" of the Garden reflected an ugly parody of this rugged, outdoor lifestyle. Gotō Reiji's lively prose takes the reader through the Fujiadian neighbourhood to reveal how goods, services, and money moved from the Garden to the streets of the poor, then into the wealthier neighbourhoods and back again. He provides biographical sketches of Garden residents who worked as pilferers, grave robbers, and beggars. He claims that pilfering alone kept this neighbourhood afloat. A pilferer's job fell somewhere between theft and recycling. A pilferer would collect or steal – often both in the same evening – everything from animal carcasses to cloth rags to pig bristles, from rat pelts to cigarette butts. He then sold the products of his labours to the various stores and small-scale manufacturers in and around the Garden, where recycled goods found their way back into the mainstream economy.[14]

Some pilferers worked in crews of three or four men. One crew Gotō studied entered factories at night to steal scrap metal and burlap bags. Each thief had a quota of at least four or five burlap bags per night. The bags could be resold for M$0.10. Some men travelled the streets carrying baskets of waste paper, a handy prop for hiding their ill-gotten gains. The pilfered materials with the highest resale value on the street were those in the underground urban fur trade.[15]

Gotō provides exquisite detail about a Manchurian fur trade that did not involve the sable pelts described by Bello. Pilferers from the Fujiadian slums collected rat corpses, cat corpses, and dog corpses. If they were lucky, the pelts of the carcasses would be glossy and fine. Each rat pelt could be sold for M$0.02. Dead cats brought in between M$0.10 and M$0.15; dead dogs could be sold for between M$0.05 and M$0.10, the price dictated by the quality of the pelt. The variety store owner, acting as middleman, sold the pelts to a tannery for between M$4 and M$10 per pound, depending on quality and colour. Tanners made the skins into eight-inch squares; two squares sewn together became a "scroll." A good tanner could create five scrolls from one pound of pelts. Tanneries could be found all along Lucky Road, near the Shandong Merchants' Association.[16]

Li Xiangchen – a thirty-nine-year-old man from Hebei in North China – owned one of the tanneries. He had invested M$200 to buy the enterprise and employed four professional tanners, who each earned M$0.10 per scroll. In one day, a good tanner could produce between twelve and fifteen scrolls. Li also accommodated four apprentices, who were not paid for the three years of their training. In addition to paying their salaries, Li fed all of his workers. He paid quarterly taxes of M$11.25.

From the tannery, the scrolls went to a dye shop. From there they went to the workhouses of fur companies, where workers used them to line hats, gloves, and scarves. These high-quality goods were then sold by high-end Harbin firms. In this way, a dead rat from the Harbin slums could make its way to the show window of a trendy shop on Kitaiskaia Street.[17]

## Police Corruption versus Professionalism

In 1932, the founders of the state of Manchukuo inherited the shreds of an old Chinese warlord policing system set up by Zhang Zuolin (1875–1928). This system included officers of dubious training and questionable loyalty.[18] The man in charge of patching together a new policing system for Manchukuo was the notorious Amakasu Masahiko (1891–1945), whose name would become associated with Manchukuo's undercover operations. Amakasu and a few of his associates hurriedly cobbled together a new policing system. Adding to what was left of the old Chinese and Russian police forces, Amakasu incorporated Japanese personnel from other colonies, *Kempeitai* officers and a slew of carpetbaggers looking for job opportunities. In the first few years, Manchukuo's cities – especially Harbin – were rife with crime. Kidnappings became a regular occurrence. The comfortable life enjoyed by many members of the expat community disappeared; elite, expatriate clubs no longer provided a protective barrier from the hardships of life in the northeastern regions. In 1932, bandits attacked a party of British sportsmen as they finished a round of golf. They barely escaped from the green with their lives. The murder of the British Mrs. Woodruff, a young bride who had only just arrived in Harbin to join her husband, followed the golf-course incident. Many Europeans sent their wives home to Europe.[19] Safety became such a concern that the American consul to Harbin carried a pistol with him at all times.[20]

One representative of the early Manchukuo police force was the head of the foreign section of the Harbin police, a Japanese man with the unlikely name of Nikolai Nikolaievich Yagi. Yagi had entered police service after a long career as a narcotics trafficker, with ties to Russian gangsters. He had taken on his Russian name after his opportunistic conversion to the Russian Orthodox faith. In 1937, during nationwide reforms, authorities in the capital Xinjing restructured the Harbin police force and removed high-ranking members from office. Yagi was rotated out of Harbin. By the time Gotō and his colleagues began the study of the Harbin slums, a newly recruited, professional group of supervising officers began to replace the likes of Yagi.

One notable supervisor was Shibuya Saburō. Shibuya hailed from a samurai family and was a graduate of the military academy – a man whose background one might expect should have prepared him for something greater than a career as a police officer. But his name was associated with the attempted coup of February 26, 1936, after which he ended up with the police force in Manchuria. Fluent in Russian, Shibuya had a reputation for being a genuinely cheerful and morally upright, a man of professionalism. In 1945, true to the samurai values of his forebears, he killed his family and then himself following the Japanese surrender to the Allied Forces.[21]

The atmosphere of coercion that permeated the puppet state of Manchukuo nurtured men like Yagi. Even when reform and consolidation brought men like Shibuya into the police organization – men who supported the kind of work done by Gotō and his colleagues – their contribution could not overcome the brutality and corruption that thrived in an unwanted occupation. Over the course of Manchukuo's brief history, the police system underwent a thorough restructuring. Katō Toyotaka, one of the authors of the *Autopsy of the Garden of Grand Vision,* called the years that followed the "golden age" of the Manchukuo police force.[22]

By 1938, the system included police-training academies, a fingerprinting bureau, and an agency called the Peace Protection Bureau. Although the name of the bureau sounded hopeful, the agency dealt with the more sinister side of police work, including the enhanced interrogation of political prisoners. As one Chinese police officer was told after he asked his Japanese superior about the bureau: "If you don't know about it, just forget it." In other words, a man in the regular police force who cared about life, limb, and career should keep his nose out of the darker side of operations.[23]

Men recruited into the Peace Protection Bureau were trained in a number of areas, including language, social customs, and electronic spying and interrogation techniques. At the "branch office" of the Peace Protection Bureau, suspects could be questioned under extreme conditions. One Chinese man who trained at the Harbin Police Academy described "interrogation" methods that included forcing hot pepper oil into body cavities, bull-whip beatings, and torture by electric current. Some prisoners of the Peace Protection Bureau ended up with the notorious Unit 731, which conducted biological experiments using unlucky captives as human guinea pigs. Gotō Reiji worked for the Peace Protection Bureau while he and his colleagues investigated social customs.[24]

The Manchurian environment supported a smuggling paradise. Rugged terrain, long borders, and proximity to the Chinese market encouraged a robust drug trade. The restructuring of the Manchukuo policing system affected the region's narcotics trade. On the Japanese home islands, laws strictly controlled narcotics use. Yet from the early years of the Japanese empire, the sale of opium was approved under a government monopoly, an institution modelled on the system used in British Hong Kong, Malaya, and Singapore, as well as in French Indo-China. Opium addiction, Japanese officials argued, was a peculiarly Chinese disease.

In the 1920s, when a political vacuum in North China set the stage for civil war and invasion, narcotics became a kind of currency, and opiates became the foundation of military adventure. After the Mukden Incident of September 18, 1931, resulted in the creation of Manchukuo, the new government began to bring the narcotics industry under its control – or at least it tried.[25] The first Manchukuo opium laws of 1932 created a government monopoly and a system of licensed dealers. Frequently, the dealers were Korean nationals who were supposed to use their profits to set up other businesses in the new state. The early monopoly aimed to reduce the number of Manchukuo addicts. On the isolated island of Taiwan, the government-monopoly system had worked reasonably well. But as the system was implemented on the Asian mainland, problems began to surface: the system could not be entirely controlled. This was especially the case in the Manchurian territory that Japan occupied after 1931. Throughout the 1920s, the competing interests of Chinese warlords, international entities, and profiteers made the area a haven for trafficking. The train routes through the north became a smugglers' highway; Japanese subjects found themselves in a multi-national, multi-ethnic arena of smuggling, production, and dealing.

In 1937, the narcotics laws were revised. State opium dispensaries appeared throughout Manchukuo. Harbin Opium Dispensary Number Thirty-Four served the Garden of Grand Vision and the surrounding neighbourhood. Koreans came to dominate the unlicensed trade in narcotics. In the Spring Garden Lodge, Korean morphine sellers rented most of the better rooms. In his report, Gotō refers to these unlicensed sellers as "ticks."[26]

A large part of Gotō's report is devoted to tracing the routes taken by this unlicensed opium as it found its way into the Manchukuo market. Gotō states that rural Manchuria was the site of an ongoing battle between government opium and smuggler opium.[27] He provides specific details about the operation of the government monopoly in the Garden, the origins of unlicensed opium, the routes taken by smugglers travelling by night from Manchukuo to the lucrative markets in North China, and the methods by which government opium leaves the dispensary and makes its way into the unregulated market.[28]

Gotō also describes the war raging in parts of Manchukuo between government buyers and smugglers. He assures his readers that for smugglers national borders are nothing more than an illusion. He is critical of the heavy-handed Manchukuo opium suppression campaigns that were carried out in the countryside.[29] He is also critical of the periodic raids that closed down smaller morphine dealers, leaving the larger dealers to continue to do business.[30] He does not mention that some of those larger dealers belonged to the *Kempeitai*. Sales from illicit narcotics operations could potentially pay for off-the-books operations that the authorities wished to keep hidden.[31]

Gotō describes a symbiotic relationship – between the slum, the government monopoly, and the Manchukuo authorities. This relationship produced a political climate in which smugglers could thrive. Though he does not point this out explicitly, at the end of the report, Gotō states clearly that the people he observed "had been pushed step by step to extinction by the pressures of life in Manchuria."[32]

## The Fate of the Dispossessed – Labour Camps

Manchukuo government officials feared that the social disorder growing inside the courtyard of buildings like the Garden of Grand Vision would migrate into the population at large. Gotō's survey coincided with a larger

Manchurian census that initiated programs aimed at enhancing military preparation and control. Arranged through the Manchukuo home ministry, the 1940 census gathered details about the residents of every city, county, and banner in the territory. Officials stated that the census was driven by the need to catalogue the many peoples of the diverse nation as well as the need to identify all able-bodied men for service to the state. More importantly, the Japanese-run military government sought information about wealth or skills that might be appropriated for the war effort.[33]

One of the stated goals of the census was reform, particularly narcotics reform. Information about the dispossessed played an important role. The restructuring immediately touched the lives of those living in the Garden. In the early 1940s, narcotics reform laws converted licensed opium shops, including the one in the Garden of Grand Vision, into opium recovery centres, complete with health clinics attached.[34] Such social engineering, however, barely disguised a pragmatic wartime reality: shortage of labour and materials. War in Asia dragged on, expanding into Southeast Asia and the Pacific. In the 1940s, reform led to the creation of local institutions variously named guidance halls or rehabilitation training facilities. In 1943, Nakaii Hisaji, head of the Police Bureau Intelligence Section, led a campaign called "Clean the Capital; Strengthen Public Safety." He set up reed huts surrounded by iron fences and called the complex a "reform training facility." These institutions began to spring up throughout Manchuria and were meant to train vagrants for some kind of meaningful work. In fact, these institutions simply housed unlucky arrestees in barracks in the countryside. The facilities were no more than slave-labour camps. These measures took society's most desperate folk – those that the police report called Manchukuo's shame in its opening lines – off the public streets, put them to work, and used the label social reform.[35]

Survivors of the slave-labour barracks remembered them as hell on earth. Inmates mined coal, cut logs, or moved rock to build roads. Work hours were long, food was scarce, and rules were strict and arbitrary. Once there, inmates lost all contact with the outside world. Electric fencing and sentry boxes surrounded the camps. One survivor recalled that inmates would say the electric fence had two doors. One door, by which they all entered, could not be used as an exit. The second door could not be used as an entrance: it was used by those who had succumbed to starvation or mistreatment and whose bodies were dispatched with as little sympathy as were those of the naked near-dead dumped into the streets outside the Garden of Grand Vision.[36]

Contact with the residents of the Garden affected the researchers themselves. One such resident was fifteen-year-old Katō, who had come to Manchukuo in 1938, joining the wave of young people looking for adventure in a new land. He was an urban version of the hopeful settlers that Annika A. Culver describes in Chapter 6 of this volume, "Constructing a Rural Utopia." Katō came alone; he settled in Harbin and became a member of the city police. Fatherless, Katō found a community in the service of which he was proud. Police chief Shibuya Saburō played an especially important role in his life. Fluent in Russian, Shibuya was the head of the Harbin school that Katō attended as a language student.

The Soviet Union began its invasion of Manchukuo on August 9, 1945, six days before Japan surrendered. Troops moved easily into the north of Manchukuo, threatening the Japanese agricultural settlements, whose vulnerability is described by Culver and Ronald Suleski elsewhere in this volume (Chapters 6 and 8). Katō shares chilling details of entire villages committing suicide, their inhabitants choosing to end their lives on their own terms rather than endure the horrors inflicted by the advancing Soviet and Chinese armies. His descriptions are brief lists of tragedy, but they leave an impression of horror similar to the horrors detailed by Ronald Suleski in Chapter 8 of this volume, "Salvaging Memories."

Katō knew some of the people stationed north of Harbin. He was of the same generation as the agricultural settlers who suddenly faced an enemy even more ferocious than the winter weather in the landscape they had made home. He describes whole villages of refugees moving south, trying to avoid the Soviet Army, Chinese soldiers, and violent roving bands of vengeful Manchukuoans. One village, whose 299 members were unable to defend themselves from attack, committed mass suicide before the Soviet army invaded. Another group from an agricultural community found its escape route to Harbin blocked. Collectively they chose poison over death at the hand of the invaders. They killed one hundred of their children before taking poison themselves. It is clear that Katō was affected profoundly by these deaths. But a greater tragedy hit closer to home: he learned of the deaths of his mentor Shibuya Saburō and Shibuya's family.[37]

After the Japanese surrendered on August 15, 1945, Katō, under orders, began burning police records while he learned of the deaths of those living in the agricultural communities to the north of Harbin. But he hardly had time to grieve before the Soviet Army came for him. On August 24, Katō and his fellow officers were arrested. They spent time locked in the basement of what had once been the Japanese consulate. In December

1945, the men were marched north, where Katō spent five years of hard labour in a series of Soviet prison camps.[38]

## Conclusion

The *Autopsy of the Garden of Grand Vision* is an eclectic document. It includes vivid descriptions of the slum, its inhabitants, its filth, and its vermin. It includes charts that detail the origins of residents and flophouse owners, the cost and purity of various drugs, the costs and profits of smugglers, and the prices paid for recycled items. It also includes pages of local slang and profanity, all neatly defined. It is at once daunting and thrilling to read. Yet only select members of the Japanese military were ever meant to read it. It was marked "top secret" and stored in a Japanese government bureau after the war. One copy is in the National Diet Library in Tokyo. It was not officially published until 2002, several years after its last surviving scholar author, Satō Shin'ichirō, had died.

The study begins by stating that the slums of Fujiadian are a shame on the glory of Manchukuo. As the study continues, its language is at times sophisticated and at times racist. For example, in a startlingly twenty-first-century tone, the authors call the prostitutes "Sex Workers." They describe the plight of these women with empathy, and note that their work was essential to the well-being of the area. At the same time, they blame Chinese men for "grotesque" sexual urges. Gotō and his collegues deride a culture that keeps women illiterate and dependent.[39] Elsewhere in the report are details about residents' background, including migration, loss of family ties, and the hard-luck circumstances that brought them to the slums.[40]

It is difficult to determine how much of the *Autopsy* is a reflection of what Gotō, Satō, and Katō really thought and how much of it reflects their need to write for a specific audience. In the last part of the study, Gotō and his colleagues tell readers that the slum's inhabitants are so exhausted by life that using drugs has become their only goal. They are not a danger to the state; they do not have the energy to revolt. There are hints along the way that Gotō had been instructed to identify spies. He found none. But he does warn that if someone were to find a way to arouse this population, there could be trouble.[41]

What Gotō and his two colleagues really felt about this study at the time is not obvious. They wrote for the Japanese military authorities, who could be vicious when criticized. And yet there are criticisms embedded within the text, such as when they take exception to the high-handed way the government treated opium farmers. They criticize the prohibition on gambling, which they state had no effect.[42]

Gotō died as a prisoner of war in a Soviet camp, probably in 1949. Katō Toyotaka survived the same camp as his supervisor. He returned home in 1950 and began writing about his experiences. Satō Shin'ichirō was captured by Chinese soldiers three times. He survived, returning home in 1946, later becoming a professor of Chinese history. Both Satō and Katō went into the Manchukuo experiment full of hope; they came away from war and prison with their own internal conflicts. Their memoirs exhibit both pride and regret. In today's understanding of post-traumatic stress and the harms that war inflicts on the human psyche, they could be diagnosed with "moral injury." The new research into the complications experienced by veterans readjusting to civilian life locates the cause of the affliction in "failing to prevent, bearing witness to, or learning about acts that transgress deeply held moral beliefs and expectations."[43] Both survivors of Manchukuo exhibit ambivalence that reflects a moral injury.

Both Satō and Katō staked a future in Manchukuo. The bilingual Satō longed for the promised land of racial harmony. In his memoirs of his Manchukuo days, he candidly states that halfway through the Manchukuo experience he became aware of the darker side of occupation policy. He accepted that he would die in China, hoping to pay back a "blood debt" for crimes against the Chinese people.[44] Young Katō sought a career. He admits the Japanese army committed horrors, yet in his extensive writing about the Manchukuo police – in multiple volumes – he is clearly proud of the organization's achievements. He states that the police system was good to him as a young man. He points out innovations in the Manchukuo police system that affected postwar police organization in Japan. Yet, he also dedicates an entire volume of his police history to the darker side of the organization – the Peace Preservation Bureau and the Garden of Grand Vision. When he did so, many of his former colleagues were furious.[45]

The opium control situation was one area where the young police officers would have their ideals challenged. Arguably, the Japanese-sponsored opium monopoly might be considered a form of harm reduction today. On paper, it suggests a humane approach to recreational drug control. In practice, however, the Japanese opium monopoly led to problems. Opium sold outside of the opium monopoly's control funded loosely organized resistance movements in remote border areas, challenging state efforts at control. At the same time, unlicensed morphine and heroin operations often funded off-budget intelligence activities. Rebels in border areas proved to be difficult to police. How could narcotics control be effective if conscientious police answered to corrupt authorities? More to the point, with an expanding battlefront, how could sufficient personnel be spared for narcotics control? The good intentions of Gotō Reiji, Satō Shin'ichirō, and Katō Toyotaka, along with men like their superior Shibuya Saburō, were buried in the rubble of a disastrous war. If anything, the men's efforts, however well intentioned, helped strengthen the environment of occupation. They became examples of the Chinese image of men who ride tigers. In spite of any regrets once on board, it was both difficult and dangerous to dismount halfway through the ride.

## Notes

*Acknowledgment:* Adapted from *Life and Death in the Garden: Sex, Drugs, Cops, and Robbers in Wartime China* by Kathryn B. Meyer (Lanham, MD: and Littlefield, 2014). By permission of the publisher.

1 Wang Yufei and Gotō Haruhiko, "Harubin shi Fujiadian Chi-iki no 'Dayuan' Shiki Kenchiku Airia ni Okeru Toshi Kūkan Keitai to Shūjū Keisei Ryōiki ni Kansuru Kenkyū" [On the relation between urban space and dwelling territory in the "Dayuan" house area of Fujiadian, Harbin, China], *Nihon Kenchiku Gakkai Keikakukei Ronbunshū* [*Journal of Architecture and Planning Transactions of the Architecture Institute of Japan*] 622 (December 2007): 113–20.

2 Hinkōshō Chihō Hōan Kyoku [Hinkō Prefecture Regional Public Safety Bureau], Keimu Sōkyoku [Manchurian Police Bureau], *Daikan'en no Kaibō* [Autopsy of the Garden of Grand Vision], *Kanminzoku Shakai Jittai Chōsa* [Investigation into the social conditions of the Chinese race], 1942, National Diet Library, Tokyo, Japan.

3 Keimu Sōkyoku, "Kichinyado no Kōsei" [Structure of the flophouses], *Daikan'en no Kaibō,* 15–21; Katō Toyotaka, "Daikan'en," in *Shōshetsu: Daikan'en* [*Garden of Grand Vision*] (Matsuyama: Ehime Tsushinsha, 1974), 173–75.

4 For posters and stamps celebrating Manchukuo internationalism, see "Five Races Under One Union (Manchukuo)," *Wikipedia,* modified February 15, 2016, https://en.wikipedia.org/wiki/Five_Races_Under_One_Union_(Manchukuo); "滿洲國" (Manchukuo), *Wikipedia,* modified April 6, 2016, https://zh.wikipedia.org/wiki/满洲国. Katō mentions the irony of the term "Kingly Way" as he describes the Fujiadian slum in Katō Toyotaka, "Daikan'en no Kaibō no Megutte" [Return to the autopsy of the Garden of Grand Vision], in *Manshūkoku no Chika Sōshiki ni Tsuite* [On the organization of Manchukuo underground] (Matsuyama: Man-Mō Dōhō Engokai Ehime-ken Shibu, 1974), vol. 2, *Manshūkoku Keisatsu Shōshi* [Short history of the Manchukuo police], 152. The daily body roundup is described in Keimu Sōkyoku, "Shakuhakumin no Matsuro" [The end of the road for the residents], *Daikan'en no Kaibō,* 259–66.

5 A photograph of one of the informants appears in the report labelled "Feng Dengdiao, The Author's Friend," *Daikan'en no Kaibō*, unnumbered photograph pages.

6 Tsukase Susumu, *Manshūkoku: "Minzoku Kyōwa" no Jitsuzō* [The true nature of Manchukuo "racial harmony"] (Tokyo: Yoshikawa Kōbunkan, 1998).

7 While 1939 may be considered the beginning of the Second World War in Europe and December 7, 1941, lives in infamy for Americans, Chinese soldiers engaged the Japanese military much earlier. September 18, 1931, the date of the Mukden Incident, looms as large in China as September 11, 2001, does for Americans. Fighting began in earnest after July 7, 1937, when a skirmish at the Marco Polo Bridge outside Beijing triggered a full-scale Japanese invasion of China, leading to all-out war.

8 Keimu Sōkyoku, "Daikan'en no Enkaku" [History of the Garden of Grand Vision], *Daikan'en no Kaibō,* 2.

9 Harbin architecture is described in detail in Nishizawa Yasuhiko, "Harbin," in *Manshū Toshi Monogatari* [Stories of Manchurian cities] (Tokyo: Kawashutsu Shobō, 1996), 12–37; see also Song Hongyan, *Dongfang Xiao Bali* [The little Paris of the East] (Harbin: Heilongjiang kexue zhishu chubanshe, 2001).

10 Keimu Sōkyoku, "Daikan'en no Enkaku," 2–5.

11 Keimu Sōkyoku, "Kichinyado no Kōsei," 21.

12 Keimu Sōkyoku, "Kichinyado no Kei'ei" [Management of the flophouses], *Daikan'en no Kaibō*, 27.

13 Keimu Sōkyoku, "Kichinyado no Kōsei," 20–23.

14 Keimu Sōkyoku, "Tōhin to Tōhin Shori" [Stolen goods and the distribution of stolen goods], *Daikan'en no Kaibō*, 178.

15 Ibid., 179.

16 Ibid., 183, 193–94.

17 Ibid., 194.

18 One man claimed that 80 percent of the police in his district could not read: Mou Jianbing, "Wei Andongxian Jingcha Tongzhi de Ji Jianshi" [A few incidents in the phony Andong County police], in *Zhimin Zhengquan* [Colonial political control], ed. Sun Bang (Jilin: Jilin People's Publishing Company, 1993), 315.

19 H.G.W. Woodhead, *A Visit to Manchukuo* (Shanghai: Mercury Press, 1932), 97–99.

20 Amleto Vespa, *Secret Agent of Japan* (Boston: Little, Brown and Company, 1938), 180.

21 Gao Yubi, "Wei Man Jingcha Gigou de Ri" [Days in the Manchukuo police organization], in *Wei Man Shi Liao Cong Shu* [Collection of Manchukuo historical materials], *Wei Man Renwu* [People of Manchukuo], ed. Sun Bang (Jilin: Jilin People's Publishing Company, 1993), 217–25.

22 Katō Toyotaka, ed., *Manshūkoku Keisatsu Juyō Shashin Bunkei Shiyosei* [A collection of important photographic resources on the Manchukuo police] (Matsuyama: Moto Zaigai Komuin Engokai, 1982), ii.

23 Wang Xianwei, "Ha'erbin Taiheqiao Duchang de Guanfang Beijing" [Official backing for Harbin's Peace Bridge Gambling Centre], in *Wei Man Shi Liao Cong Shu, Wei Man Shehui,* 519.

24 Makuuchi Mitsuo, *Manshūkoku Keisatsu Gaishi* [An unauthorized history of the Manchukuo police] (Tokyo: San Ichi Shobō, 1996), 57–62; Katō Toyotaka, *Manshūkoku Keisatsu Shoshi* [A short history of the Manchukuo police], vol. 1, *Manshūkoku Kenryoku no Jittai ni Tsuite* [Concerning the actual condition of power in Manchukuo] (Matsuyama:

Man-Mo Dōhō Engokai Ehime-ken Shibu, 1968), 10, 98–99, 139; Yue Ruheng, "Bayan Xiang Weijing Wuke Neimu" [The inside story of the Bayan County police] in *Wei Man Shi Liao Cong Shu, Zhimin Zhengquan* [Colonial political control], ed. Sun Bang (Jilin: Jilin People's Publishing Company, 1993), 346–47.

25 Kathryn Meyer and Terry Parsinnen, *Webs of Smoke; Smugglers, Warlords, Spies, and the History of the International Narcotics Trade* (Lanham, MD: Rowman and Littlefield, 1998); John Jennings, *Opium and Empire: Japanese Imperialism and Drug Trafficking in Asia, 1895–1945* (Westport, CT: Praeger, 1997); Norman Smith, *Intoxicating Manchuria: Alcohol, Opium, and Culture in China's Northeast* (Vancouver: UBC Press, 2012).

26 Cabot Coville to State Department, Consular Monthly Report, March 1934, RG 59, Internal Affairs of China. 893.00P.R.HARBIN/65; Keimu Sōkyoku, "Ahen no Mitsuyu" [Opium smugglers], *Daikan'en no Kaibō,* 136–37.

27 Keimu Sōkyoku, "Ahen," 115.

28 Ibid., 117–24.

29 Ibid., 139.

30 Ibid., 115–44; "Sennin no Mohi Mitsubaisha" [Korean narcotics dealers], *Daikan'en no Kaibō,* 144–56.

31 Yamauchi Saburō, a heroin manufacturer, described this arrangement in his memoirs, "Mayaku to Sensō: Nitchū Sensō to Himitsu Heiki" [Narcotics and war: The Sino-Japanese war and its secret weapon], *Jimbutsu Ōrai* [Affairs of eminent men] (September 1965): 165–69.

32 Keimu Sōkyoku, "Seikaku" [Character], *Daikan'en no Kaibō,* 292.

33 Hao Rangxian, "Heishan Xiang Linshi 'Guoshi Diaocha' Jilu" [Recollections of the Heishan County provisional census], in *Wei Man Shi Liao Cong Shu* [Collection of Manchukuo historical materials], *Wei Man Shehui* [Manchukuo society], ed. Sun Bang (Changchun: Jilin renmin chubanshe, 1993), 271.

34 Lu Shouxin, "Haerbin de Yapian Yindu" [Opium and narcotics in Harbin], in *Wei Man Shi Liao Cong Shu, Wei Man Shehui,* 444–46.

35 Shen Wenming, "Weiman Xinjing Gengsheng Xunliansuo Neimu" [The inside story of the Manchurian reform training centres in Shinkyō], in *Wei Man Shi Liao Cong Shu, Wei Man Shehui,* 147–57; the comment comes from the document's introduction, Keimu Sōkyoku, "Jo ni Kaete," *Daikan'en no kaibō,* i.

36 Xiao Chongguang, "Renjian Diyu–Fulang Ying" [Hell on earth – the vagrant barracks], in *Wei Man Shi Liao Cong Shu, Wei Man Shehui,* 139–47.

37 Katō Toyotaka, *Manshūkoku Keisatsu Shōshi* [Short history of the Manchukuo police], vol. 3, *Manshūkoku no Kaitai to Keisatsu* [The dismantling of Manchukuo and its police] (Matsuyama: Man-Mō DōHō Engokai Ehime-ken Shibu, 1976), 136–38.

38 Ibid., foldout insert after 201.

39 Keimu Sōkyoku, "Seiyoku Rōdōsha [Sex workers]," *Daikan'en no Kaibō,* 83.

40 Ibid., "Seikaku," 290–93.

41 Ibid., "Seikaku," 293.

42 Keimu Sōkyoku, "Ahen," 139; Keimu Sōkyoku, "Tobaku, Buraikin" [Gamblers and villains], *Daikan'en no Kaibō,* 100-1.

43 Shira Maguen and Brett Litz, "Moral Injury in Veterans of War," *PTSD Research Quarterly* 23, no. 1, http://www.ptsd.va.gov/professional/newsletters/research-quarterly/v23n1.pdf, 1.

44 Satō Shin'ichirō, *Satō Shin'ichirō Senshū* [Satō Shin'ichirō selected writings] (Tōkyō: Satō Shi'ichirō Senshu Kankōkai, 1994), 213.

45 Katō Toyotaka, *Manshūkoku no Chika Sōshiki ni Tsuite* [On the dismantling of Manchukuo and its police], vol. 3, *Manshūkoku Keisatsu Shōshi* [Short history of the Manchukuo police] (Matsuyama: Man-Mō Dōhō Engokai Ehime-ken Shibu, 1976), 213.

# 8

# Salvaging Memories: Former Japanese Colonists in Manchuria and the Shimoina Project, 2001–12

*Ronald Suleski*

In September 1931, imperial Japanese forces of the Kwantung Army (*Kantōgun*) occupied all of Northeast China, known as Manchuria. In 1932, they established the puppet state of Manchukuo (*Manshūkoku*). It was a new territory to add to Japan's growing empire, a vast land of mineral and agricultural riches. In order to secure its control over the new territory and to take advantage of the region's fertile fields, the Japanese government began to seek Japanese farmers to send into Manchuria. Recruiting agents visited villages in Japan, and government notices were sent out to many rural communities, promising assistance to those who decided to sign up for the program. The Japanese farmers would be given tools, some provisions, and, most importantly for them, the promise that when they had met certain conditions, they would receive title to the land.[1] In Manchuria, the farmers would create farming communities and work the fields surrounding their hamlets. Eventually, the Japanese government hoped, Manchuria would be populated by Japanese families working the land. The families would be loyal subjects of the Japanese emperor and would produce more than enough food to feed the people of the home islands, where agricultural land was insufficient for the expanding population.[2]

By the end of the Second World War, over a million Japanese were living in Manchukuo. They came from many areas of Japan, with a large number hailing from Nagano Prefecture (*Nagano-ken*), a mountainous area south of Tokyo on the main home island of Honshū. Although life in Manchuria was not easy because of the extreme climate (short, intense summers and long, frigid winters), the farmers were used to a hard life of outdoor work and few material comforts. Japanese villages were built among the villages of Chinese peasants who originally worked the land. Many Chinese resented the arrival of the Japanese, and relations between the Japanese settlers and their Chinese neighbours were not always cordial. But the Japanese military and the Japanese-led police forces controlled the area; in general, the Chinese and the Japanese farmers each went about their own business, working to survive the challenging climate of Manchuria.[3]

When Russian forces streamed across the border in August 1945, the lives that the Japanese agricultural colonists (*nōgyō imin*) had known were shattered. Those who could not escape were slaughtered by the Russian invaders. Those who could ran ahead of the Russian forces, heading to the south of Manchuria, where they hoped to find a boat to take them back to Japan. Thousands of Japanese were killed or taken prisoner by Russian troops. After the war ended, those Japanese not taken to Siberia were left to roam south Manchuria, begging for food and turning to prostitution, or doing any kind of work they could find. From August 1945 until they were finally able to return home (in some cases not until the 1950s), these displaced Japanese lived in extreme conditions, unwanted and often mistreated by the Chinese. The Japanese government, operating in Tokyo under the American occupation, was not able to give much assistance.[4] This period of Japanese history is full of dramatic tales of survival. Much has been written in Japan about these remarkable stories, but relatively little is known in the West.

The agricultural colonizers who managed to escape Manchuria and return to a war-impoverished Japan tried to pick up their lives as best they could. Through persistent hard work in the 1950s, and then with the improving Japanese economy in the 1960s, their lives gradually became stable. Most remained silent about their experiences in Manchuria. In the 1970s and 1980s, however, a number of accounts of the trials and tribulations of the former colonists in Manchuria were published in Japan. Many of these were in the form of memoirs; some accounts were written by professional writers who recognized the power of the stories. In these accounts, several themes are prevalent. First is the sense of victimhood; colonists were victims of the wartime Japanese military, victims of Chinese anger following the Japanese defeat, and victims of circumstance. Second was the near-universal acceptance of the value system in place in Japan in the 1930s and 1940s. Very few of the writers questioned closely their previous absolute loyalty to the emperor, or the legitimacy of the Japanese military's actions, or the right of the Japanese to move into Chinese territory and take land away from the peasants living there.[5] Third, few of the accounts are detailed or graphic in the portrayal of events. Specific names, places, and numbers are given, but despite the mention of rape, wanton death, and extreme hardship, the accounts remain somewhat removed in the telling from the gritty, bloody, and dirty reality of what took place. The writers of these accounts are interested in telling their dramatic stories, and of presenting themselves as survivors, martyrs, or victims, but they do not present themselves as conflicted figures willing to expose the totality of what they witnessed. In fact, underlying many of the earlier accounts is a sense of adventure or heroism – a sense that the writer is retelling an important moment in Japanese history.[6]

To be sure, after the war there were also more critical views of the role of the Japanese farmers in Manchukuo, often presented in well-documented studies by academics who examined a wider range of the materials available and severely questioned conduct in the colonization efforts in Manchuria. But despite the soundness of the research, these were minor voices overpowered by the higher volume of laudatory or nostalgic works and memoirs of the former Japanese farmers and officials who viewed Manchukuo as a great and worthwhile experiment.[7]

## The Shimoina Project

Accounts of the Japanese agricultural colonizers in Manchuria continued in this vein until after the turn of the millennium. In early 2001, researchers in the city of Iida in Nagano Prefecture began to collect the stories of local residents who had been to Manchuria as Japanese farmers in the 1930s and 1940s. The region, especially the hilly country of Shimoina surrounding the city, had seen a great number of local residents, sometimes almost entire villages, move to Manchuria as colonizers. Teams of interviewers facilitated by the Iida city government and Kyoto University visited with willing residents and recorded their stories. Researchers took photographs of the citizens they interviewed and began to publish their accounts in the series *Shimoina no naka no Manshū* (*Shimoina and Manchuria*), within a series called *Kiki gaki hōkokushū* (*Collection of Oral and Written Reports*) published by the Iida City Institute of Historical Research (*Iidashi rekishi kenkyūjo*).[8]

These accounts were unlike those that had appeared in the preceding decades since the end of the war. In the Shimoina accounts, former colonizers often spoke frankly and in graphic detail about their experiences in

Manchuria. They no longer seemed to feel the need to be constrained by presenting a positive, grand narrative that cast a patriotic glow over what they had experienced. They no longer dismissed the Chinese inhabitants in terms of "us" and "them" but instead saw conflicted human beings on all sides. While many continued to emphasize the themes present in the earlier accounts and spoke glowingly about the cooperation between Japanese colonizers and Chinese farmers (because there *had* been instances of cooperation and many cases of individual friendships between Chinese and Japanese), they also related stories that could not be found in the earlier published accounts. The new narratives described elements such as the haughty manner with which some Japanese farmers would steal from the itinerant Korean merchants who showed up at their gates to offer goods for sale, or how the Japanese felt at liberty to severely beat Chinese peasants that they suspected of stealing from them. In these accounts, the narrators were willing to omit the wartime discourse about nation building (*kenkoku*) and service to the august emperor (*tennō heika no tame ni*) in order to tell their story as they chose. Suddenly, their accounts gave us a version of history that had been denied or ignored in almost all the previously published Japanese-language books about the Japanese presence in rural Manchukuo.[9]

As the project began to publish interviews with former colonizers, it was not clear what had brought about this dramatic change in the tenor of the accounts. It could be that the informants, now elderly, were no longer afraid to tell the truth as they saw it and decided there was simply no longer any purpose in telling a sanitized version of their story. It could be that the thoughtful and genuine questions of the researchers elicited a like response from those interviewed. A conscious editorial decision was made to preserve in the published interviews the regional dialects and expressions of the Nagano people interviewed. This made the stories all the more vivid and compelling, providing a clear sense of the former colonists' provincial origins. Retaining the regional dialects also revealed clearly the personalities of the individuals as they related their stories.

The narrative below is the account of one of the Japanese farmers in Manchuria from the end of the war in 1945 until his return to Japan in 1948. He was interviewed in 2003, and his long account (of forty-nine pages) was published in 2005 in the third volume of the series. A shorter version of this account appears in the book *Manshū yimin: Iida Shimoina kara no message* (*Manchuria Immigrants: A Message from Shimoina and Iida*).[10] In this story, the life-and-death imperatives facing the Japanese colonizers in the days immediately following the end of the war in August 1945, and the conflicting values the colonizers had to confront, are presented in a straightforward and matter-of-fact narrative. The account reflects the new honesty that appeared in the published interviews.

The project started in 2001 when Saitō Toshie, a librarian working with the Iida municipal government, received a visit from Professor Araragi Shinzō of Kyoto University.[11] The professor brought with him some graduate students and wanted to study the Manchuria experiences of Nagano citizens. Ms. Saitō had long been interested in the stories of the Japanese recruited for service in Manchuria in the 1930s and 1940s, and she had been collecting published accounts for the city's library holdings. Together they organized the Society for Narratives about Manchuria and Mongolia Colonization (*Man-Mō kaitaku o katari tsugu kai*). Local newspapers announced the early seminars and, to the delight of Saitō and Araragi, many local people, some of them former agricultural colonists, expressed eagerness to learn about the project.[12]

Willing local citizens who responded to the newspaper ads were visited by a team of (usually) two interviewers. Interviews were recorded on tape, and consisted of short, polite but direct questions followed by the more lengthy replies of the interviewees. Most of the former colonists were interviewed on several occasions over a period of weeks. After each individual interview session, they were given the opportunity to review the written transcript and to suggest changes or corrections. Interviewers also circulated the transcripts among themselves and suggested follow-up questions and clarifications. The youngest interviewees were in their sixties, while the oldest former colonists offering comments on the project were in their nineties. Many were able to provide old photos of their youth in Nagano or their days in Manchuria.

Professor Araragi's goal was to collect the life histories of former colonists in order to build an archive of oral histories as narrated by the colonists themselves. He was not especially interested in the master narratives about the Manchuria colonization scheme that circulated in postwar Japan and which reflected Japanese imperialism or emperor worship. Such interpretive frameworks had the effect of slotting the narratives into a victimization paradigm. Araragi simply wanted people to tell their stories as best they could, in as complete a manner as possible.

In many respects, it is not surprising that active efforts to collect and record the memories of Japanese agricultural colonists in Manchuria were most successful in Nagano Prefecture. Of the 1.5 million Japanese in Manchuria at the end of the war, more than 220,000 were farmers living in Japanese-organized hamlets. In the 1930s and 1940s, when recruiters sent by the Japanese government had visited most of the rural prefectures in the home islands, Nagano had provided the highest number of citizens willing to join in the effort to populate Manchuria with Japanese farm families. According to the best statistics now available, Nagano Prefecture sent at least forty groups of colonizers to Manchuria. Most of these were family groups, totaling 31,264 adults. The prefecture also provided 6,595 teenage boys who went as members of the Manchuria Youth Corps. At 37,859 colonists in total, Nagano easily provided the highest number of its citizens, of all regions in Japan, for the Manchuria colonization effort.[13]

Government recruiters made big promises to those who signed up for the patriotic exercise. Colonists would receive a subsidy from the government to cover most of their costs; they would find living accommodations and be able to farm fertile fields. After a few years of working the land, they would be given title to the property. What has become clear, especially since the Shimoina Project began to publish its records, is that the recruiting process was neither benign nor subtle. Recruiters spoke loudly about the importance of moving to Manchuria as part of the national effort to safeguard Japan's newest imperial territory. When farmers demurred, recruiters resorted to threats and bullying, calling out the reluctant people as traitors, intimating they might be investigated or arrested by the police because of their uncooperative and suspect attitude. The coercion on the part of government agents still rankles the Nagano people, who remember the pressure they felt to give in to government demands.[14]

As the war hysteria in Japan increased into the 1940s, it became very difficult for ordinary citizens even to imply they had doubts or questions about the recruiter's message. In the personal accounts, former farmers relate how the conditions they encountered in Manchuria were often far different from what the recruiters had promised. The houses made available to them (often the huts of the original Chinese owners, who had been evicted from the land) were poor and shabby, reflective of the widespread poverty among most of the Chinese peasants in Manchuria. In some cases, the Japanese had to construct their own living quarters with inadequate tools and construction materials, in regions where running water was hard to obtain and electricity was non-existent. The bitterness of Manchuria's harsh climate was life-threatening; cold Siberian winds blew across the flat plains, and the ground was frozen by the end of November.[15]

The area of Shimoina and Iida City, which supported the Shimoina Project with great fervour, had sent a high proportion of its population to Manchuria as agricultural colonists. According to the records, 8,379 colonists from the immediate area were sent to Manchuria; 4,199 of these managed to return after the end of the war. Each of those who returned had suffered the loss of friends and family members, often including their own children.[16]

Once the Shimoina Project was underway, it was discovered that a large number of elderly people living in the area had experienced life in Manchuria during the war years.[17] Many of those caught in the traumatic events at the end of the war had horrendous stories to tell, memories that were still vivid to them nearly sixty years later. The first volume of *Shimoina no naka no Manshū* (*Shimoina and Manchuria*) was published in 2003 and included five accounts by local citizens. In total, the ten published volumes, with the tenth volume appearing in the summer of 2012, contain eighty-five extended interviews.[18]

Kubota Isamu, the main narrator of the story in this chapter, is now in his mid-eighties and still lives in the same region in Japan where he grew up. His general good health as a teenager likely helped him to survive life

FIGURE 8.1 Kubota Isamu's speech in 2012.
*Source:* Web photo from a Japanese Google images site

in Manchuria and in the horrific days that followed Japan's surrender in August 1945. Like many people who have experienced extreme life situations, Kubota hardly ever spoke about those dark, almost surreal events after the surrender. He managed to survive the early years with his health and his spirits intact, and then was preoccupied with rebuilding his life after returning to Japan.

In 2003, Kubota agreed to be interviewed by the team sponsored by the local government in Iida. He told his story in five interview sessions in 2003 and 2004. He was straightforward and honest in recounting the story. Former Japanese colonists had generally tried to create a positive image when recounting their time in Northeast China, but it was no longer necessary to distort or ignore the darker side of events of those days. Still, it was not an easy story for Kubota to recount. The government interviewers noted in their reports how Kubota's emotions would surface as he told his story. His story centred on the mass suicide of August 1945. Underlying his account was his decision, made silently to himself and then followed with dogged persistence, to live.

## A Japanese Village in Manchuria

In 1945, a group of Japanese farmers were living in Manchuria in a hamlet they had built and which they had named Kawano Village, after a village in their home prefecture of Nagano. Japan was increasingly desperate that spring in 1945, as the tide of the Pacific War unalterably turned against the imperial forces. Cities in the home islands were bombed daily by American planes. All of Japanese society was mobilized for defence in support of the war effort. When the Soviet Union declared war against Japan on August 9, 1945, the adult, able-bodied men in Kawano Village in Manchuria were mobilized and left the village to report for duty at Shinkyō, the nearby capital of Manchukuo.[19] Of the few males remaining in the village, the oldest was seventy-three, followed by the village leader, Tsutsui Aikichi, age sixty-six. Nakagawa Kōichi was twenty-two, and hearing-impaired. The youngest, Kubota Isamu, was fifteen years old. All the other inhabitants of the village were either women or children under the age of eleven.

It was somewhat unusual for a boy as young as Kubota to be living on his own in a village of colonists, but Kubota had been restless at home in Japan. He was not the eldest son in his family, so he would not inherit the family's property, nor would he be responsible for his parents in their

FIGURE 8.2 Kubota at the age of fourteen in 1944 with his family in Nagano; Kubota is standing on the extreme left.
*Source:* Photograph supplied by Mr. Kubota to the author

old age. After he graduated from his compulsory schooling in 1944 at the age of fourteen, his options were either to enter a defence plant or to join the military. Being too young for military service, he thought that a stint in Manchuria working in a government-sponsored farming village would provide adventure. He left for Manchuria that year.

### *The War Abruptly Ends*

News spread quickly through Manchuria on August 15, 1945, that the emperor had announced Japan's surrender in a radio broadcast. The Japanese villagers were worried about how the neighbouring Chinese would react to this news. The Japanese presence in Manchuria had been forced on the Chinese. The Japanese government had taken Chinese land in order to build Japanese villages, paying paltry sums for farmland and taking away the livelihoods of many Chinese farmers. Worries about the Chinese reaction were well founded. Not long after the surrender was announced, belligerent crowds of local Chinese began gathering in a threatening manner just outside Kawano Village. Several hundred

Chinese were yelling, "Japan is defeated! Defeated!" When someone in the crowd fired a rifle shot, the gathering turned into a riot. Chinese farmers entered the Japanese village and opened barn doors to lead the farm animals away. With hoe handles, they broke into houses, grabbing up clothing.[20]

The terrified colonists ran away from the village towards the fields of tall corn and *gaoliang* (sorghum/millet – a tall grass, widely grown in Manchuria, used to make dough and alcoholic drinks), heading for the low mountains nearby. It didn't take long for a number of the Chinese rioters to catch up with the women. They tore at the women's clothing and demanded their belongings. Meanwhile, other groups of Chinese were dismantling doors and window frames from the village houses for use in their own homes. Amidst the total confusion, the residents of Kawano Village split up into several groups, hoping to hide in the fields until nightfall.

Village head Tsutsui, along with Kubota, Nakagawa, and some of the women, headed towards nearby Jilin City, where they hoped to find police protection. Some of the Chinese farmers caught up with the group and demanded clothing and items carried by the Japanese colonists. The colonists refused to hand over what they carried, saying, "We will have nothing left."[21] In his encounters with the angry Chinese farmers in these violent days, Kubota was to lose his work shirt, his tall boots, his watch, and his wallet. The crowd then attacked and beat the three males in the group – the village head, the hearing-impaired Nakagawa, and fifteen-year-old Kubota himself – but they took nothing more from the colonists and went back to their own homes.

Tragedy for the colonists of Kawano Village really struck after the Chinese mob left. Dusk was falling on the evening of August 16, the day following the first mob attack on the village, when the villagers (the women, the children, and the three males who had been attacked the previous day) moved out into the *gaoliang* and cornfields, planning to hide there through the night. The women especially felt completely vulnerable, fearing robbery and rape.

Tsutsui was in bad condition following his beating by the Chinese mob earlier that day. He had been badly pummelled, was very bruised, and likely had many broken bones. He was having great difficulty breathing and finally told the women and children, "I can't breathe. This is too hard and I can't take it anymore. You young people should try to survive this war and tell others what has happened." Then he repeatedly begged to be killed, saying, "Please let me die."

At first, no one was willing to act on his wish. Then one of the women raised her voice and said, "Can't you see how he is begging to die? Shouldn't we let him die to relieve his suffering?" Others in the group agreed that death would bring the man relief from his extreme pain. Two or three of the women put their hands around his neck and began to choke him, wishing him farewell, saying, "Sayōnara, sayōnara." His eyes closed and he took his last breath. No one knew the exact time of his death, but it was after nightfall on the evening of August 16, 1945.

### *Death Stalks the Villagers*

The death of their leader was only the beginning of the tragedy that awaited the group. The women began to discuss what they ought to do next. One woman reflected the thinking of many patriotic Japanese of the time and said, "We all know we won't be able to return to Japan. Our husbands have already been drafted into the military, and we need to understand that they won't be coming back. In war, the losers are not allowed to live, so either they have lost their lives or they have killed themselves. It's only a matter of time until the Chinese find us and do whatever they want to us. It would be a shame on all Japanese women if we let that happen. It would be better if we take our fate into our own hands and end our lives now."[22]

Fifteen-year-old Kubota heard the women's discussion and understood they had decided to die. He was frightened of what awaited all of them. He sat there hopelessly, trying to think of some way to avert the looming tragedy. One of the women saw him sitting there and said, "Kubota! What are you doing? Hurry up. You need to help us, too. We don't have much time. The sun will be up soon and the Chinese will come back. We have to do it before they get here." Kubota saw that some mothers had already started choking their children.

Infants were choked to death first; the small children were next. Things got more difficult when the women tried to kill their school-age children because those children were physically mature enough to resist being killed. One boy, about eight years old, cried, "I don't want to die. Don't kill me, mom." His mother replied, "Don't be so difficult. Your dad has already died in the war, and we are going to where your dad is. Kneel down on the ground and hold your hands in prayer. There is nothing to be scared of. Your dad is waiting for you." The mother persuaded the child to listen to her and eventually, as the boy tried to sit straight as if reciting a Buddhist prayer, his mother choked him to death. One of the women, who was a nurse, said, "I can't bear to watch these children die a slow death. The young ones should die quickly, before they realize what is happening."

One by one, the mothers continued to strangle their children. Kubota, who was watching in confusion, has said the nurse told the adults how to kill each other without prolonging their suffering. She said using a knife would be a better way, but since no one had a knife, they should use their cloth belts or sashes instead. But the women were not strong enough to tighten the sashes sufficiently to cause strangulation. So the nurse suggested they put an ear of corn between the sash and the back of the neck and twist the ear of corn, tightening the cloth around the child's neck. The children tried to sit up straight with their hands in prayer while being killed, and then they fell over on the ground. Once they fell on the ground, Kubota was told to kick them in the stomach to make sure they were dead. Years later, he reported that he killed at least twenty children by kicking them violently in the stomach after they had fallen to the ground. Kubota's classmate and the small daughter of twenty-two-year-old Nakagawa were killed in this manner. First the younger children, then the middle-aged women, then the younger women were all killed. In total, seventy-three women and children from Kawano Village died between the fall of darkness on August 16 and the morning of August 17, 1945, in a cornfield in central Manchuria, not far from the capital city of Manchukuo.

## *A Time to Die*

Finally, in the darkness before dawn, only Kubota and Nakagawa remained alive. Nakagawa thought that if he could bite through his wrist he would be able to bleed to death. He bit repeatedly and blood began to flow, but it did not seem to him that there would be enough blood loss to cause his death. Kubota remembered that he had heard that being hit on the head between the eyebrows would cause death. The two began to search for rocks large enough to use for this purpose. They stood facing each other and took turns hitting each other in the forehead with rocks. Blood began to flow as they hit each other, and they became dizzy and grew weaker.

The next thing Kubota remembers was that the sun was high in the sky, and everything was damp. A heavy rain had fallen. Kubota slowly lifted his head to look around, and saw that Nakagawa, lying nearby, was trying to do the same. They then realized that all of their clothes were gone; both of them wore only their undershorts.

The scene before them was truly a hell on earth. Scattered all around were the dead bodies of the women and children. Most of the bodies were completely naked. Some of the women's bodies still had underwear or *haramaki,* a woolen sash that was wrapped around the stomach as a way to keep body temperatures constant. The sun shone down on this scene of carnage. Likely sometime after dawn, some of the Chinese farmers had returned to the field, wanting to know what had become of the Japanese hiding there. Being poor themselves, they had probably taken the clothing from the bodies, as cloth was difficult to obtain during wartime conditions. In Kubota's mind, he still saw those women and children as living beings; when he recalled the dreadful scene nearly sixty years later, he still remembered them this way and remarked, "Those women and children who were choked to death never came back to life again."[23]

A few local Chinese came by to look at the carnage. They told the boys to try to get to one of the rail lines, where they might catch a train heading south towards the large port of Dairen in southern Manchuria, where there might be a ship to take them back to Japan. Weak and disoriented, Kubota and Nakagawa were led away from the field. They were directed to the headquarters village of the Japanese colonists in the area. There they hoped to find the headman of the Chinese workers (the Japanese called them "coolies") employed to work on the Japanese farms. Since the Chinese headman could speak Japanese, the two young men were hopeful that he would be willing to help them.[24] The youths were weak and unsteady on their feet. They decided to avoid the roads, hiding in the fields during the day and moving only at night. They found a burlap bag large enough to pull over their shoulders to provide some protection while sleeping during the day or while walking through the fields at night, since even the August nights could be chilly. Their only refreshment was the water from muddy puddles.

## *A Time to Survive*

On August 23, 1945, after six days of making their way through the fields, the boys found the headquarters village and the home of the Chinese headman. On seeing their nearly naked, muddy bodies covered with dried blood, the headman's wife was visibly upset. The headman sombrely concluded, "You boys must have been beaten." The boys told their story, then were given some ears of corn to eat and were shown a place to lie down and sleep. The next morning, the headman's wife used cooking oil and salt to make fried eggs for the boys. Waking that morning to the sound of eggs frying in oil was a memory that, even fifty-eight years later during his interview for the Shimoina Project, brought tears to Kubota's eyes. The Chinese farmers were poor people whose main diet at that time of year was corn and corn soup. Cooking oil, salt, and eggs were luxury items, as was the white rice the youths were offered during their stay at the headman's house.[25]

After a week of rest at the headman's house and after having been provided with shirts and trousers by the Chinese family, the boys were taken by the headman and several of his relatives to the region's former capital city of Shinkyō to a refugee centre that had been set up there for the Japanese. The headman had asked his brothers to accompany them on the journey in order to protect the two young Japanese boys in case they met hostile Chinese on the way.

So began three years of struggle for Kubota as he wandered about in south Manchuria, searching for food and shelter, avoiding the authorities as much as possible, taking on odd jobs for money or food when he could.[26] Russian soldiers swarmed over Manchuria in 1945 and 1946. Kubota and Nakagawa stayed together for that year. In August 1946, Nakagawa suddenly developed a high fever and died at the age of twenty-three. That left Kubota on his own. Sometimes he met and travelled or stayed for some time with other Japanese refugees, but he always survived by his own wits. In 1947, the Russians pulled out of Manchuria and the Chinese civil war between the communists and the nationalists reached Northeast China. At that time, Kubota often encountered soldiers of the Communist Eighth Route army, the army that would eventually win the war.

To survive, Kubota was willing to do heavy labour for the Russian troops and, later, the Chinese. He worked as a stretcher-bearer for a time, and then did some stringing of telephone lines. Sometimes he found a Christian group that offered meals to the homeless. One enterprising Japanese refugee began a small wine shop that served light meals; Kubota discovered he was a good short-order cook, popular with the other customers, who were mostly stateless Japanese like himself. Through handouts and hard work, Kubota managed to survive. South Manchuria was in a state of chaos at the time, as defeated nationalist soldiers hoped to get out of the area and increasingly confident communist troops moved in; the defeated and stateless Japanese were more or less ignored. Like Kubota, they worked for money or food when they could find a job, peddled whatever they could get their hands on, and carried on even though every day was a struggle for survival.[27]

Kubota kept his ear out for news of ships taking refugees back to Japan from the southern Manchurian port of Dairen. He was finally able to board one of those infrequent ships and made it back to Japan in July 1948. When he stepped onto the land of his home village after three years of rough living, he was just eighteen years old.

FIGURE 8.3 Kubota at twenty years of age, two years after returning from Manchuria. *Source:* Photograph supplied by Mr. Kubota to the author

The farm to which he returned had little to offer Kubota, and his family had their own tales of struggle from the bleak war years. But the farm grew potatoes, and Kubota devised a method to slice and bake the potatoes into a salty cracker, *senbei*. The inexpensive and nourishing snack was welcomed in the postwar years; Kubota opened a *senbei* shop and operated it for thirteen years, from 1948 until 1961. He married a local girl in 1954 and they had a long marriage. After a typhoon destroyed his shop and business in 1961, Kubota worked in his hometown for thirty years as a carpenter until he retired in 1991 at the age of sixty-one after suffering a stroke.

## Conclusion: Salvaging Memories

In the summer of 2012, the twelve-year oral history project to collect and publish the accounts of the Nagano people who had joined in the Japanese agricultural colonization effort in Manchuria ended. Among the project's distinguishing features was the fact that it documented and made public its methods of collecting, verifying, and editing the oral histories. Each published volume explained the project's history and operating procedures. As a citizen-run project, it strived for transparency, giving both the interviewees and the interviewers an opportunity to react to and/or edit the information provided. The typed interview manuscripts were circulated among the interviewers, who edited them for grammar and indicated where further clarifications were needed. Each interviewee had a chance to read over his or her interview, several times if desired, before it was published. There is no record of any of the interviewees having lodged a complaint against the project after his or her interview was published.

In the spring of 2012, the Society for Narratives about Manchuria and Mongolia Colonization received an award from the Nagano edition of the *Mainichi Newspaper*, the Shin-Mai Prize (*Shinmai shō*; the historical name for Nagano was Shinano, used as the "Shin" in the prize title). It was awarded in recognition of the project's contribution to local history and public knowledge about the region's past.[28]

The accounts published in the Shimoina Project reflected an honesty and forthrightness not usually seen in the many volumes of former colonists' recollections published in Japan in since the end of the war. In that sense, the project has been one of the most successful wartime-memory salvage operations yet. What might account for this new openness in these stories? Why, after all these years, was there a willingness to reveal a complete account of those events and the conflicting views and feelings of those who experienced them? What motivated these elderly citizens to step forward with such stark and brutally honest accounts of what life was like in Manchuria during the years of the war, the harrowing years that followed imperial Japan's collapse, and the long struggle for survival afterwards?

Kubota Isamu has written that, around the time the Shimoina Project was beginning, he was surprised by the fact that so many people were interested in hearing his story. The more he told of his life experiences (in his visits to a number of schools in the area), the more he found that young people raised questions as they tried to understand what had happened to him. One university student in particular asked a number of probing questions at the end of a presentation, and the questions convinced Kubota that he ought to reveal as much as he could about what had happened. Later he wrote that he was telling his story in part to encourage those who had a desire for world peace.

Professor Araragi has published insightful analyses of the challenges faced by project participants in approaching former colonists. He has also written about the hesitation and fear that former colonists had to overcome in order to recount events that were often emotionally painful, characterized by inner conflict. The elderly interviewees had managed to rebuild their lives in postwar Japan, and they had done so in part by internalizing what they had experienced and by emotionally moving on. Many had experienced numerous tragedies. They had seen or suffered rape; they had seen loved and respected members of their community gunned down by Soviet troops or by Chinese bandits. They had seen Japanese children, sometimes their own siblings, sold to Chinese farmers in exchange for food and in the hope that the farmers would feed and clothe the children. They themselves had stolen food or bartered away meagre possessions in their efforts to stay alive in a world that had turned violent and seemed heartless. Many also struggled with the knowledge that at times they had cruelly mistreated their Chinese or Korean neighbours.

As Professor Araragi has pointed out, in telling their stories, the former colonists confessed in a way to their past transgressions, revealing the most horrendous events they had ever witnessed. In making these confessions, they were asking for forgiveness and for understanding of the circumstances that had caused them to act in violation of all the social and moral values they held dear. The interviews could not have been easy for them; the resurrection of buried memories forced each interviewee to confront the "truth" of what had happened. Many of the interviewees wondered why they had managed to live to enjoy their old age while so many people they had known and loved lost their lives brutally or had suffered even more profoundly than they had.

Professor Araragi also suggested why most of the former colonists had hesitated so long and with such resolve to publicize what they had experienced: their silence was due to their sense of shared involvement and responsibility. They had all been members of transplanted villages, networks of rural communities that had local leaders whose social status allowed them to mobilize, lead, and inspire fellow colonists. Each interviewee had made a commitment to a larger group; criticizing the leader of a group was in essence to criticize all the other members of the group. Group social values called for everyone to accept his or her suffering in silence rather than to place blame on another.[29]

Onitsuka Hiroshi, a social commentator and scholar who contributed to the book that includes the account of Kubota Isamu used for this chapter, has written that the elderly colonists are no longer afraid of the local power-holders – those people who had persuaded them to join in the Manchuria colonization effort. These local leaders and village bosses were now long dead, their wartime exhortations completely repudiated by Japanese society, which now condemns wartime values and calls for peace. According to Onitsuka, many of the former colonists feel the Japanese government misled them. Colonists were dumped in Northeast China and then suffered extreme hardship. They had once felt they were part of a nation-building effort, but today they think of themselves as victims of the national policy of the Manchuria emigration scheme. They are very old now, and do not fear telling their stories.[30]

Onitsuka also feels that the warmth and sympathetic concern displayed by Saitō Toshie did much to help interviewees overcome their reservations and their reticence. Her kindness towards the elderly colonists set the tone for the project, allowing the colonists to share their (frequently painful) memories of life in Manchuria.[31]

Saitō Toshie, who was with the project for the duration and was one of its driving forces, feels that the former colonists now realize they are in the final stages of their lives. They want to tell their stories so that the truth of what they experienced – the entire truth – will not die with them. It is important for their children and grandchildren, and equally important for all those whose lives were sacrificed, that what the colonists tried to do and why – and in what ways they were misled or wronged or made to suffer – is recorded for all time. A number of the interviewees make the same point in their published recollections. In that sense, telling their stories is a way for them to live on even after their death.[32]

The published interviews – the salvaged memories – provide an unvarnished record of the lives of the former colonists. Because of their honesty and forceful narratives, these accounts will likely accomplish what many of the former colonists hoped they would: they will give us pause for thought as we remember the lives and struggles of those who were caught up in the Manchuria colonization effort. The former colonists managed to survive the great drama and tragedy of the sudden end of the war, but then they had to live with the memories and strong emotions embedded in their hearts. The losses they suffered in 1945 sprang to life because of the interview project. Yet we must never forget that participants in the Shimoina Project were volunteers, repeatedly asked if they wished to restate or withdraw their stories. At each stage of the process, they reaffirmed the importance of revealing their stories.

For Kubota, the rekindling of old memories came as something of a surprise. He never expected that young people, this new generation that had not experienced the traumas he lived through, would want to understand what it was like to be forced into the environment of mass suicide where women and children were the victims. By telling his story at local schools and community meetings, Kubota was able to let go of the burdens of the suicides, to liberate his own spirit from the pains of the past. The burdens of those memories were allowed freedom to weaken and dissipate in the air of the present-day.

In terms of laying out the historical record, the interviews in the Shimoina Project in general, and Kubota's story in particular, finally begin to clear away the officially sanctioned and strongly manipulated memories of those times that dominated the historical record for sixty years after 1945. The wartime rhetoric of willing service to the emperor, of selfless service in the name of patriotism, of the comradeship between occupying Japanese colonists and accepting Chinese farmers, has been allowed to crumble. It was not true at the time, and now, from this documentation of past times, it can be seen to have been fabricated by officials of the imperial Japanese government. At each step of the way, the fabrication was reinforced and ushered along by common citizens: by the agents recruiting people for work in Manchukuo, by the local village leaders, and by the beleaguered Japanese residents of Nagano Prefecture who were forced to believe it and who also wanted to believe it because everyone else seemed to believe it.

The Shimoina Project has salvaged the memories of Japanese agricultural colonists in Manchuria. The published interviews are authentic recollections and are documented as carefully as possible. Henceforth, all of the published accounts of Japanese life in the rural areas of Manchukuo must be read in reference to the stories published in the pages of these reports. As these stories slip away from living memory with the passing of the people who experienced them, the accounts brought to light by the Shimoina Project must be acknowledged as among the most authentic records we have of those places and those times.

## Notes

*Acknowledgments:* I am grateful to several people who helped me in the production of this work. Saitō Toshie's commitment to the project has been unwavering. Onitsuka Hiroshi, who assisted in the project for a time and is now an independent scholar and commentator, facilitated communications with interview subjects and gave me good counsel. Kubota Isamu, whose story is recounted here, is a singular figure now in his eighties, whose narrative touches us all. Finally, Noji Kaeko was my reliable research assistant at Suffolk University.

1 What the Japanese farmers were promised, and what they actually received from the government, is discussed in Greg P. Guelcher, "Paradise Lost: Japan's Agricultural Colonists in Manchukuo," in *Japanese Diasporas: Unsung Pasts, Conflicting Presents, and Uncertain Futures,* ed. Nobuko Adachi (New York: Routledge, 2006), 71–84. An overview of the Manchuria colonization efforts in the 1930s and 1940s is presented in Kevin McDowell, "Japan in Manchuria: Agricultural Emigration in the Japanese Empire, 1932–1945," *Eras Journal* (November 2003), http://artsonline.monash.edu.au/eras/. The total population of villages did not move to Manchuria in their entirety. Instead, a figure of less than 20 percent of a village's population signed up for the Manchuria colonization effort. In that sense, the Japanese villages in Manchuria might be called "hamlets." Onitsuka Hiroshi has written that, "even in the case of Seinaiji and Kamihisakata villages, which sent the largest portion of villagers, the ratio of the emigrants against the whole villagers [was] 18.9%." See *Manshū yimin: Iida Shimoina kara no message* (Manchuria immigrants: A message from Shimoina and Iida (Iida: Historical Research Section of Iida City, 2007), 4. Personal communication from Onitsuka-san, email of August 2, 2012.

2 An overview of Japanese policy towards Manchuria colonization, issues involving Chinese peasants as well as Koreans in the region, is presented in Chapter 6 of Okabe Makio and Kogawa Tsuneko, "Nongye imin" [Agricultural immigrants], in *Wei Manzhouguo de zhenxiang; ZhongRi xuezhe gongtong yanjiu* [The truth about Manzhouguo; collaborative research by Chinese and Japanese scholars] (Beijing: Shehui kexue wenxian chubanshe, 2010), 135–63.

3 The usual figure is about 1.5 million Japanese in Manchuria by the end of the war, but this figure included government bureaucrats, businesspeople, traders, railway workers, military, and the spouses and children of all these people. Numbers for the agricultural

colonists are discussed in the concluding section of this chapter. Manchuria was also a destination for Chinese immigrants. An overview of numbers of Chinese immigrants in Manchuria is in Ueda Takao, "Tōhoku Ajia ni okeru Chūgokujin yimin no hensen, 1860–1945" [Changes in Chinese immigration patterns into Northeast Asia from 1860 to 1945], in *Nippon teigoku o meguru jinkoyidō no kokusai shakaigaku* [Population movements under Japanese imperialism as part of international sociology], ed. Araragi Shinzō (Tokyo: Fuji shuppan, 2008), 158–205.

4 Many of the abandoned Japanese children who survived the ordeal of the Soviet invasion remained in Manchuria, where they married, had children, and made their lives. These people began to return to Japan only in the 1980s and 1990s. Some went to Japan in those decades and met their long-lost relatives, then decided they preferred to continue living in China. For consideration of these points, see Mariko Asano Tamanoi, *Memory Maps: The State and Manchuria in Postwar Japan* (Honolulu: University of Hawai'i Press, 2009); and Mayumi Itoh, *Japanese War Orphans in Manchuria: Forgotten Victims of World War II* (New York: Palgrave Macmillan, 2010).

5 Many of the Japanese farmers who went to Manchuria indeed never even thought about how the land they worked in China came to be owned by Japan. These recollections were stated in *Shimoina no naka no Manshū* [Shimoina and Manchuria], vol. 5, in the series *Kiki gaki hōkokushū* [Collection of oral and written reports] (Iida City: Institute of Historical Research, 2007), 159, 204. Typical examples of the pro-imperialist wartime rhetoric can be seen in *Manshū kaitaku monogatari: Manshū kaitaku no chichi Tōmiya taisa ni shagu* [The story of Manchuria colonization: In praise of Colonel Tomiya] (Shinkyō: Manshū jijō annaijo, 1941); also see *Takushoku* [Colonizing] (Tokyo: Jistugyō kyōkasho kabushikigaisha, 1942).

6 The bulk of memoirs published in postwar Japan about the wartime experiences of Japanese agricultural colonists in Manchuria in the postwar years contained some or all of these characteristics. Criticisms of wartime government policy were usually circumspect. Among the many examples that could be given of this genre of work, see Kobayashi Kōji, *Manshū yimin no mura: Shinshū Yasuoka mura no Shōwa shi* [A village of Japanese immigrants: A history of Yasuoka Village from Shinshū in the Showa era] (Tokyo: Chikuma Shobo, 1977); *Nichitakami kaitakudan shi* [Record of the Nichitakami Colonization Corps] (n.p.: mimeographed in 1979); Fukada Shinshirō and Fukada Makoto, *Aro ronsan* [Two-dragon mountain] (Hakuzaki: Hakuzaki nippōsha, [1970] 1998). The sense of victimization is mentioned in Yeeshan Chan, *Abandoned Japanese in Postwar Manchuria* (New York: Routledge, 2011), 55.

7 Examples of the better-documented and more academic studies about Japanese colonization in Manchuria are Asada Kyōji et al., *Nihon teigoku shugi shita no Manshū yimin* [Japanese immigrants under Japanese imperialism] (Tokyo: Ryūkei shosha, 1976); Yamada Shōji, *Manshū yimin* [Manchuria immigrants] (Tokyo: Shinjinbutsu ōraisha, 1978); Nishimura Shigeo, *Chūgoku kindai tōhoku chi'ikishi kenkyū* [Research on the history of modern China's northeast] (Kyoto: Hōritsu bunkasha, 1984); Enatsu Yoshiki et al., *Kindai Chūgoku Tōhoku chi-ikishi kenkyū go shinshikau* [New perspectives about historical research on China's northeast] (Tokyo: Yamakawa shuppansha, 2005).

8 The project began in 2001, and between 2003 and 2012 it published ten volumes of interviews, one each year. Not all the researchers were from Iida City. Some of the researchers who participated in the project were residents of towns and villages in Shimoina County and the nearby Kamiina area as well.

9 The case of mistreating the Korean peddler is in *Shimoina no naka no Manshū,* vol. 5 (2007), 158; the case of the beating of a Chinese peasant is in ibid., 74–75.

10 Historical Research Section of Iida City, 2007. Five individuals involved with the Iida City project cooperated to produce this volume. Kubota's story is on pages 115–23. The longer version of Kubota's story appeared in *Shimoina no naka no Manshū,* vol. 3 (2005), 9–58.

11 He now teaches at Sophia University in Tokyo.

12 For Ms. Saito's recollections of the earliest days of the project, see *Shimoina no naka no Manshū,* vol. 1 (2003), 157–60.

13 Numbers vary for all the categories of Japanese who were sent to Manchuria. A reliable source is *Nagano-ken Manshū kaitakushi* [A history of Nagano Prefecture's colonization of Manchuria], edited by the prefectural government's Nagano-ken kaitaku jikyōkai Manshū kaitakushi kankō, 1984. Various sources give a total figure of 240,000 to 300,000 Japanese colonists sent to Manchuria. The publication cited above places the total numbers of rural farmers sent to Manchuria at 220,255 colonists and 101,627 youth corps boys, for a total of 321,882 persons. A chart covering all Japanese prefectures is on page 309. I thank Onitsuka Hiroshi for his guidance on statistical issues.

14 In the early months of 2012, from February to April, the regional Nagano edition of the *Mainichi Shimbun* published a series of articles based on interviews and comments from Nagano citizens talking about government-civilian relations in the period of war hysteria. The articles detail how schoolteachers passed along government propaganda to their students, urging the boys to join the Manchuria Youth Corps; how the government moved to stop left-wing potentially anti-government political activity by arresting and rounding up people, then sending them out to Manchuria; how farm girls were told to train to be good housewives because they, too, were to be sent to Manchuria – to marry male colonists and youth corps boys. The picture that emerges is of forceful and intolerant government recruiting efforts, a story that has almost never been told in Japan. Anger towards the Japanese educational authorities who imposed government policies on young students continues to this day, as shown in the comments of a former youth corps member who accuses the wartime Shinano Educational Association (Shinano kyōikukai) of never having honestly reflected on its wartime misdeeds. See *Shimoina no naka no Manshū betsatsu kirokushū,* 108 in the series *Shimoina no naka no Manshū* [Shimoina and Manchuria], in *Kiki gaki hōkokushū* [Collection of oral and written reports] (Iida City: Institute of Historical Research).

15 The series of *Mainichi Shimbun* articles mentioned above also tell of a village head who opposed sending his fellow citizens to Manchuria; when he stepped down from his position after heated controversy, the transfer of villagers began in earnest. A further episode tells of a villager who went to Manchuria to see some of the colonists' villages there, then returned to Nagano saying things were not so great in Manchuria – in contradiction to all of the government propaganda of the time that praised the opportunities for agriculture in Manchuria and the chances to do heroic deeds for the Japanese nation. See the article "Nihonjin no ibari ni–fuan" [The uncertainties of Japanese bravado], *Mainichi Shimbun, Nagano Edition,* April 12, 2012. This article also states that of the 33,700 colonists and youth corps boys going to Manchuria, at least 14,900 – nearly half – died there, never making it back to Japan. Note that the 33,700 figure given here is lower than the 37,859 figure for total colonists sent to Manchuria, cited above in *Nagano-ken Manshū kaitakushi,* 309.

16 See *Shimoina no naka no Manshū,* 1: 157.

17 A group of residents from the Yasuoka Detached Village (*Yasuoka bunson*) in Manchuria, all of whom hailed from Shimoina in Nagano Prefecture, published a volume of recollections of their wartime experiences in 1979. They were among the first of the former colonists to recount their wartime experience in a transparent manner. Of the 984 villagers who went to Manchuria, 613 died, 57 did not return to Japan, and all traces of 50 others had been completely lost. Professor Araragi reissued this volume as *Manshū Yasuoka bunson; Nanajūnen no rekishi to kioku* [Yasuoka Detached Village in Manchuria: History and recollections of seventy years] (Tokyo: Fuji shuppan, 2007). A review of the book by Shimajima Tetsurō appears in *Shūkan Dokushojin* [*Weekly Reader*], no. 2721 (March 2008): 6. I thank Toda Yasushi of the Harvard-Yenching Library for alerting me to this reissue.

18 The end of the Shimoina Project was announced in the summer of 2012. See the *Asahi Shimbun* digital edition of July 27, 2012, http://www.asahi.com/ajw/.

19 This was the *Nekosogi dōin* [Stripping out the roots call-up]. See Ozawa Chikamitsu, *Hishi, Manshūkokugun* [Secret history, the Manchukuo army] (Tokyo: Kashiwa shobō, 1976), 193; Mayumi Itoh, *Japanese War Orphans in Manchuria,* 18; Yeeshan Chan, *Abandoned Japanese in Postwar Manchuria,* 164. Many units of the Kwantung Army had already been transferred out of Manchuria to protect the home islands or been sent to the South Pacific as reinforcements for decimated Japanese units, so this final mobilization ordered all the remaining able-bodied Japanese males in the villages to report to their nearest major city for active duty. The villagers in Kawano Village soon came to realize that this was the final attempt to defend the Japanese empire, which becomes clear as the story unfolds. In the account that follows, I have paraphrased the narrative as given by Kubota, while direct translations from the published report are in quotation marks.

20 The animosity between Chinese peasants and the Japanese colonizers because of the land issue is explored in Ronald Suleski, "Northeast China under Japanese Control: The Role of the Manchurian Youth Corps, 1934–1945," *Modern China* 7, no. 3 (1981): 351–77.

21 The phrase used by the colonists, as recalled by Kubota in local dialect, was "*yaran, hadaki ni natchaou*" (but then, we'll be naked).

22 According to many of the published Japanese-language accounts cited by me in the notes here, this chain of reasoning was widespread among the Japanese colonists in Manchuria following the abrupt end of the war.

23 Obtaining cloth or wearable clothing was difficult for the Chinese peasants in the countryside, so it is understandable that local peasants took whatever they could once they came upon the dead bodies. Many males at the time wore traditional loincloths (*fundoshi*) as underwear, so I asked Kubota-san, via correspondence with Onitsuka-san, why he and Nakagawa were wearing "boxer shorts" (*pantsu*). He said both types of male underwear were common at the time, and he had been issued the boxer style of underwear of the type also given to boys in the Youth Corps (*Seishōnen giyūgun*). I also asked him why, if many bodies were completely naked, both he and Nakagawa were still wearing shorts. He said that probably as the Chinese were removing the clothing from both boys, they realized their bodies were still warm and that both were still alive, so they left the shorts on. For an overview of the Manchuria Youth Corps, see Ronald Suleski, "Reconstructing Life in the Youth Corps Camps of Manchuria, 1938–45: Resistance to Conformity," *East Asian History* 30 (2005): 67–90. The most useful single volume on the Manchuria Youth Corps is Kami Shōichirō, *Man-Mō kaitaku seishōnen giyūgun* [The Manchuria Youth Corps] (Tokyo: Chūōkoronsha, 1973). Mr. Kami was helpful to me during my initial research

on the Youth Corps. Kami accepts figures showing that Japanese agricultural colonists in Manchuria numbered 222,970, of which 72,000 died and 11,000 went missing. He holds that 24,200 Youth Corps boys died in Manchuria (75, 177–78). These figures vary somewhat from those discussed in the volume *Nagano-ken Manshū kaitakushi.*

24 There existed language barriers among the Japanese colonists, the Chinese peasants, and the Koreans who lived in many parts of Manchuria. Most Japanese learned very little Chinese and conducted all of their affairs in Japanese, so the Chinese who could communicate in Japanese were relied on by the Japanese, in this case to act as bosses over the other hired Chinese workers. One interesting case, where a Chinese "coolie" actually lived in a simple dwelling with a Japanese family, is given in *Shimoina no naka no Manshū,* vol. 9 (2011), 134–36. This interview discusses the use of language and indicates that the Japanese made no special effort to learn Chinese or Korean.

25 The state of agriculture in Manchuria at the time, particularly for poor farmers, is discussed in *Nihon teigoku shugi shita no Manshū yimin,* 397-411. The diet of typical farmers in Manchuria today is similar to farmers' diets in 1940s, except that the quality and quantity is better and high-protein meats are now easily available. See the comments on food given in Michael Meyer, *In Manchuria: A Village Called Wasteland and the Transformation of Rural China* (New York: Bloomsbury Press, 2015).

26 One of the more revealing accounts of the panic among Japanese in the postwar days in Manchuria is Ikeuchi Hajime, *Anjyaku: Dairen no rakujitsu* [Warming stone: The setting sun on Dairen] (Osaka: Dasō shobō, 1980). This description of events is focused on the Japanese who were living in Dairen, as recounted by a Japanese woman who was a young lady then. She describes the fears of the Japanese community over the impending occupation of the city by Soviet forces. She describes how Chinese citizens were stopping any Japanese they encountered in the street and taking from them anything they desired – watches, jewellery, and so on. The Japanese began hiding such precious objects in their socks and shoes. There were cases of rape, so most of the young women went into hiding for months at a time, living in small rooms and fearing each knock on the door.

27 Many detailed descriptions of how the Japanese refugees managed to survive during the chaos following the end of the war have been published in Japanese. For example, see Matsui Hitoo, *Kotaribe no Manshū* [Narratives of Manchuria] (Nagano: Ginga shobō, 1980); and Imai Yakichi, *Manshū nanmin yuki* [Wanderings of Manchurian refugees] (Tokyo: Tsukiji shokan, 1980). To the extent they could, some Japanese tried to offer relief to the wandering refugees. See Morimoto Shigeru, *Kōryō no nubari: Chūgoku Tōhoku chiku Nihon nanmin kyūsai no kiroku* [A needle in the gaoliang: A record of relief work for Japanese refugees in northeast China] (Osaka: Dasō shobō, 1979).

28 See *Shimoina no naka no Manshū,* vol. 10 (2012), 350.

29 Professor Araragi's thoughtful analytical comments are in *Shimoina no naka no Manshū betsatsu kirokushū,* 162–86.

30 Personal communication from Onitsuka-san, email of August 1, 2012.

31 Ibid.

32 Personal communication from Saitō-san, letter of August 13, 2012. An interesting point to note is that Saitō Toshie, and virtually all of the people interviewed in these volumes, are people of few words. They convey their meaning and then stop speaking or writing. Professor Araragi and Onitsuka-san, both of whom have written and published analytical studies, are people for whom expressing themselves in lengthy exposition comes easily. In spite of this, Saitō-san and her interviewers managed to compile ten volumes of precious and revealing information.

# 9

# Exile to Manchuria: Stories in the Qing and the PRC

*Wang Ning*

China has a long tradition of internal banishment – exiling various offenders (common criminals, disgraced officials and scholars, as well as other disfavoured personnel) to border regions, in the south and the north. Although legal systems changed over time, exiling convicts in lieu of execution remained a constant practice and a feasible option, due to rulers' professed "benevolence" and their reluctance to execute any but the most serious offenders. In much of the imperial era, however, the distance of the banishment was considered to reflect the seriousness of penalty. Banishments typically ranged from 2,000 to 4,000 *li* (1,000 to 2,000 kilometres).[1] Common exile destinations included the regions of Guangxi and Yunnan in the Qin Dynasty (221–206 BCE), Hainan in the Song Dynasty (960–1279), and Liaodong in the Ming Dynasty (1368–1644). Being exiled to one of these areas was considered a punishment second only to capital punishment in severity.

In the Qing Dynasty (1644–1912), Manchuria became the favoured exile locale, before Xinjiang was incorporated into the empire in the mid-eighteenth century.[2] The goal was to punish offenders while at the same time populating Manchuria and providing labour for banner troops stationed there. Rulers did not send exiles to the furthest frontier lines (such as beyond the Amur River), however, due to the region's uncertain borders with Russia, the lack of adequate conditions for human habitation, and the difficulties of administration. Instead, exiles were dispatched to places where there was at least some human habitation, where administrative systems had been established, or where Manchu garrisons needed labour.[3]

The communist government of the People's Republic of China (PRC) inherited this imperial legacy. During regime consolidation in the 1950s, millions of Chinese fell victim to political campaigns and persecution, and many were exiled to Manchuria, where they were forced to work in army farms, labour camps, mines, and other types of labour reform settlements. In the 1960s and early 1970s, many millions of people were forcibly relocated to rural areas of the northeastern provinces for the purpose of alleviating population pressure on urban areas and removing social undesirables (the unemployed or vagrants, for example), or for other political reasons. According to Thomas Bernstein, 3.24 million urban youths (including ex–Red Guards) were sent to the rural areas of Liaoning, Jilin, and Heilongjiang from 1968 to 1975.[4] The vast Manchurian countryside became a state-preferred destination for the involuntary migrants.

This chapter provides a sketchy but vivid picture of Chinese exiles in Manchuria in the early and mid–Qing Dynasty and in the early PRC period. Specifically, it compares the experiences of Qing exiles (chiefly, disgraced scholars and officials) in several banishment destinations from 1644 to 1759, with the experiences of PRC exiles in army farms and labour camps in the Heilongjiang borderlands in the 1950s and early 1960s – their struggles and suffering, as well as their adaptation to the environment. I argue that while the rich natural resources in the northeastern borderland provided various means of survival, unfavourable physical and environmental conditions (remoteness from "China proper," long and cold winters, and an inhospitable environment) were compounded by state oppression and inevitably gave rise to hardship. Yet the Qing exiles fared much better than their counterparts from the PRC, in terms of material conditions, physical and psychological treatment, and social circumstances; they successfully built influence and prestige through their educational and commercial activities, while contributing to cultural and economic change in the Manchurian borderlands. Notably, in the Qing, women were banished as part of a family, not due to their own actions; in the PRC, on the other hand, women (writers, journalists, and so on) were sent into exile due to their own actions and critical commentary. There were substantial differences between Qing exile and exile during the PRC, but what was common to both was the use of the Manchurian environment as a penal regime to deal with opponents, critics of the state, and the socially undesirable.

## Stories from the Qing

Banishing offenders to Manchuria accorded with both frontier colonization (*shibian*) and the need to move undesirable elements to remote regions. With tens of thousands of bannermen migrating into China proper in the early years of the Qing Dynasty, the depleted population in Manchuria was in need of replenishment. During the Shunzhi reign (1643–61), Qing policy encouraged a population influx into Manchuria, including the immigration of poor northern peasants to Liaodong in southern Manchuria and the distribution of convicts to various prefectures and settlements further north.[5] Although this "encouragement" policy was rescinded in 1668, early in the Kangxi reign (1661–1722), the practice of banishment continued, with operations ranging from sending individual offenders into exile separately to the group banishment of dozens or even hundreds at a time – including convicts, their families, and those implicated in their cases.[6] At the end of the Yongzheng reign (1722–35), there were an estimated one hundred thousand exiles in Manchuria.[7]

The Manchurian borderlands, especially Heilongjiang, had long been viewed in China proper as the "furthest frontiers, with bitter cold" (*jibian kuhan*).[8] Records from the seventeenth century indicate that ground frost often appeared during the seventh lunar month (roughly, August), and that snow was everywhere during the eighth lunar month. The Amur River started freezing in the tenth month, and people in Qiqihar would tell stories of individuals' ears occasionally falling off during the coldest period at the end of the year. In many areas of Heilongjiang, snow could last until the fourth lunar month the following year.[9] Before the Chinese exiles arrived, people living north of Shenyang had not planted vegetables or worn cotton clothes; food was coarse – chiefly sorghum, millet, soya bean, and edible barn grasses – even though arable land was abundant and soil fertile. For the state, the harsh physical environment and unpleasant living conditions in Manchuria made the region the perfect practical destination for banished offenders and those who had fallen out of favour; on an abstract level, however, the idea of Manchuria could also generate considerable fear. Records indicate that many of the offenders and their families were visibly shaken and distraught on hearing the imperial edict of banishment to Manchuria.[10]

The first Qing exile to Manchuria (in 1645) was the monk Hanke, who was found guilty of composing eulogistic biographies of Ming loyalists, and who suffered physical abuse while in custody in the prison of the Ministry of Punishment. However, Hanke's life of exile was not as hard as one might imagine. His designated destination was Shenyang, the Qing Dynasty's secondary capital, which had a well-established civil administration, material amenities, and a degree of intellectual life. With the help of sympathetic gentry, Hanke quickly adapted to local life and began to preach Buddhism. He was able to raise funds to construct seven Buddhist temples and was then able to recruit more than six hundred disciples before he died peacefully at the age of forty-four.[11] Not all exiles had such good fortune, however. In the following years, thousands were sent further north to Shangyangbao and Tieling, places that were physically and materially far removed from Shenyang. Other exile locales included Chuanchang in Jilin, and, later, Ninguta, Qiqihar, Mergen, Beduna, and Aigun in Heilongjiang. Of these, Shangyangbao, Ninguta, and Qiqihar received the greatest influx of exiles.[12]

In the Qing era, exiles to Manchuria could be categorized loosely as either criminal offenders or non-criminal (that is, political) offenders. The criminal offenders included robbers, bandits, counterfeiters, and smugglers. Traditionally such crimes ranked among the most serious, but not as serious as rebellion and treason. According to Joanna Waley-Cohen, "many of these exiles had originally been sentenced to death but had had their sentences commuted."[13] These exiles were usually enslaved at their exile destinations – awarded to bannermen as family slaves, to government farms (*guanzhuang*) as forced labourers, or to garrison troops as servants.[14]

Non-criminal exiles overwhelmingly comprised disgraced scholars and former officials, as well as their families and relatives. The reasons for their banishment were multifarious. Some non-criminal exiles, such as the famous scholars Wu Zhaoqian (1631–84) and Sun Yang, were blamed for irregularities or fraud allegedly committed at civil service examinations. Others, such as the brothers of Zha Siting (d. 1726), were victims of the Qing literary inquisitions (*wenziyu*). In such cases, principal offenders were condemned for advocating the restoration of the Ming Dynasty or for defaming Qing rulers by circulating seditious materials or anti-Qing allegories. Even after such "offenders" died, their descendants and relatives could be banished by virtue of collective responsibility.[15] Still others were banished for offending the emperor by lodging criticisms or for giving "inappropriate advice." These included Minister of Rites Chen Zhilin, who argued that Manchu officials who committed offences be punished, and Ji Kaisheng, who argued against the Shunzhi emperor's visit to Yangzhou to select women for himself.[16]

Lastly, some individuals were banished for being peripherally involved in anti-Qing rebellions. The ringleaders of these rebellions were already dead – having died of illness or by execution – and the state exiled those who played an accessory role. For instance, after the Rebellion of the Three Feudatories (1673–81) was put down, more than one hundred rebellion advisors and associates were sent to Shangyangbao. All literati and merchants accused of collaboration (real or alleged) with Zheng Chenggong (Koxinga) and Zhang Huangyan in their assault on southern Jiangsu in 1659 were banished to Ninguta, including Zheng's father, Zheng Zhilong, who had long before surrendered to the Qing.[17]

For many exiles, especially those from the south, living in (what they perceived to be) primitive and desolate Manchurian borderlands was a difficult experience. Those who were not enslaved to banner families had to start their lives over. On their arrival, they "cut wood to make shelters, piled earth as sleeping platforms, and sold belongings for food."[18] An imperial edict of June 1682 described such exiles in Ninguta and Jilin as being deprived of proper shelter, lacking the means to engage in farming, and burdened with government orders to provide labour services.[19] In 1676, in order to prepare defences against incursions by Russian Cossacks in the Amur region, several thousand families of exiles were sent to Chuanchang, a major dockyard on the Sungari River, to build boats and serve as sailors.[20] For the banished scholars and officials who had never performed such manual labour, life in exile was indeed difficult.

Exile Fang Gongqian provides a window on Ninguta life around 1659–60:

> There is no person of leisure in Ninguta, and women are the busiest of all. The beating of cloth for pasting over the windows in lieu of paper, and the cutting of hemp and mixing it with chaff for lighting are all done by women. Grain is not husked by grinding but by pounding, which goes on day and night. Polished grain requires five or six poundings. Recently a mill was built, but the grain has to be carried to the mill. Soon after the pounding of grain comes the task of drawing icy water. The well-site is as slippery as a hill. Barefooted and thinly garbed, those who weep and lament while shouldering their burdens are all descendants of the rich and honourable families of China. How heartrending a sight it is![21]

Physical punishment was also part of the exile's life. Shi Songqiao, a high-level official who was banished to Qiqihar, appeared late for the roll call and was flogged in public, a punishment rarely inflicted on officials in China proper. Unable to bear such humiliation, Shi fell ill and died soon afterwards.[22]

The exile life in Qing Manchuria was not necessarily only suffering and misery, however. The experience of exiles depended on many factors, such as their socio-political and financial status before banishment, the personalities of local officials, and the conditions of the exile settlements. For instance, exiles who were not enslaved and not assigned to serve in military garrisons were free to make their own living. As Diana Lary and others demonstrate in this volume, the vast territories of Manchuria boasted fertile soil and rich natural resources. For those with modest resources and a willingness to adapt to the new environment, making ends meet was possible. One could relatively easily acquire a plot of land to plant grain and vegetables; fish in the bountiful streams; collect mountain products such as nuts, wild fruits, mushrooms, and ginseng; or hire natives to hunt wild animals for the fur trade. They could even avoid forced labour by making donations in lieu of work.[23] For example, except for once having to run an official errand in a blizzard, exile Wu Zhaoqian lived a relatively comfortable life in Ninguta for most of his exile years with his wife and son, Wu Zhenchen, author of *Ninguta jilüe* (*Notes on Ninguta*). They owned a thatch-roofed house in a big, treed courtyard, with a wooden fence flanked by a flower garden. With his savings and his income from teaching, Wu was free from having to do farm work except for growing vegetables and mushrooms in his courtyard – and was still able to support his family. His financial position even allowed him to make a donation towards repairing a local government granary, thus exempting himself from short-term, military-related service.[24] In *Ninguta jilüe,* Wu Zhenchen recounts in some detail a joyful childhood spent collecting nuts, wild fruits, and rose flowers (to make jam), as well as practising mounted archery.[25] These descriptions indicate that some exile families, at least, did not live in abject poverty, and that they benefited from making physical and mental adjustments to their new life in Manchuria.

It would be a mistake to assume that enslavement necessarily led to hardship. In Qing Heilongjiang, the families of bannermen and related ethnic groups (such as Dahurs and Solons), rather than officials or government employees, were eligible to take on exiles as slaves. Slaves were awarded based on an individual or family's merit and distinction, or due to their destitution.[26] However, these people were generally considered to be simple-minded, rough, and rugged by nature;[27] many were not capable of effectively employing or dealing with the sophisticated and somewhat shrewd exiles coming from China proper. Two cases found by Fu Yu, the military governor of Heilongjiang, reveal the easy life of some slaves and the challenge faced by their masters. One slave assigned to a banner family often refused to get up in the morning, so the members of the master family had to do the farm work themselves. In another case, a banner soldier carried water himself rather than using his slave, who was a scholar sent from Jiangxi Province. When he was asked the reason for this, the soldier replied, "The scholar cannot do such a work." In both cases, Governor Fu intervened, flogging the "lazy" slaves and replacing them with more "tamed and kind ones."[28] It is difficult to ascertain whether such examples were common, but these stories suggest that the "mastership" of the bannermen did not guarantee effective control of their slaves: master families could be reluctant or unable to drive the slaves hard, and slaves were unlikely to be punished for their "laziness" unless their inaction was discovered by officials. In some cases, the banner families simply could not support their slaves and asked for them to be freed or transferred elsewhere.[29]

Exiles were nominally under the surveillance of Manchu officials or, in the case of slaves, the bannermen. In reality, however, slaves were not always watched carefully. In the Shunzi and early Kangxi reigns, for instance, supervision of exiles was rather lax on the frontiers. Roll call, an important method of monitoring, was done once every five, ten, or even fifteen days, depending on location.[30] According to Yang Bin, who travelled to Ninguta to visit his father, "local officials have not been keeping a strict lookout over the exiles; many of them were granted furloughs to cross the Shanhaiguan pass (to visit home), and the cases of sudden night escape did happen at times. There was neither local checks nor official certificates needed when crossing the pass."[31] Wu Zhenchen indicated that native people were reluctant to report cases of fleeing. Even when an escapee was caught, punishment was unlikely: people were inclined to explain to the authorities that the escapee had returned voluntarily.[32] Such "benevolence" and lax management may be due in part to the tolerant and easygoing nature of the local people, but it was also likely that the need for convict labour in the area was limited, as was the bannermen's ability to effectively command convict labourers. Consequently, many labourers managed to escape. For instance, exiled scholar Fang Gongqian was able to flee to his native home in Jiangsu in 1670 and then joined Wu Sangui's rebellion in Yunnan. Another exile managed to transport the wife of his friend back to his hometown.[33] It was not until 1690 that official papers were required at check posts to prevent such escape.[34]

For the exiles who were not enslaved, or for those who had committed less serious offences, a variety of job opportunities existed. Exiles could work in military-run agricultural colonies, government farms, postal stations, or dockyards. Those serving in garrison units could work as river patrollers. Since the Qing government had not intended to exercise close control over the life of exiles, nor had it envisaged systematic planning of their work, the exiles were allowed to take on various freelance opportunities, including work as land-holding farmers, medical doctors, craftsmen, and petty traders. Some exiles in Qiqihar even ran gambling houses for profit.[35] These occupations provided exiles with the opportunity to establish themselves economically and socially.

The experiences of the descendants of Lü Liuliang (1629–83) are telling. Lü Liuliang was a Ming loyalist scholar who composed a number of volumes of anti-Manchu writings. Forty-seven years after his death, in 1730, some of his writings were discovered, resulting in a brutal literary inquisition. The corpses of Lü and his son were disinterred and dismembered, and one of his surviving sons was put to death. All the other members of Lü's extended family, in total 111 people, were exiled to Ninguta. At first they were enslaved to bannermen; after being freed in 1737, some members of the family engaged in commercial activities and eventually became successful businessmen. One traded fox and sable furs, ran a hot water shop in the village, lent money at 20 percent interest, and owned several hundred acres of land. Another Lü operated a medicine business and also sold grain and salt; his properties included a 28-room thatch-roofed house, a medicine shop, and a cache of three hundred silver taels. Several others held substantial wealth in the area as well. Two Lü family members were even able to purchase *jiansheng* status, legally forbidden to convicts. Subsequently, in Qiqihar, the Lü families again managed to establish themselves as the most valued teachers in the town.[36]

The story of the Lü family is merely one example of exiles establishing themselves as successful businessmen. The first trader in Ninguta (perhaps the first resident Chinese trader in Manchurian frontiers) was Chen Jingyin, who would go on to become a prosperous merchant. As he recounted to Yang Bin:

> I was exiled to Ninguta in the twelfth year of the Shunzhi reign [1655], when no Han lived here. The rich Manchus mostly wore hemp winter coats stuffed with hemp wadding, and the poor all put on roe deer furs, having no idea of cotton clothes. Such things became known to the locals only after I arrived. I once traded some cloth for three and a half *dan* of

grain [175 kg]. An official got from me some white cloth to make a coat; when wearing it on New Year day he was greatly admired by others. Nowadays the well-to-do people in Ninguta all wear silks and satins in summer and long coats made of goatskin, and furs of lynx and wolf in the winter, whereas the poor wear cotton clothes.[37]

This narrative suggests how the cotton trade initiated by the exiles might have influenced the clothing materials used by Ninguta residents. Further description of commercial activity can be found in a letter penned by Wu Zhaoqian:

After the Dingsi year of the Kangxi reign [1677], businessmen and traders have flocked to Ninguta; 60–70 percent of precious goods originating from the south can be found here, with shops and stalls burgeoning and the lights of cargo wagons flashing. The exiles are good at doing business, all conduct trade in ginseng and sable furs, and have thus earned hundreds even thousands of taels of gold ... I myself, however, cannot give up my "pen and wine" taste to strive for that kind of profitable work, as I am content with a simple and easy life.[38]

The commercial boom set in motion by the exiles changed the social landscape of Manchuria's frontier region and helped to establish exiles' influence and prestige in local communities. Yang Bin explains how commercial importance was translated into social prominence for the exiles:

All the merchants of the eastern and western boroughs were Chinese. The Manchu officers and soldiers were poor. Their clothing and food were all received on credit from merchant acquaintances[,] whom they paid when their monthly stipends arrived. Consequently, they were extremely diffident in their daily intercourse, being afraid that the merchants might withhold their credits. Furthermore, the merchants were all prominent and scholarly men among the exiles. They associated with the military governor and his staff on a basis of equality. The older ones among them even greeted the military governor and his associates as younger brothers, to say nothing of those with lower ranks.[39]

When the Manchus gained sovereignty over the Han people in China proper through military conquest and the ensuing political dominance, a reversing trend started in their homeland as Chinese merchants gained soft power through commercial dominance.

With increasing prosperity, some scholar-merchants involved themselves in local affairs, especially charities. After prospering through business (chiefly with other Han exiles), Yang Ancheng engaged in a number of charitable ventures – buying the freedom of slaves from official organizations, supporting poor elders and orphans, and providing financial aid to those unable to pay for marriages and funerals. According to one observer, "the local rich were impressed by his righteousness, and thus jumped forward to assist him, feeling shameful when not joining in." The result was that "no inhabitant ever starved or suffered from cold. Whoever was hungry or cold was given food and clothing donated by all."[40] Such philanthropic undertakings can be seen as the starting point of Chinese exiles' assumption of leadership roles in the frontier society.

The influence and prestige of the exiles, however, derived not only from their economic strength and philanthropic contributions. Teaching was among the most esteemed careers that the banished scholars could engage in, and it was perhaps the most suitable job for the literati trained in the Confucian classics. Their pupils included Manchu officials and their children, as well as the general public. Mastery of the Chinese classics was the key to writing a successful civil service examination and to personal advancement. For the frontier Manchu elites, a command of Chinese language and literature was an indication of cultural refinement. High-level Manchu officials in Jilin and Heilongjiang were eager to hire exiled scholars as tutors for their children and themselves, and patronized the men of letters.[41] Xi Qing has shown that the Heilongjiang exiles "who commanded certain knowledge of literature and writing all took teaching as the means of self-support."[42] In Qiqihar, the best-known figure was Wang Tinglin, who taught classics and literature at a community school for the children of bannermen. In Ninguta, Wu Zhaoqian tutored two sons of the military governor Bahai; his own son, Wu Zhenchen, was granted the chance to learn with them. The books that the members of the Wu family brought with them into exile included *Wu jing* (the Five Classics), *Shi ji* (*Historical Records*), *Han shu* (*The Book of Han*), and *Li Tai Bai quan ji* (*The Complete Works of Li Bai*). This was the first time these Chinese classics had appeared in the frontier communities.

The annual income from teaching ranged from ten to thirty taels of gold, which ensured that the tutors were financially secure.[43] However, some did not take payment for all of their services. Yang Ancheng, for instance, provided basic-level classes in Chinese language and literature for local ordinary people, free of charge.[44] In this way, native Manchus and other locals were able to access education in the Han Chinese classics, which helped to spread Confucian norms such as loyalty, filial piety, and propriety, and other aspects of Chinese culture, such as protocol and etiquette, through the frontier society. Commenting on the experience of Qing exiles, Robert Lee argues that the former social status of these elites was reproduced in a somewhat distorted form on the frontier: "It was distorted because the exiled scholar-officials, having been deprived legally of their political privileges, played a purely intellectual, and sometimes commercial role in the frontier society."[45]

Given their important social role, the exiled scholars won considerable esteem and admiration from Manchus, officials, and commoners. Natives in Ninguta respectfully addressed the exiles who had been awarded imperial degrees as *hafan* ("officials" in Manchu), together with other local scholars.[46] Wu Zhaoqian's prestige was such that he was able to persuade Bahai to free his son, Zhenchen, from labour service, which allowed Zhenchen to focus on scholarly work. Another military governor in Wula invited Wu to be his secretary and the manager of postal stations under his jurisdiction, an offer that became moot when amnesty was granted to Wu by the court.[47] Zhang Runan (a former high-level Gansu official and a *jinshi* imperial degree holder) repeatedly refused to be enlisted as a secretary to a local Qiqihar official, as he felt it did not appropriately reflect his status and importance.[48] For the Lü families, their double disgrace – as the relatives of an anti-state culprit and as lawbreakers in their own right for having illegally purchased *jiansheng* status – did not inhibit their growing prestige in the exile settlement. When Zhang Binglin (1868–1936) visited Qiqihar in the early twentieth century, he was surprised that the "Lü families, though classified as convicts, were all too often invited as tutors by local residents, and banished officials on arrival would always pay courtesy visits to their house. The native people had no reason to despise them; neither did they assume any self-debasing status themselves."[49]

As time went on, some exiles (primarily scholars and ex-officials) wished to resume their previous gentrified lifestyle and re-establish their previous gentrified lifestyle in the exile settlements. They sought each other out and formed literary clubs and other, similar cultural societies. Monk Hanke organized the Poetry Club of the Ice World (*Bingtian shishe*) with thirty-three fellow exiles in Shenyang, and Wu Zhaoqian established the Society of Seven Exiles (*Qizhe zhihui*) in Ninguta. They met regularly in gardens, to enjoy viewing the flowers and to drink wine, composing and chanting poems and verse, through which they could vent, implicitly or explicitly, their grievances over their personal misfortunes and their distress over the loss of the Han homeland. In this way, they seemed to have consoled each other to some degree. Some, such as Zhang Tangong and Li Jianru, owned slave singers and entertainers in exile.[50] Having been unsuccessful in their political or social endeavours in China proper, they attempted to acquire in exile some pleasure and meaning of life in a lifestyle of "poetry, wine, and beauty."

Another meaningful aspect of exile life that is often ignored by historians concerns the contribution of Qing exiles to agriculture in Manchuria. The local peoples, especially Manchus, had developed certain techniques in the cultivation of sorghum, millet, soya bean, and wheat. Regular hunting of bear, roe deer, river deer, elk, fox, and pheasant (in the most favoured season, the tenth lunar month) provided precious sources of meat. This diet was supplemented with wild herbs. However, the local population knew little about cultivating vegetables, except for bok choy. In the mid-seventeenth century, exiles in Jilin and later in Heilongjiang began to cultivate a variety of vegetables, including celery, leaf mustard, chives, spinach, lettuce, eggplant, radish, cucumber, pumpkin, green onion, green pepper, and garlic.[51] In Ninguta, the family of Wu Zhaoqian planted "squashes and vegetables" in their big courtyard; "every [exile] household did the same farm work, as we could not buy vegetables from others."[52] In Qiqihar, when exile families could not consume all the vegetables they grew, they sold the surplus at markets. Thus, the technologies of vegetable cultivation spread.[53] Yet despite the pioneering efforts of the Qing exiles, because the growing season in north Manchuria was short and frosts came early, the cultivation of many types of vegetables was simply not possible.

Banishment to Manchuria gradually came to an end during the Qianlong reign. The Qing rulers had long seen Manchuria as their homeland, "birthplace of the dragon" (*long xing zhi di*). For cultural, strategic, and economic reasons, they wished to preserve the Manchus' traditional values and lifestyle, and to maintain government monopolies over the lucrative trade in ginseng, pearls, and fur. Their concerns became increasingly grave on seeing steady Han penetration into Liaodong, Jilin, and even Heilongjiang; exile families in the northeast – in Heilongjiang in particular – began to outnumber the garrison troops stationed there. At the same time, Han migrants began to form loosely organized lawless populations on the frontiers.[54] Thus, after Xinjiang was conquered and incorporated into the empire in 1759, the Qing government began to send the majority of offenders to the northwest instead. There was still residual and small-scale banishment to Manchuria, however, and descendants of many of the early exiles continued to live there for generations.

## PRC Exiles in Heilongjiang

After the 1949 takeover of China, the governing Chinese Communist Party (CCP) waged a series of campaigns to eliminate real or potential opponents, consolidate the regime, and establish a new political and socio-economic order. Initiatives ranged from land reform and the Campaign to Suppress Counterrevolutionaries (1950–52) to the Anti-Rightist Movement of 1957. Millions of victims were targeted, and many former Nationalist Party (GMD) soldiers, counterrevolutionaries, Rightists, landlords, and members of secret societies were banished to Manchuria, overwhelmingly to the Heilongjiang borderlands, which had 141 labour camps and army farms in the 1950s and early 1960s (and many existed beyond the 1960s).[55] The vast majority of exiles became agricultural labourers and lumber workers, except those delegated to the mines. PRC leaders believed that sending offenders to difficult regions was instrumental to their rehabilitation as well as to building up areas where voluntary migration was difficult to achieve.[56]

Several characteristics stand out in the PRC banishment of exiles in the 1950s and early 1960s. One feature is the intensive use of forced labour and strict physical control over the exiles in the designated labour camps. Unlike their imperial antecedents, who, except for those in enslavement, were given a variety of job options or were simply left to their own devices, all PRC exiles to Heilongjiang were put into labour camps, army farms, and other kinds of labour reform settlements, under the rigorous supervision and monitoring of camp cadres or farm officials. They had to provide various compulsory labour services in agricultural, lumbering, and hydraulic projects.

After the 1955 Campaign of Liquidating Counterrevolutionaries, for instance, more than ten thousand convicts were sent to the Xingkaihu area of eastern Heilongjiang, to work in infrastructure construction, building a large labour reform complex. In March 1958, more than twelve hundred so-called Rightists were sent from Beijing to a number of army farms in Mishan and Hulin, close to the Soviet border. There, they engaged primarily in land reclamation, as well as lumbering work and charcoal making.[57] Put into well-organized labour teams, the inmates (Rightists and convicts) cultivated the land, harvested crops, built reservoirs, cut wood, excavated earth, made bricks, erected barracks, built roads, and raised livestock.[58] Work goals were set; those who failed to fulfill the goals, individually or collectively, would be scolded and even punished – denied food or sleep. Dai Huang, a former journalist with Xinhua News Agency, recalls that during the Great Leap Forward (1958–59), army farm cadres frequently withheld food if tasks were not completed, or used the threat of withholding food to promote the accomplishment of work goals. One of his fellow Rightists in Farm 850 was denied food by farm cadres for inadequate work and subsequently died.[59] References to such brutal treatment and disciplinary measures are not found in the records relating to the Manchurian exiles during the Qing.

For the convicts in labour reform camps (*laogai*), the work day was officially eight hours in length, with nine hours as a maximum. This rule was largely followed, except during the busy sowing and reaping seasons, because the soldiers escorting the convicts worked on a fixed shift. However, for those not formally sentenced but sent by police to re-education through labour camps (*laojiao*), and for those sent to army farms for "labour under supervision" (such as the Rightists), the length of the work day was determined according to the need or whim of managerial officials, often ranging from ten to twelve or even fourteen hours a day.[60] Since Heilongjiang is the most northeasterly province of China, and local time is fixed by Beijing, a summer dawn could be as early as 3 a.m. The exiles had to wake up at the sound of the cadre's whistle, leave the barracks, rush to the fields with their meals, and work collectively until dark. During the busy seasons, and in the years of the Great Leap Forward, there was no rest even on weekends.[61] The exiles were often driven to physical exhaustion. Peter Liu recalls his experience during that time:

> It was 6 p.m. … We had to work one or one and a half extra hour each day since the launching of the Great Leap … I felt all my senses were blind, my shoulders benumbed under the load, and my legs stiff except that they could still move back and forth with me and the load. The real ordeal was the grinding pain in my joints when I lay down on the *kang* after a day's leap, but more painful was the knowledge that the leap was to continue on the morrow.[62]

The state's use of intensive forced labour to achieve its goals amounted to physical torture for many exiles, especially when labour camp and farm cadres were career-oriented and did not hesitate to overwork the exiles in order to ensure their own advancement.

Another characteristic that stands out in the PRC banishment of exiles to Heilongjiang is that the exiles suffered harsh physical and environmental conditions. Three hundred years had passed since the Qing ruling house had begun to send political offenders to Manchuria in the mid-seventeenth century – but the weather conditions had remained the same. In northern and eastern Heilongjiang, where many of the labour camps and farms were established, weather conditions were formidable: flurries and small snow squalls heralded the arrival of winter – in September. With the arrival of winds from the north, the soil hardened quickly. In November, the temperature could drop to –20°C; in January, –30°C degrees or lower, with wind-chill, was not unusual.[63] Such temperatures posed extreme hardship for anyone working outside. Before the communist takeover of the area, rural Northeasterners customarily stayed indoors during the winter months, drinking, gambling, gossiping, and so on. Though at times local inhabitants hunted roe deer in the snow, "hiding from the winter" (*mao dong*) was a common practice.

The communist government, like the earlier leaders of Manchukuo (see Chapter 5 in this volume, by Norman Smith), wished to see the winter months productive; hydraulic projects and lumbering conducted at this time of year were seen as positive signs of progress under the CCP. Large numbers of exiles were mobilized for such winter enterprises. Survivors of the Heilongjiang banishment have left many records that detail the building of dams to make reservoirs, the digging of relief channels to discharge water from marshes in preparation for the summer months, and winter lumbering.[64] Yin Yi could not forget the months that he and his fellow Rightists worked to break the frozen ground with pickaxes and shovels in temperatures of –30°C, the biting wind piercing through their cotton-padded clothes, right to their bones.[65] Work efficiency was high, ironically, because people exposed in the wilderness had to work fast: it was too cold to be slow or to stay idle. Only when blizzards raged and outdoor work was impossible might the labourers be kept in their barracks and allowed to take a break. Getting lost in a blizzard or fleeing the camp in the winter meant certain death. In the Hailun labour farm, three of Huang Zhan's campmates died in an unpredicted snowstorm in the winter of 1956 after getting lost when returning from fieldwork.[66]

Other environmental factors also challenged the constitution of the exiles. In the 1950s and 1960s, most of the labour camps and army farms of Heilongjiang were established in the plains north of Harbin and Yilehuli Mountain, in the Three River Plain (an alluvial plain formed by the Amur, Ussuri, and Sungari rivers), and in the Mudan River Plain. Often collectively called *Beidahuang* (the Great Northern Wilderness), these areas were still virgin land, bleak and desolate.[67] A large part of the area was a wide stretch of grassy marshland and swamp. During the rainy season (roughly from May to August), the region could be waterlogged, when thousands of square kilometres flooded. (In the next chapter, Sun Xiaoping provides vivid descriptions of this watery world that hundreds of thousands of forced labourers were expected to alter with their bare hands.) In the early spring, Chen Fengxiao recalls, the Xingkaihu convicts had to stand in icy mud, digging trenches to drain swampland and to level the ground, with hardly any protective equipment.[68]

Such a harsh environment inhibited any plans for escape. In the early stages of land reclamation, the marshes consisted of thriving *wula* grasses, which would entomb anyone who got entangled in them during the wet season. Even if one could manage to get out of the guarded camp, the dangers of the swampland made escape virtually impossible. Peter Liu recalls that a Xingkaihu inmate had managed to flee from the encampment after careful preparation. Armed with a map, a compass, and food he had stolen from a camp kitchen, he marched into the depth of ripened *wula* grass before an autumn dawn, slowly wading through ankle-deep water in what he thought was the right direction. He walked for the whole day, only to find himself trapped in neck-deep water and silt. In a panic, he shouted for help. A peasant collecting wild-duck eggs in a rowboat rescued him and returned him to the camp.[69] Selecting such formidable sites for labour camps and exile settlements suggests the state's intentions, both to alter the harsh landscapes of the natural world and to prevent the flight of exiles.

Mosquitoes were another vexing problem. The insects bred in swamps and grassy wetlands, and were larger than other breeds elsewhere in Manchuria; they tortured everyone in the Heilongjiang borderlands. One exile recalled: "Once wakened from grass, these extremely vicious creatures would swarm those of us who worked in the field, and would try to get into our noses, eyes, ears and mouths … They were hard to repel when we were working. People would get mad in this situation."[70] Exposed skin led to instant bites. They had no insect-proof suits, so the exiles in Xingkaihu labour camp covered their entire faces except for their eyes, tied their cuffs and the bottom of their trouser legs, and worked without stop – trying to hold the mosquitoes at bay. For many, the mosquitoes were one of the worst aspects of exile.[71] Mosquitoes were even used as a form of physical abuse in the labour camps. In Xingkaihu, a group of inmates wanted to punish an unpopular fellow inmate. They stripped him, tied him up, and left him outdoors for the night, thinking that a night with the mosquitoes would teach him how to behave. The man howled in agony at first and then was silent. The next morning, his persecutors went out to release him, only to find that his body was swollen beyond recognition, he was having difficulty breathing, and he was covered with feasting mosquitoes and gadflies.[72]

Most of the large wild animals noted by Qing exiles were hardly mentioned by their counterparts in the PRC's Heilongjiang – except for wolves. Wolves posed a threat not to those congregating in settlements or working outside in groups but to individuals who travelled alone in the wilderness. Huang Zhan, who had been a cartographer before he was banished, was once tracked by three wolves one late afternoon. He was lucky enough to find a telegraph pole to climb. After a lengthy stalemate, the wolves were diverted by an escaping convict, who was quickly hunted down and devoured. A group of camp guards pursuing the escapee arrived on the scene in time to save Huang, if not the hapless convict.[73] Others were not so lucky. At Farm 853, a female Rightist mentally broke down after her young son was killed by a pack of wolves while he was travelling by himself to a nearby market town.[74]

Aside from hardship due to physical and environmental elements, food shortages in the famine years and mass deaths due to starvation and disease in the exile settlements were common experiences for the Heilongjiang exiles. In principle, food supplies were guaranteed for the exiled labourers in Heilongjiang, not only because they worked in rich agricultural areas, which usually saw bumper harvests, but also because the labour farm authorities had to secure the physical strength of the exiles for heavy workloads. "Upon arrival at the farm," recalls Liu Meng of Farm 850, "Rightist laborers were allowed to eat as much as they could because the farm needed strong land reclamation workers," although some complained about unpleasant tastes and coarse food such as corn gruel. Xingkaihu inmate Wu Ningkun also recalls, "The standard staple was still corn flour here, but of edible quality. On a day of rest or a satellite-launching day, we would be fed rice or steamed bread made of wheat flour, along with vegetables that we had grown and fish we had caught. Later on, we raised pigs, and there was even a little pork."[75]

An unforeseen misfortune for the Heilongjiang exiles was the fact that their banishment coincided with three years of bad famine from 1959 to 1962; the labour camps and farms were obligated to submit large quantities of grain to the national granaries. Food supplies for the farm labourers decreased dramatically as a result. Starting from the winter of 1959, for instance, monthly grain supplies for each labourer dropped to 40 *jin* (20 kg) at Farm 850, 30 *jin* (15 kg) at Xingkaihu, and 21 *jin* (10.5 kg) at Farm Hailun; during 1960, the worst famine year, a Rightist at Farm 853 received only 8 *jin* (4 kg) of coarse food per month.[76] Such stringent monthly rations (also subject to the embezzlement of canteen and kitchen staff and the manipulation of labour farm cadres) were provided primarily in the form of hardtack corn buns supplemented with watery vegetable soup, without a trace of dietary oil or protein. Such a meal could hardly have alleviated the hunger pangs of those forced to perform heavy manual labour, and was far from adequate for subsistence needs.

Vast stretches of the Heilongjiang borderlands provided a variety of wild foods for the famished exiles – edible herbs, wild fruits, plant roots, mushrooms, bird eggs, and algae – before land reclamation had a significant impact on the environment. Wild game included pheasants, hares, snakes, and field rats. These wild foods, together with vegetables and soya beans left unharvested in the fields or damaged by early frost, provided opportunities for the exiles to help themselves.[77] Recollections of survivors suggest that whenever they had the chance, exiles would scour the fields for anything edible, even searching rat burrows for caches of grain stored by the rodents underground. In Farm 850, Wu Yongliang and his fellow Rightists were sent to dig rat holes for soya beans, and their six days' work produced more than ten bags of soya beans – an important addition to their farm's food supply. In Farm Hailun, Huang Zhan and his friends put hundreds of field rats out of their homes by digging up their holes and seizing 300 *jin* (150 kg) of corn.[78]

Small animals also became the targets of food hunts. Sources show that Heilongjiang exiles focused particularly on field rats and snakes, which were plentiful and easy to catch. They dug the ground to catch field rats, together with the grain they had stored; for many, rodents became a normal dietary supplement and an important source of nutrition.[79] Dai Huang describes how he and his fellow exiles devoured captured voles in hunger: "After work that day, we began to prepare these voles … We put water in a washbasin, mixed these voles with salt and chili powder, and boiled them. When they were cooked, I devoured them very fast … When we could not find voles, we started to pick up raw corn to eat."[80] Camp cadres sometimes frowned on the search for rodents, however. Yin Yi recalls that one of his fellow exiles caught rats to cook but was condemned by farm cadres as "trying to discredit socialism."[81]

Commanded by their empty stomachs, the exiles also tried "to entice snakes out of their lairs" in order to add them to their diet. Tang Zunwen, a former opera performer, was popular among his fellow Rightists in Farm 853 for his technique in capturing snakes: "At times in that summer he went out to the riverside and came back with a couple of snakes. He took out their poison gland (in the case of venomous snakes), washed them, cut them into pieces, and cooked them in an iron barrel with a bit of salt to make a tasty snake soup ... Even those who hated snake could not help but come to try."[82] During the years of famine, when meat and cooking oil were scarce, wild animals became precious sources of protein for the exile community.[83] However, some labour farm cadres did not allow the exiles to leave the barracks to collect wild food. According to Yin Yi and his campmates, an important reason for this was the cadres' need to maintain discipline and avoid showing leniency to the exiles (generally considered to be "opponents to the Party and socialism") as a form of self-protection in the post-1957, anti-Rightist political climate.[84]

With the forced labour system and adverse managerial policies in place, wild game turned out to be only a minor and temporary relief from hunger. Beginning in 1959, hunger overwhelmed the majority of exiles, eventually leading to a massive number of deaths due to starvation. Chen Fengxiao recalls that of the seventy-five convicts sent to his labour team at Xingkaihu in April 1959, only twenty-nine were still alive in January 1967; the majority had died during 1960 and 1961. Huang Zhan recalls that of about fifteen hundred inmates in his branch farm of Hailun, 193 died in the winter and spring of 1960–61.[85] According to Yang Congdao, who worked as a secretary at Farm 850, from May to November 1960, twenty-nine Rightists in Farm 850 died of starvation and related diseases, such as edema and tuberculosis, as a result of prolonged malnutrition.[86]

Though the figures provided by the survivors of exile may be distorted by individual memory over time, they are valuable as anecdotal evidence; official records and statistics for Heilongjiang labour camps are lacking or largely evasive.[87] Overall, it is difficult to estimate the total death toll of exiles in the famine years: we do not have even anecdotal evidence from most of the labour camps, official records (if they exist at all) are incomplete, and labour camp and farm authorities have never formally provided death figures. Accurate death tolls will probably never be available to the public. Regardless, it is not presumptuous to conclude that the severe starvation and high death tolls in the PRC labour camps and exile settlements stand in sharp contrast to the experiences of exiles in the Qing era.

A final notable characteristic of the PRC exile is the mandated ideological "remoulding" that the exiles underwent. Imperial China often resorted to Confucian-based moral teaching, providing individuals with moral examples to follow in order to "civilize the masses,"[88] and convicts were expected to repent of their offences and appreciate the benevolence of the throne, but "thought reform" as it was practised during the PRC exile was unheard of during the Qing exile. The imperial authorities had neither the intention nor the resources to implement it. It was the exiled scholars who came to "Sinicize" the local Manchus, as discussed above.

The experience of the PRC exiles was very different. On taking over China, CCP leaders were confident that they could "remould" all Chinese persons into "socialist new men" with Marxism-Leninism-Mao Zedong thought. They also believed they could successfully reconfigure the subjective world of the convicts (while at the same time having them "change the natural world"), transforming the exiles into pliant tools of the Party.[89] Ideological re-education, which went hand in hand with forced labour, became an important regimen, at least in principle, for convict rehabilitation. Consequently, exiles in Heilongjiang underwent considerable ideological remoulding, though variations in this program existed from camp to camp.

The most common method of thought reform was the political study session. In the labour reform camps of Xingkaihu, political study sessions were well organized. *Renmin ribao* (*The People's Daily*) and *Xuexi tongxun* (*Study Report*), an official newspaper, was compulsory reading, as was the *Renmin ribao* (*The People's Daily*) and *Xuexi tongxun* (*Study Report*), an internal publication for inmates. After a day of strenuous labour, inmates had to attend the study session for one or two hours. During the session, publications would be read by the camp cadre, or by a literate inmate. Every inmate was required to say something, to repeat or to rephrase materials from the newspapers, to praise the CCP's policies on certain issues, or to eulogize economic achievements.[90] Inmates were also required to conduct a self-examination (of their daily thoughts and work productivity), and to expose those of their fellow inmates whom they felt had performed inadequately. After study sessions, those who had said something inappropriate were ordered to write "thought examination" (*sixiang jiancha*) as a form of self-critical reparation, and to swear to make improvements.[91]

In addition to such "thought work," labour camp cadres also attempted to cultivate the "crime consciousness" of the inmates, using *xunhua* (rebuking) as admonishment. They reminded inmates of their "crimes" and of the CCP's "benevolence." Peter Liu recalls that one camp cadre preferred to "remind us that we Bourgeois Rightists were guilty of very grave offense against the Party, the state and the people, that we were here to receive a penalty far lighter than we deserved, and that we ought to be very grateful and do our very best to reform ourselves."[92] Chen Fengxiao recalls that his political instructor reminded him earnestly that the CCP had given him an opportunity to receive a school education but he had ignored this kindness and committed serious political crimes.[93] The purpose of these lectures was to generate inmates' gratitude to the Party, and make them satisfied and content with the situation they found themselves in, as well as scare them into complying with Party rules.

In some labour farms, such political lectures were predominantly *xunhua*. For instance, in Farm 850, whenever cadres found that the Rightists had failed to fulfill designated work quotas, or when they perceived a less-than-positive work attitude among inmates, *xunhua* would be meted out, likely followed by the denial of food or the imposition of extra work hours. The language used was insulting, including phrases such as *cunzhu* (stupid pigs), *laomianyang* (old sheep), and *tufei* (bandits).[94] In other situations, labour farm cadres used *xunhua* as a regular means of pressuring the Rightists. Topics would include the Rightists' "bourgeois taste," their ignorance of agriculture, and their lack of physical strength for work. The Rightists were reminded constantly of their inferiority to "the working people" and their need to engage in long-term labour reform.[95] In this way, political lectures were an effective way of stripping exiles of their self-esteem and convincing them that they would never receive any respect from labour farm officials.

Notably, not all the local authorities and camp cadres shared the same enthusiasm for ideological remoulding. Some Rightists found that as long as work assignments were completed, farm cadres would be satisfied; many officials were not rigourous about arranging political study sessions or providing ideological re-education. At times, the Rightists were asked to submit "thought examinations" that addressed their ideological change through political study: certain forms needed to be filled out by the Rightists to demonstrate their self-appraisal and ideological growth. But, ironically, these materials, which the Rightists had taken pains to compose, were often used by the cadres as wrappers or toilet paper.[96] Over time, more and more exiles came to realize that the labour farm authorities were actually more interested in using them for forced labour rather than attending to their "ideological renewal." As Yin Yi puts it, Rightists in these camps realized that the goal of "thought reform" was nothing more than a camouflage, and that punishment through labour was in fact the whole point of exile.[97] While ideological remoulding was an important aspect of the experience of many exiles, including those in Heilongjiang borderlands, remoulding practices were not uniformly implemented, and "educators" were not necessarily uniformly interested in reprogramming the minds of their labourers. It is likely that not many exiles were truly transformed into the "new socialist men," although many may have learned the rhetoric in their political study sessions.

## Conclusion

Both the imperial rulers and the communist government of China used exile to the northeastern borderlands as a form of punishment for both political and criminal offenders. The region's geographical remoteness from China proper and its physical and environmental challenges, as well as the state's need to colonize the frontier, were among the reasons for implementing banishment in this way. In both the Qing and the PRC, the exiled struggled to adapt to their new physical environment while playing an important role in land reclamation, landscape transformation, and economic development.

Yet there were significant differences between the Qing and the PRC in terms of how the state implemented banishment and the actual lived experience of the exiles. Strict physical control of the exiles, as well as an emphasis on indoctrination and ideological remoulding, were characteristic of PRC practice. These elements were generally lacking in the Qing, except in its treatment of the small number of slaves (who were physically restricted but not indoctrinated). Modern totalitarianism, with its tight physical control and ideological indoctrination of state subjects, is a twentieth-century development. In contrast, the rulers of the Manchus, an ethnic minority in China, had not formulated clear and effective measures at this early stage of their rule to supervise its offenders in exile. They were unprepared to deal with the sophistication of (often intellectual) exiles, who were overwhelmingly from the Han majority. Locals at exile settlements also tended to be of an easygoing nature, displaying generosity and even admiration for the well-educated and resourceful "guests" now resident in their homeland. These circumstances contrasted sharply with those during the communist era. The cadres of the PRC labour camps, imbued with a "class consciousness," were eager to show their commitment to the communist state.

During the Qing, there was ample opportunity for exiles to manipulate their status. They worked as teachers, traders, medical doctors, even secretaries to officials. With their involvement in the education of locals, their commercial success, and their contributions to charity, Qing exiles not only facilitated the spread of Chinese culture in Manchuria and advanced the region's economic development but also increased their own prestige, winning respect from local communities. They were considered the living embodiment of Chinese-gentry values. Far from experiencing physical or intellectual deprivation, many Qing exiles commanded informal power on the frontier. In contrast, PRC exiles in Heilongjiang were, for the most part, forced into manual labour and stripped of their social status. Although many were intellectuals, they had little chance to apply their expertise or exert social influence as their imperial predecessors once had. Considered opponents of the state, or "socially undesirable," they suffered various abuses as part of their "legitimate" punishment, and received little or no respect from the camp cadres. During severe famine, they suffered starvation and a high death toll. Thus, while exile to Manchuria as a form of punishment was constant in both the Qing and the early PRC state, the experience of the exiles was not.

## Notes

1 Joanna Waley-Cohen, *Exile in Mid-Qing China: Banishment to Xinjiang, 1758–1820* (New Haven, CT: Yale University Press, 1991), 53–54.
2 For a detailed discussion of exile to Xinjiang, see ibid.
3 Xie Guozhen, *Qingchu liuren kaifa dongbei shi* [A history of colonization of the Northeast by exiles in the early Qing] (Taibei: Kaiming shudian, 1969), 3–5; Yang Bin, *Liubian* [Notes from the Willow Palisades], in *Xuxiu siku quanshu,* vol. 731 (Shanghai: Shanghai guji chubanshe, 2002), 247–48.
4 Thomas Bernstein, *Up to the Mountains and Down to the Villages: The Transfer of Youth from Urban to Rural China* (New Haven, CT: Yale University Press, 1977), 26.
5 Robert Lee, *The Manchurian Frontier in Ching History* (Cambridge, MA: Harvard University Press, 1970), 78.
6 In 1709, for example, fifty-nine of those implicated in the case of "Prince Zhu San" were banished in a group to Ninguta. Xie, *Qingchu liuren kaifa dongbei shi,* 79.
7 Lee, *The Manchurian Frontier in Ching History,* 80.
8 Xi Qing, *Heilongjiang waiji* [An alternative record of Heilongjiang] (Taibei: Wenhai chubanshe, 1967), 165.
9 Ibid., 23, 29; Wu Zhenchen, *Ninguta jilüe* [Notes on Ninguta], in *Xuxiu siku quanshu,* 731: 606.
10 Xie, *Qingchu liuren kaifa dongbei shi,* 9.
11 Ibid., 12–13.
12 Lee, *The Manchurian Frontier in Ching History,* 79; Waley-Cohen, *Exile in Mid-Qing China,* 57; Xi, *Heilongjiang waiji,* 165–66.
13 Waley-Cohen, *Exile in Mid-Qing China,* 57.
14 Xi, *Heilongjiang waiji,* 165–66; Xie, *Qingchu liuren kaifa dongbei shi,* 84.

15 Xie, *Qingchu liuren kaifa dongbei shi,* 100. A similar case included banishing the descendants of Lü Liuliang.
16 Ibid., 37, 74–75.
17 Ibid., 41; Xi, *Heilongjiang waiji,* 85.
18 Xie, *Qingchu liuren kaifa dongbei shi,* 45.
19 Lee, *The Manchurian Frontier in Ching History,* 80, citing *Jilin tongzhi,* 1: 8.
20 Ibid., 80.
21 Ibid., 81, citing Fang, *Jueyu jilüe* [Notes on a remote land]. *Xiao fang hu zhai yu di cong chao* (The little square vase studio geographical series) (N.p.: 1894).
22 Xi, *Heilongjiang waiji,* 220.
23 Lee, *The Manchurian Frontier in Ching History,* 80; Xi, *Heilongjiang waiji,* 228–29.
24 Xie, *Qingchu liuren kaifa dongbei shi,* 24–25; Wu, *Ninguta jilüe,* 606; Yang, *Liubian jilüe,* 430. Wu Zhaoqian's release in amnesty in 1680 was also mainly due to the collective donation made by his friends and patrons on his behalf. Xie, *ibid.,* 29.
25 Wu, *Ninguta jilüe,* 606–7.
26 Xi, *Heilongjiang waiji,* 168.
27 Xie, *Qingchu liuren kaifa dongbei shi,* 45; Xi, *Heilongjiang waiji,* 185–86.
28 Xi, *Heilongjiang waiji,* 168–69.
29 Waley-Cohen, *Exile in Mid-Qing China,* 58.
30 Xie, *Qingchu liuren kaifa dongbei shi,* 3; Xi, *Heilongjiang waiji,* 166; Waley-Cohen, *Exile in Mid-Qing China,* 224.
31 Yang, *Liubian jilüe,* 468
32 Wu, *Ninguta jilüe,* 609.
33 Xie, *Qingchu liuren kaifa dongbei shi,* 100; Yang, *Liubian jilüe,* 468.
34 Yang, *Liubian jilüe,* 468.
35 Xi, *Heilongjiang waiji,* 198.
36 Xie, *Qingchu liuren kaifa dongbei shi,* 77–78; Waley-Cohen, *Exile in Mid-Qing China,* 224–25.
37 Yang, *Liubian jilüe,* 418–19.
38 Xie, *Qingchu liuren kaifa dongbei shi,* 88.
39 Yang, *Liubian jilüe,* 420; Lee, *The Manchurian Frontier in Ching History,* 82.
40 Xie, *Qingchu liuren kaifa dongbei shi,* 45–46; Yang, *Liubian jilüe,* 420.
41 Lee, *The Manchurian Frontier in Ching History,* 82.
42 Xi, *Heilongjiang waiji,* 222; Wu, *Ninguta jilüe,* 606.
43 Yang, *Liubian jilüe,* 421.
44 Xie, *Qingchu liuren kaifa dongbei shi,* 90.
45 Lee, *The Manchurian Frontier in Ching History,* 82–83.
46 Yang, *Liubian jilüe,* 421.
47 When Wu Zhaoqian was about to return to Beijing, governor Bahai arranged an elaborate escort service all the way to Shenyang. Wu, *Ninguta jilüe,* 612.
48 Xi, *Heilongjiang waiji,* 221.
49 Xie, *Qingchu liuren kaifa dongbei shi,* 78.
50 Ibid., 86–87; Yang, *Liubian jilüe,* 467.
51 Xi, *Heilongjiang waiji,* 229.
52 Wu, *Ninguta jilüe,* 605, 610.
53 Xi, *Heilongjiang waiji,* 230. The Han residents also taught the locals techniques for raising bees and collecting honey.

54 Waley-Cohen, *Exile in Mid-Qing China,* 23 (citing Fang, *Longsha jilüe,* 58).

55 *Heilongjiang shengzhi guoying nongchang zhi* [Gazetteer of Heilongjiang Province: The state farms volume] (Harbin: Heilongjiang renmin chubanshe, 1992), 6, 77.

56 See Mao's speeches, respectively, on October 9, October 13, and December 8, 1957, in *Mao Zedong sixiang wansui* (Long Live Mao Zedong's Thought) (A collection of important speeches of Mao Zedong) (N.p.: 1967).

57 Xingkaihu changshiban, *Xingkaihu nongchangshi* [A history of Xingkaihu Farm] (Mishan: n.p., 1988), 41; Chen Fengxiao, *Mengduan Weiminghu: Ershier nian laogai shengya jishi* [Broken dreams at Weiming Lake: True stories of my twenty-two years of Laogai experience] (Washington, DC: Laogai Research Foundation, 2005), 87; Bawuling nongchangshi bianxie bangongshi, *Bawuling nongchangshi* [A history of Farm 850] (Hulin, Heilongjiang: n.p., 1986), 30; Bawusan nongchangzhi bianshen weiyuanhui, *Bawusan nongchangzhi* [Gazetteer of Farm 853] (Beijing: n.p., 1986), 2–3; *Heilongjiang shengzhi guoying nongchang zhi,* 76.

58 Narratives of their labour can be found in Ni Genshan, *Chensi ji* [On contemplation] (Hong Kong: Tianma chuban youxian gongsi, 2005), 69; Dai Huang, *Jiusi yisheng: wo de youpai licheng* [A narrow escape from death: My experience as a rightist] (Beijing: Zhongyang bianyi chubanshe, 1998), 115–16.

59 Dai, *Jiusi yisheng: wo de youpai licheng,* 239–40.

60 Ibid., 120; Wu Yongliang, *Yuxue feifei: Beidahuang shenghuo jishi* [Floating rain and snow: True stories of life in the Great Northern Wilderness] (Beijing: Zhongguo xiju chubanshe, 2002), 7.

61 Dai, *Jiusi yisheng: wo de youpai licheng,* 120; Wu, *Yuxue feifei: Beidahuang shenghuo jishi,* 64; Chen, *Mengduan Weiminghu,* 92–93.

62 Peter Liu, *Mirror: A Loss of Innocence in Mao's China* (Philadelphia: Xlibris, 2001), 224.

63 Huang Zhan, *Yongyuan de Beidahuang* [The everlasting Beidahuang] (Washington, DC: Laogai Research Foundation, 2004), 66.

64 Chen, *Mengduan Weiminghu,* 88–89; Huang, *Yongyuan de Beidahuang,* 76–77; Yin Yi, *Huishou canyang yi han shan* [The setting of the sun over the mountain] (Beijing: Shiyue wenyi chubanshe, 2003), 46–47; Dai, *Jiusi yisheng,* 126.

65 Yin, *Huishou canyang yi han shan,* 47.

66 Huang, *Yongyuan de Beidahuang,* 69–70.

67 Mei Jimin, *Beidahuang* [The Great Northern Wilderness] (Taibei: Shuifurong chubanshe, 1975), 7.

68 Chen, *Mengduan Weiminghu,* 93.

69 Peter Liu, *Mirror,* 240.

70 Wang, "Menghui huangyuan," in *Lishi zai shenpan* [Retrial by history], ed. Liu Meng (Chengdu: Sichuan renmin chubanshe, 1996), 338.

71 Chen, *Mengduan Weiminghu,* 98. Many survivors of the Heilongjiang banishment see mosquitoes as among their worst experiences there. See Yin, *Huishou canyang yi han shan,* 33; Wu, *Yuxue feifei,* 60; Zheng Jiazhen, *Zhongguo dongbeijiao* [The northeast corner of China] (Harbin: Heilongjiang renmin chubanshe, 1998), 67.

72 Liu, *Mirror,* 239.

73 Huang, *Yongyuan de Beidahuang,* 64–65.

74 Huang Wu, *Mahuatang waiji* [Additional collection of Mahua Hall] (Guangzhou: Guangdong wenhua chubanshe, 1989), 32.

75 Liu Meng, interview, Beijing, October 4, 2002; Wu Ningkun, *A Single Tear: A Family's Persecution, Love and Endurance in Communist China* (New York: Atlantic Monthly Press, 1993), 96.

76 Dai, *Jiusi yisheng,* 157; Chen, *Mengduan Weiminghu,* 104; Huang, *Yongyuan de Beidahuang,* 87; *Bawusan nongchang zhi,* 388.

77 Yin, *Huishou canyang yi han shan,* 91–92; Liu Meng, *Chuntian de yu qiutian qing* [The whisking of rain in spring, the clearing skies in autumn] (Beijing: Zhongguo gongren chubanshe, 2003), 140–41.

78 Wu, *Yuxue feifei,* 112–13; Huang, *Yongyuan de Beidahuang,* 88.

79 There are many recollections by Dai Huang, Liu Meng, and Yin Yi about their experiences devouring field rats. Liu, *Chuntian de yu qiutian qing,* 141; Yin, *Huishou canyang yi han shan,* 75–76; Dai, *Jiusi yisheng,* 242.

80 Dai, *Jiusi yisheng,* 242.

81 Yin, *Huishou canyang yi han shan,* 76.

82 Huang, *Mahuatang waiji,* 25.

83 Famished exiles who gorged on wild species risked being poisoned. In *Xingkaihu nongchangshi,* the author admits that accidental death was frequent when inmates devoured poisonous wild herbs: *Xingkaihu nongcangshi,* 404. According to Chen Fengxiao, thirteen convicts in Xingkaihu Branch 4 died of poisonous wild herbs in the spring of 1960. Chen, *Mengduan Weiminghu,* 105.

84 Yin, *Huishou canyang yi han shan,* 93–94.

85 Chen, *Mengduan Weiminghu,* 111; Huang, *Yongyuan de Beidahuang,* 99.

86 Yang Congdao, "Bawuling Yunshan xumuchang liuren feizhengchang siwang mingdan" [The list of the abnormal deaths in Yunshan Branch of Farm 850] (Unpublished note, Nanjing). Huang Zhan gives an account of how he was entrusted to handle the dead in his labour farm in a cold winter: "A discarded seed house behind the horse stall was used to stock corpses. When I sent in Li Fucai, it was almost full. I had set up … the bodies into five rows, A, B, C, D and E, each of which had eight bodies piled together … Each of them had a number on the bottom of his shoes. Number A6 was now assigned to Li. In the cold temperature of minus 20–30 degree Celsius, corpses would never get putrescent … When families came to claim the bodies, two of my assistants would use iron hooks to pull them out. As the bodies got frozen solid, it was easy to get this done." Huang, *Yongyuan de Beidahuang,* 94.

87 For example, *Xingkaihu nongchangshi* documents in detail the natural calamities and difficulties that Xingkaihu confronted, in terms of the amount of crops destroyed, financial deficits the camp suffered, and even the death toll of pigs and chickens; however, the author mentions only in passing that ten persons died of eating poisonous herbs (*Xingkaihu nongchangshi,* 403–4).The deaths of many inmates were apparently omitted from records.

88 See Philip Williams and Yenna Wu, *The Great Wall of Confinement: The Chinese Prison Camp through Contemporary Fiction and Reportage* (Berkeley: University of California Press, 2004), 8–9, for detailed discussion of the imperial regimen of "civilizing the masses."

89 For instance, Mao Zedong declared in June 1949 that the CCP would "remould" Chinese reactionaries through "forced labour" and "propaganda" into "new men"; Mao Zedong, *Mao Zedong xuanji* [Selected works of Mao Zedong], vol. 4 (Beijing: Renmin chubanshe, 1967), 1476–77.

90 Chen, *Mengduan Weiminghu,* 110, 118, 125.

91 Ibid., 125. According to Wu Yongliang, newspaper reading and political talks given by demobilized soldiers were the main forms of political study in his labour farm,

but the Rightists were not required to "examine" or criticize themselves. Wu, *Yuxue feifei*, 70–72.

92 Liu, *Mirror*, 213.

93 Chen, *Mengduan Weiminghu*, 94–95.

94 Dai, *Jiusi yisheng*, 131; Yin, *Huishou canyang yi han shan*, 51; Wu, *Yuxue feifei*, 66.

95 Li Hui, *Renzai xuanwo: Huang Miaozi yu Yu Feng* [People in the eddy: Huang Miaozi and Yu Feng] (Jinan: Shandong huabao chubanshe, 1998), 305; Wu, *Yuxue feifei*, 66–67.

96 Wu, *Yuxue feifei*, 97–98.

97 Yin, *Huishou canyang yi han shan*, 63.

# 10

# "War against the Earth": Military Farming in Communist Manchuria, 1949–75

*Sun Xiaoping*

*Extraordinary PLA officers and soldiers, heroes!*

*You are the elite of the six hundred million people.*

*Under the leadership of the Party and the guidance of Chairman Mao,*

*You have smashed the alliance of imperialism, feudalism and bureaucratism.*

*[ . . . ]*

*Now many of you comrades are putting down the armor and returning to the land.*

*No, you are only changing the battleground to wage war against the Earth.*

*[ . . . ]*

*Immediately start the war against the Earth, comrades.*

*Even to the ends of the Earth, charge!*

– *"War against the Earth," Guo Moruo, 1958*

"War against the Earth" ("Xiang diqiu kaizhan") was written at the request of the first minister of the Ministry of Agricultural Reclamation (*Nongkenbu*), Wang Zhen (1908–93), and originally published in the *People's Daily* (*Renmin ribao*), the most widely read official newspaper, in April 1958.[1] Guo Moruo (1892–1978), a household name in modern Chinese literature but also a controversial figure who gained political patronage through expressing his loyalty to the Chinese Communist Party (CCP) in poetry, composed the poem in order to boost morale for a new wave of military farming in the recently established People's Republic of China (PRC). In three months of intensive mobilization from March to May of 1958, about 100,000 veterans of the People's Liberation Army (PLA) from central and eastern China were deployed in the "wilderness" of Heilongjiang, the northeastern border province, where army farms already dotted the landscape.[2] In fact, from February 1949, when the first veteran farm was established, to the end of 1975, when military leadership of the Heilongjiang Production and Construction Corps was withdrawn from state farms, over 140,000 discharged officers and soldiers from other parts of China were moved to the border region for farmland reclamation work.[3]

The land reclamation area in Heilongjiang is much better known by its nickname, *Beidahuang*, the Great Northern Wilderness. For many Chinese, the name evokes a nostalgia for the lost adolescence of the half-million urban youth who worked and lived on Beidahuang farms during the Cultural Revolution (1966–76). It also induces a collective lament for the sufferings of the hundreds of thousands of political exiles banished to Beidahuang farms from the 1950s to the 1970s. Indeed, as Wang Ning shows in Chapter 9 of this volume, Heilongjiang borderlands had served as an internal-banishment destination since imperial times. Even the Chinese characters that form the name "Beidahuang," in their conventional sense, convey the intimidating qualities of distance, cold, bleakness, and desolation.

In the 1950s, however, the reputation of Beidahuang was quite different. In fact, Beidahuang first gained its fame through the propaganda aimed at mobilizing veterans for land reclamation in border regions. Extensive demobilization after the civil war and the Korean War aggravated the predicament facing the new regime, which was recovering from the devastating impact of these conflicts and from unemployment and rampant food shortages. Under the state's "grain first" policy (*yi liang wei gang*), Beidahuang demonstrated an extraordinary capacity to absorb surplus labour and offered powerful promises to develop socialist agricultural modernity – eventually transforming Beidahuang into *Beidacang*, the Great Northern Granary.

The creation of the myth of Beidahuang was an essential component in the transition from putative wilderness to granary. The myth simultaneously highlighted Beidahuang's inexhaustible agricultural promise and its intimidating challenges that only "extraordinary heroes" could overcome. Rhetorically reversing the emasculating transition from military to agrarian service, this myth turned demobilized veterans into re-masculinized heroes. This hero narrative allowed ex-soldiers who had laid down their swords to take up the plough, turning farming tools into weapons in the battle to conquer the wilderness. Thus, through the making of Beidahuang, the state and veterans together turned the pressure of massive demobilization into the fanfare of war mobilization, something at which the CCP excelled. Although the remote region was already populated, the state's emphasis on the emptiness of Beidahuang further legitimized state-mobilized migration and accentuated the heroics of the intrepid explorer of *terra incognita*.

Warriors could conquer and occupy, but they could not settle Beidahuang's "emptiness" alone. While veterans fought to "dress up" the virgin land as a "new bride" (*xin jianiang*) in "the most fashionable attire" of "waves of golden wheat,"[4] thousands of young women arrived on army farms to domesticate the warriors of the wilderness in stable marriages. The women's less visible work and life on army farms complicated the masculine narrative of nature taming, making possible the long-term perpetuation of human domain over Beidahuang.

The "war against the Earth" was economically victorious but ecologically devastating. By the early 2000s, Heilongjiang had become the largest commercial grain production base in China, with the highest concentration of state farms in the country. Yang Shaopin, Director of the Bureau of State Farms and Land Reclamation under the Ministry of Agriculture, stressed the importance of Beidahuang to the nation: "In 2006, the reclamation area in Heilongjiang Province produced a total of 6.66 billion kg [of grain], which is enough to sustain the total population in the four municipalities of Beijing, Shanghai, Tianjin and Chongqing and all Chinese army soldiers for one year."[5] However, the irretrievable loss of wetlands to farmland has become the leading cause of soil erosion in recent years, has increased and intensified floods and droughts, and has resulted in disappearing wildlife habitats and decreased biodiversity.[6]

This chapter takes military farming in Beidahuang as an entry point to examine how the PRC looked to the border region for solutions to the predicaments facing the new regime. Certainly, the PRC was not the first to look to the borderlands for this purpose. Manchuria had served as the "life line" for the Japanese empire up until August 1945, soon before the CCP took over.[7] However, the few English-language studies of the region during the PRC era have focused mainly on political exiles and urban development.[8] Chinese-language research, although abundant, predictably follows the Party line, reinforcing the myth of Beidahuang and celebrating the glorious achievements under the CCP's leadership. This chapter unpacks the politics inherent in the discursive construction of Beidahuang to reveal the strategies and impacts of the multi-layered transitions – from war among humans to war against non-humans, from warrior to farm worker, and from "empty" wilderness to populated farmland.

## Creating the Myth of Beidahuang

Oh, this piece of land, it has always been in deep, sound sleep!
No human voices for hundreds of miles, only the roar of wolves,
bears and tigers.
Oh, this piece of land, it has always been in deep, wild grass!
No human figures in dozens of days, only the blue sky, green water and the
red sun round like a wheel.
[ ... ]
... Inherit it, our offspring for generations to come!
It is everlasting property – renewable for tens of thousands of years.
... Keep cultivating it, masters of the future world!
It is miraculous land – ever to be found on Earth or in heaven.
– "Engraved onto the Land of Beidahuang," Guo Xiaochuan, 1963[9]

These verses by prominent poet Guo Xiaochuan (1919–76) epitomize the creation of the myth of Beidahuang in the 1950s and 60s, painting the region as vast, unoccupied land with boundless resources, patiently waiting for communist heroes to conquer it and for their descendants to inherit it. The name "Beidahuang" refers specifically to the Songnen and Sanjiang Plains of Heilongjiang. In the late 1940s, the dominant landscape of Beidahuang was of wetlands, forests, and meadows. The Sanjiang Plain in particular, extending to the easternmost corner of the country, is often characterized in Beidahuang literature as *gen'gu huangyuan,* an "eternal wasteland" that has been populated and cultivated only since the late 1940s, when communist-led land reclamation began.

Historical records, however, tell a very different story. Recent archaeological findings have confirmed human activity in the area 175,000 years ago.[10] Between 300 BCE and 400 CE, for example, about 100,000 people lived on the Sanjiang Plain. The sites of more than 750 early cities and villages have been identified – roughly the same number that exist today.[11] Chinese historians usually attribute the lack of development of the region in the intervening centuries to the policies of the Qing Dynasty (1644–1912), which forbade Han Chinese from settling in the Manchu homeland. Although Han migration never completely stopped during the ban, the lifting of the ban in 1857 ushered in an era of unprecedented migration. Millions of Han Chinese from densely populated provinces such as Shandong and Hebei flooded into Manchuria. However, as James Reardon-Anderson's book title suggests, Han migrants to Manchuria in the late nineteenth and early twentieth centuries were "reluctant pioneers" who were "rarely driven by a quest for fortune or adventure or by a religious, political, or ideological calling."[12] Instead, their agenda was to serve the families at home, their mission to Manchuria being "to go, to earn, and to return."[13] While southern and central Manchuria were gradually settled by Han agricultural migrants, northern Manchuria remained a frontier of "few people, much land" (*renshao diduo*). According to Reardon-Anderson, in the 1930s, fully 68 percent of the arable land in Heilongjiang remained uncultivated, in contrast to neighbouring provinces, where this percentage was significantly lower.[14]

During the existence of the Japanese state of Manchukuo (1932–45), the northern frontier of Manchuria was developed as an agricultural base for the Japanese empire. According to Ronald Suleski, at least 270,000 Japanese agricultural colonists were living in Manchuria by 1945, and more than 86,000 Japanese farm boys, recruited through the Manchurian Youth Corps, served in Manchuria from 1934 to 1945.[15] Most of the colonists and Youth Corps camps established sites in the northeastern section of Manchukuo, namely, the Sanjiang Plain.[16] As Suleski shows, although the 1939 "Outline of the Basic Policy for the Colonization of Manchuria" required the colonists and Youth Corps to seek unclaimed but fertile land, the Japanese often took top-quality farmlands from Chinese owners by forcing them to sell their land at a price well below its market value. Nevertheless, despite the colonial propaganda that depicted happy Japanese farmers enjoying their new life in Manchuria, as Annika Culver demonstrates in Chapter 6 of this volume, many Japanese colonists could not withstand the harsh environment and deserted their land to venture into the cities. The eventual exodus of Japanese settlers after Japan's defeat and the ensuing civil war in the region left most of the land fallow. By 1949, only 7.22 percent of the Sanjiang Plain was cultivated; forty-five years later, in 1994, farmland had already become the dominant landscape, covering 42 percent of the area.[17]

The achievement of transforming a "wasteland" (*huangdi*) into a state granary attests to the remarkable human effort in the war against the

Earth. Within the three decades spanning 1949 to 1978, millions of people were moved from China south of the Great Wall to Beidahuang, including tens of thousands of peasants, veterans, convicts, Rightists, and both rural and urban youth. Getting people to the wilderness was one thing; keeping them there was quite another. The vast majority of Rightists and urban youth returned to the cities in the late 1970s. According to Thomas Gottschang and Diana Lary, in the 1950s the return rate of Shandong rural migrants was as high as 50 percent:[18] "even revolutionary hyperbole could not make [Beidahuang] more attractive than its god-forsaken name."[19]

In fact, the official adoption of the "god-forsaken name" Beidahuang is a relatively recent development. Some Chinese historians trace the origin of the term back to the *Shanhaijing* (*Classic of Mountains and Seas*), a fourth-century BCE text of geography and mythology, suggesting that the term has been commonly used since then. It is hard to imagine, however, that the politically Sinicized Manchu emperors would use this term to characterize their homeland as an uncivilized wilderness, and in this way negate their efforts to legitimize their rule over the Han Chinese. Similarly, Japanese propaganda avoided using the term; instead, it portrayed Manchukuo as a utopian paradise to entice agricultural settlers, as Annika Culver argues in Chapter 6 of this volume. "Beidahuang" itself indicates a Sino-centric point of view, wherein the vast wilderness of Chinese civilization's northern periphery is uncultivated and uncivilized and therefore in need of Chinese settlement to cultivate and civilize it. The idea of a northern "wilderness" itself legitimized communist intrusion into the region.

Communist propaganda about Beidahuang emphasized the region's abundant resources in order to attract migrants from heavily populated regions, as evidenced in Guo Xiaochuan's poem above. The most commonly mentioned advantage of farming in Beidahuang was its long daylight hours in summer compared with those in the south. However, unlike the typical booster literature of settler migration schemes in the boreal wilderness of early-twentieth-century Canada,[20] official Chinese literature stressed the hardship of the natural environment in order to excite revolutionary enthusiasm: only heroes could conquer such a wild world. Though Chinese officials certainly knew of the infrastructure left behind by the Japanese Manchurian Youth Corps, and even directed veterans to take advantage of the remaining tools, machines, and buildings – even the interned Japanese mechanics[21] – for obvious ideological reasons, the Japanese colonial legacy is rarely mentioned in Chinese literature.

The Japanese settlement was referenced only to highlight what ruins the communists had to reconstruct, not the foundations on which they could build. Even now, years after the successful transformation of the region into productive farmland, the Chinese discourse about "god-forsaken" Beidahuang retains its hold on the popular imagination. Art genres such as Beidahuang literature (*Beidahuang wenxue*) and Beidahuang woodblock painting (*Beidahuang banhua*) emerged from the military farming culture and flourished into the twenty-first century. Beidahuang has been written into the identity of the region's settlers, especially the veterans who established army farms in the 1950s and worked on the farms their entire lives. Enthusiastic or reluctant, these veterans carry on the political glory imposed on them as the first generation of *Beidahuangren* (Beidahuang people) who overcame environmental hardship and set heroic examples for farmers and migrants in the establishment of a modern socialist agriculture.

### Making Heroes

Thank you Mr. Guo for your praise,
We are launching war against the Earth.
[ ... ]
Give our red hearts to the Party,
Heroes put down the armor but return to the battleground.
[ ... ]
The hand that used to hold the gun now picks up the hoe,
To force the land to surrender grain.
[ ... ]
Heroes put down their armors but never their guns.
– "Never Put Down the Gun," Xu Xianguo, 1958[22]

First published in the *People's Daily* on May 7, 1958, "Never Put Down the Gun" was lauded by Minister Wang Zhen as the "heartfelt voice of hundreds of thousands of people."[23] Xu Xianguo, who penned the poem, was a propaganda assistant (*xuanchuan zhuliyuan*) with the rank of second lieutenant in the Ideological Department (*Zhengzhibu*) of the Xinyang Infantry Academy in Henan. He was one of the hundred thousand military personnel mobilized and transported to Beidahuang in the spring of 1958 to establish military farms. As this section will illustrate, the making of heroes was not merely due to a top-down state mobilization; it also involved willing and eager rank-and-file participation. Relationships between state and individuals, on the one hand, and between central and local governments, on the other hand, were dynamically intertwined. In fact, the central demobilization strategy of systematically turning military divisions into production units emerged from local experiments in damage control to accommodate discharged veterans during and after the civil war, since the late 1940s. More importantly, production units organized along military lines were entrusted with the responsibility of modelling collective farming for peasants (*shifan nongmin*) and protecting border regions from external threats.

Ironically, the first veteran farms in Beidahuang resulted from the inability of the communist government to deal with the social and political crises caused by discontented ex-soldiers. Heilongjiang, the first area in the Northeast under CCP control, had provided crucial logistical support for the civil war since late 1946. While newly established local governments were preoccupied with enlistment and mobilization to reinforce the frontlines in other parts of China, tens of thousands of wounded and disabled PLA soldiers, who were granted the glorious title *rongjun* (honoured soldiers), flooded into Heilongjiang. Except for those who attended the newly established *rongjun* schools and who were placed in *rongjun* factories and construction teams, most received no placement at all. They wandered the streets, robbed local shops and residents, and even violently attacked local government offices. In the fall of 1948, some *rongjun* broke into the offices of the provincial government in Qiqihar, the provincial capital of Heilongjiang until 1954, and detained the chairman and vice chairman of the province, forcing the government to shut down for a few days.[24] Clearly, something had to be done.

The idea of establishing *rongjun* farms to remove the threat to urban order came from Hao Guangnong, a *rongjun* who had sustained permanent injury to one of his arms and lost an eye during the war before becoming the director of the Ideological Department of a *rongjun* school in Qiqihar. The provincial *Rongjun* Committee leader initially dismissed Hao's proposal in light of the challenges inherent in organizing disabled ex-soldiers for land clearance and farming. Undeterred, Hao gathered twenty-eight willing *rongjun* and, with a small fund from the school to purchase farming tools and seeds, established a farm in April 1949 in the unclaimed area of Dongping, 135 kilometres from Qiqihar.[25] Together they cleared 250 hectares and cultivated 100 hectares in two months.[26] Impressed by their achievements, the provincial *Rongjun* Committee moved to support their efforts and made farming a "primary solution" for *rongjun* placement in September 1949.[27] By April 1951, nine veteran farms had been established in Beidahuang, absorbing more than two thousand *rongjun* and fourteen thousand former Guomindang veterans.[28] While a few *rongjun* gained honour as distinguished labour models, the conversion of ex-soldiers into producers established a new hero narrative: veterans "in the past bled for the Revolution, in the present sweat for Construction" (*guoqu wei geiming liuxie, xianzai wei jianshe liuhan*), and "refused to rest on their laurels but continued to provide meritorious service to the people."[29]

Although veteran farms emerged as a local remedy for social crises, undeniably the evolution of military farming in Beidahuang had its roots in the tradition of self-sufficiency among communist troops. It had been a consistent policy for communist soldiers to carry multiple responsibilities in addition to combat, including production, construction, and ideological work. The famed 359 Brigade served as the ultimate exemplar of this ideal. During the War of Resistance against Japan (1937–45), the 359 Brigade led by Wang Zhen not only achieved self-sufficiency through farming, converting wasteland in Nanniwan into farmland, but also contributed its surplus to the survival of the communist base in Yan'an. Under the command of Liu Zhuanlian, the 359 Brigade liberated the first large city, Harbin, in April 1946. Once the civil war was over, the Central Military Commission (*Zhongyang junshi weiyuanhui*) issued an order in December 1949 to emphasize the double roles of the PLA in national defence and production. It stipulated that, starting from the spring of 1950, the PLA should engage in production and construction to "improve its standard of living and reduce costs to the state."[30] Mao Zedong (1893–1976) issued another order in February 1952 that required certain military divisions to "save weapons for combat and [instead] pick up the weapons for production and construction."[31] According to Yitzhak Shichor, by 1959, 7 million troops had been demobilized and 5 million of these had been resettled in the countryside, usually at their place of origin.[32] However, as Neil Diamant points out, there were also "uprisings" in rural areas such as in Shandong, led by returning veterans who lacked land and housing. Veteran protests in cities like Harbin and Dalian forced government offices to occasionally "close for business" in the 1950s.[33] The government began to move troops from densely populated areas to the border regions in order to protect and develop them. The 97th Division of the PLA became the Second Agricultural Construction Division (*Nongjian ershi*) and moved from Shandong to Beidahuang in September 1954. There it established three farms and named them after their unit codes: Farm 290, Farm 291, and Farm 289, which later amalgamated with a local farm to become Tieli (Iron Force) Farm.[34] As troops of the Second Agricultural Construction Division were originally from

Shandong, their relocation was also intended to encourage "necessary preparations for rural migration from their native places to the Northeast."[35] Although discussion of rural migration is beyond the scope of this chapter, it is important to note that military farms were intended not only to exemplify for local peasants the modern efficiency of collective farming but also to serve as exemplars of frontier settlements to entice other would-be migrants.

Further development of military farming in Beidahuang can be attributed to Wang Zhen, who gained fame first as the commander of the 359 Brigade in Nanniwan during the Great Production Drive in the early 1940s and then established the Production and Construction Corps for military farming in Xinjiang in the early 1950s. His renowned contribution to agricultural reclamation in Beidahuang began with his career as the chief commander of the Railway Soldiers (*Tiedaobing*).[36] He saw the potential of Beidahuang to accommodate demobilized troops when he visited the 5th Division of Railway Soldiers, which had returned from the Korean War and was engaged in railway construction in a forested area in Heilongjiang in 1954. Under his command, the first Railway Soldiers farm, Farm 850, was established in January 1955 in Hulin, close to the eastern border with Russia.[37] Encouraged by the success of Farm 850, Wang Zhen submitted a report to the Central Military Commission in August 1955, proposing to establish farming and livestock enterprises and reclaim farmland from the wilderness.[38] By 1957, ten farms in the Sanjiang Plain employed about twenty-three thousand Railway Soldiers veterans, most of them from densely populated provinces such as Sichuan, Shandong, and Henan.[39]

The intensive opening of Beidahuang in the early and mid-1950s, together with other border-region development programs, led to the creation of the Ministry of Agricultural Reclamation in May 1956, with Wang Zhen as its first minister. According to the former director of the Chinese Agricultural Reclamation Research Institute, Xu Renjun, in the 1950s and 1960s the Ministry of Agricultural Reclamation had the highest concentration of top-ranked military officers in the State Council, while the majority of farm workers in land reclamation areas along the Chinese border were PLA veterans.[40] This arrangement highlighted the strategic significance of military farming in veteran employment, frontier development, and border security.

The most glorified veteran migrants in Beidahuang were among the wave that arrived in 1958. These 100,000 veterans (including about 20,000 veteran families) migrated during unprecedented propaganda campaigns, exemplified by the poems of Guo Moruo and Xu Xianguo. Wang Zhen had requested that Guo boost enthusiasm for land reclamation with his poetic compositions, but the central government, too, restated the importance of military farming in March 1958.[41] At the local level, the response to mobilization for border-region development was mixed. For example, among the 480 Xinyang Infantry Academy cadets and officers mobilized to go to Beidahuang, more than 80 were discharged because of "political mistakes" (such as Rightist remarks, undesirable family origins, and inappropriate overseas connections).[42] Regardless of the "motley crew" (*zapaijun*) nature of the "volunteers," Xu, the propaganda assistant at the academy, was ordered to write a resolution statement to express the migrants' ambitions at a farewell gathering. Xu apparently outperformed his duties by writing an additional poem, "Never Put Down the Gun" ("Yong bu fangxia qiang"), which proved immensely effective in encouraging mobilization for land reclamation.

The story behind "Never Put Down the Gun" reveals the importance of propaganda in the opening of Beidahuang. The poem was published in the *People's Daily* in May 1958 and edited three weeks later by Minister Wang Zhen, who recommended that the poem be set to music so it could be sung. This reflects the CCP revolutionary tradition of using songs as effective propaganda tools. However, Wang Zhen was not satisfied with the original poem, which read:

> Let blood-stained uniforms
> be rewarded with engine oil and dirt.
> Let bullet-penetrated scars
> turn red and shine on the black soil.

In order to achieve a "more harmonious and more cheerful mood," Wang suggested that "blood-stained uniforms" (*xueji jinran de junzhuang*) be substituted for "glorious and victorious uniforms" (*guangrong shengli de junzhuang*), and that "bullet-penetrated scars" (*zidan chuantou de shangba*) be changed to "strong and powerful arms" (*jianqiang youli de bibang*).[43] By editing out the battle sacrifices of soldiers, Wang reinforced the hero narrative that was beginning to emerge as *rongjun* were placed on veteran farms and encouraged continued industriousness on the part of settlers. As Wang expected, the poem was sung in more than ten tunes in various parts of Beidahuang, and was used as lyrics for the theme song of a documentary film, *Heroes Conquered Beidahuang* (*Yingxiong zhansheng*

*Beidahuang*), which was shown throughout the country in 1959.[44] In September 1958, the revised and adapted poem appeared in supplementary Grade 9 and 10 textbooks in provinces such as Heilongjiang and Hubei.[45] Many young people responded to the propaganda campaign, writing to newspapers, military farms, and even Xu Xianguo himself to express their aspirations to join the land reclamation efforts.[46] By the time his poem was published, Xu was already working on a farm in Luobei, across the Amur River from the Soviet Union.

Veteran farm workers were quick to use official propaganda to their own advantage. Inspired by the power of Wang Zhen's letter, Xu's co-workers asked Xu to seize the "opportunity" and write to Wang in the hope that Wang would send the letter to the *People's Daily* so "people throughout the country would know their enthusiasm and good deeds."[47] Xu's letter was published by the paper on July 7, 1958, with the editorial comment that the letter represented "heroes with ambitions higher than the sky" (*zhiqi bi tiangao de yingxiong haohan*).[48] The campaign reached new heights in September 1958 with the publication of *War against the Earth – an Anthology of Beidahuang Poetry*. The volume included nearly two hundred poems by veterans working in Beidahuang. The book takes its main title from Guo Moruo's poem – a nod to veterans' "imposing manners" (*haomai qigai*) in the war against nature.[49]

While land reclamation and grain production remained the priority of military farming, Sino-Soviet border conflicts in the 1960s pushed the veteran farms in Beidahuang to the front lines of national defence as well. Following the precedent set by the Xinjiang Production and Construction Corps (*Shengchan jianshe bingtuan*), an armed production unit led by the PLA,[50] more than ten thousand demobilized officers were dispatched to the farms and local agricultural reclamation bureaus in Heilongjiang in 1966 in order to organize the Heilongjiang Production and Construction Corps. This project was disrupted briefly by the outbreak of the Cultural Revolution. In 1968, however, the project was resumed because border tensions between China and the Soviet Union had intensified and the Cold War was perceived by Chinese leaders as an international threat. On January 5, 1968, the Sino-Soviet skirmish over Qiliqin Island in the middle of the Amur killed four Chinese militiamen, who, together with more than three hundred Chinese troops armed only with Little Red Books of selected quotations from Mao and sticks, faced two Soviet platoons outfitted with machine guns.[51] By June 1968, about five thousand military officers were deployed through the Heilongjiang Production and Construction Corps to farms throughout the region, including those previously run by provincial and local governments.[52] From 1968 to 1971, there were twelve production and construction corps and three production and construction divisions established along the Chinese border, all with highly specialized production priorities (grain in Heilongjiang, cotton in Xinjiang, livestock in Inner Mongolia, rubber in Hainan and Yunnan, and sugar in Guangdong and Guangxi).[53] Given the border tensions along the Amur, the Heilongjiang Production and Construction Corps was the most heavily armed; twenty-two thousand farm workers were equipped with weapons and artillery, including 130 anti-tank canons and 320 bazookas.[54] Although the Heilongjiang Production and Construction Corps did not directly participate in combat over Zhenbao Island in March 1969, it was credited for its logistical support. However, ineffective farm management by military officers led to the cancellation of the Production and Construction Corps project, with the last corps in Heilongjiang decommissioned by the end of 1975.[55]

### Taking Root

On our army farm,
Arrived five city girls.
Wow, their well-tanned skin,
So healthy, so beautiful!
[ … ]
Casting off their school uniforms,
They select seeds and transplant seedlings.
[ … ]
All of them are actually *guniang,*
But they all say they are engaged.
When being asked whom they are engaged to,
They say:
"It is the gorgeous Beidahuang."

– "City Girls Fell in Love with Beidahuang," Chen Di, 1958[56]

The narrative of the Beidahuang settlement has been told as a masculine story. Not only did former military personnel regain their masculinity through their "war" against nature but they also performed this masculinity by taming the "wasteland" of Beidahuang and establishing farmland. The "betrothal" of young women to Beidahuang in Chen's poem reveals an even deeper layer of the politics of settling the "wilderness." The poem refers to the women as *guniang,* a term that suggests youth, virginity, and sexual availability. The women were recruited to join border-region development programs as equal participants (epitomized in the glorified Maoist adage "Half the Sky" [*banbiantian*]). They performed various tasks, from seed selection to road construction. The implied objective of this mobilization rhetoric was to bring young women to the farms so that they could become army brides. They were instrumental in taming the newly unleashed masculinity of single, young veterans within the parameters of conventional marriage, thus enabling the "bare sticks" (*guanggun*, unmarried men) to "take root" (*zhagen*) in Beidahuang and ensuring that the virgin land was fruitful – in terms of both agricultural production and human reproduction.

While economic production was one of the primary reasons for venturing into the wilderness, the state's ambition to populate the "no-man's land" reflected its desire to consolidate the frontier. Despite the massive migration program that moved millions of rural households into Beidahuang in the 1950s and 60s, military farms (usually at some physical distance from civilian villages) presented particular challenges. Although a few officers brought their families along, most veterans arrived as *dan'ganhu* (single-person households). Authorities perceived the disproportionate number of young male settlers on these farms to be a potential threat to local order; the move to attract young female settlers to the region was an attempt to establish greater social stability.

Wang Zhen, the mind and force behind military farming, was particularly keen to use the prospect of marriage as a way to encourage productivity and settlement. As early as 1950, when Wang was still commander of the PLA troops who liberated Xinjiang, he had already arranged to bring in a large number of women recruits from Hunan province, mostly "18 to 19 years old, educated, unmarried girls," to go to Xinjiang to become "good workers, good wives, and good mothers."[57] His recruitment efforts continued in Beidahuang. He was known to have encouraged ex-soldiers to get married in their hometowns and take their brides back to Beidahuang; he also arranged the recruitment of a great number of young women from Shandong and Sichuan to Beidahuang army farms.[58] Official communist discourse often paid particular homage to his matchmaking manoeuvres, honouring him as the "greatest marriage broker [*hongniang*] of the twentieth century."[59] Most noteworthy was the way he openly used young women as a lure for migration to Beidahuang. He told veterans on various occasions that young women would come once they had developed the frontier.[60] On April 12, 1958, during a period of intensive migration of one hundred thousand veterans to Beidahuang, Wang addressed newly arrived veterans at the train station in Mishan, the last major town before the borderlands. In his speech, he acknowledged that most of the young men were still *dan'ganhu* but reassured them that this was a "solvable" problem:

> Many girl graduates from high school and junior high wrote to us that they want to build Beidahuang. When they arrive, they are *dan'ganhu* too. In a couple of years, how could you not "collectivize" [*hezuohua*]? But, *guniang* loves heroes. They love labour models. If you want to find a good wife, you have to demonstrate your revolutionary motivation through great accomplishments in work.[61]

Without using the word "marriage," Wang turned the political tide of agricultural collectivization into a metaphor for living happily ever after. However, that promise of a happy ending came on the condition that veterans find a legitimate outlet for their renewed masculinity – in socialist agricultural production. This reinforced the hero-making discourse that underpinned the state settlement of the frontier.

Farm leaders took on similar paternalistic responsibility for veteran settlers. The arrival of young female recruits was usually advertised prominently at local theatres, often on big banners: "Beidahuang welcomes Sichuan *guniang*."[62] Women were sometimes strategically assigned to posts that would facilitate their contact with veterans. Minister Wang Zhen recommended "a couple of years" of romantic courtship (to encourage the free-choice marriage championed by the New Marriage Law), but local leaders often did not wait that long to exert their influence. Once relationships began to develop, farm leaders eagerly staged group weddings for new couples – sometimes within days of their acquaintance – in order to seal the "betrothal."[63] By officiating at such mass marriages, farm leaders effectively turned personal events into public sermons; newlyweds became powerful symbols of a promise fulfilled, and single veterans were further encouraged to follow suit and "take root" in the wilderness. This evidence of a "happily ever after" reportedly "stabilized veterans' moods" (*qingxu wending*) and reduced delinquent activities.[64]

While the myth of Beidahuang promised veterans a renewal of their masculinity, it also offered young women the prospect of empowerment. Coming of age in the heyday of Maoist gender equality – "whatever men can do, women can do, too" – many young women saw building Beidahuang as the opportunity to demonstrate their qualifications as socialist workers. Many did not wait until veterans had completed the wilderness-to-granary conversion but arrived alongside the men to help establish farms. For example, the first group of thirty-two young women from Sichuan arrived in May 1956 at Farm 852, even before the June announcement of the farm's establishment, to "participate in farm construction."[65] The following year, nearly four hundred more young women from Sichuan, Hebei, and Mudanjiang joined Farm 852. Together, they constituted 9.6 percent of the farm's workforce.[66] Another three thousand young women from Shandong arrived in 1959.[67] Combined with about twenty-three hundred veterans' wives who arrived as dependents (*jiashu*), women made up 33.1 percent of the workforce of Farm 852 in 1960.[68] They earned compliments for challenging the traditional gendered division of labour: they "no longer ground embroidery needles, but sharpened sickles instead and displayed heroism [*cheng yingxiong*] in the shock harvesting [*qiangshou*] of wheat."[69] Dong Xueqin, a young woman who had arrived from Harbin in March 1958, became the first woman Party member on Farm 852.[70] A *guniang* platoon led by Dong set the record of soya bean sowing per person per day on Farm 852 – a record that was unrivalled by men.[71] When the entire farm was mobilized to finish road construction in time to celebrate the anniversary of the establishment of the Party, on July 1, 1962, Dong marshalled women's teams and personally carried sod on her back (against her doctor's advice); she died a few days later of meningitis.[72]

There was significant differentiation between women's populations on the farms, however. Single, young women embodied the state's vision of "good workers," independent and competent participants in socialist construction. Women who arrived on the farms as dependent wives of verterans were often depicted as subservient "good wives and good mothers." Although army dependents were incorporated into the workforce during the Great Leap Forward, they soon constituted the majority of Beidahuang's reserve army of labour. From 1961 to 1965, of all the army farms and state farms in Beidahuang, over sixty thousand married women were laid off due to their "family burdens" (*jiating tuolei*) and were hired as temporary workers only during busy seasons.[73] Despite the promises of Maoist emancipation, marriage in socialist rural China still placed major constraints on women's economic independence and social equality.

And though matches were made and families grew, taking root in Beidahuang remained a constant test of the resilience of army migrants. The harsh living conditions were intimidating even to seasoned "heroes." In the early days of army farms, veterans and farm workers had to share saddle-shaped shelters, *majia,* which were made of thatch and tree branches. In 1958, Farm 859 built over 360 *majia* in about twenty days to accommodate 3,475 newly arrived veterans.[74] Veterans joked about having the largest amount of land per capita but the smallest living space per person in China, as each had a single bed made of thatch and branches in the communal *majia*. Newlyweds and veteran families usually curtained off their beds with boxes and cloth for privacy. Even when they did manage to find some personal space, they faced other challenges. A veteran who had built a small *majia* for himself (so as not to disturb his comrades with his loud snoring) was bitten by a wolf at night.[75] One couple lived in a deserted fishing hut, but the husband was called back from felling trees in the forest to guard his newborn against wild rats.[76] Moreover, exhaustion on army farms was routine. Workdays were arduously long, usually more than ten hours; Sundays did not become a rest day until 1963. One kitchen staff member at Farm 859 was so exhausted that he slipped and fell asleep on his way to fetch water.[77] And despite the promise of fairytale endings, desertion was rampant. In one case, twelve veterans from a single platoon deserted.[78] In order to quiet complaints and to set a model for others, Wang Zhen moved his sister's family and her adult children from rural Hunan to Farm 852 as common workers.[79] Overall, most veterans and young women did take root in the wilderness and, together with their offspring, made it the largest commercial grain production base in the country.

## Seeing Off "Beidahuang"

If future generations want to know what Beidahuang is,
Please go to the history museum.

– "Seeing Off 'Beidahuang,'" Qiang Xiaochu, 1959[80]

In the wake of propaganda about the Great Leap Forward over land reclamation, there appeared a collection of poems written by Heilongjiang provincial leaders. Entitled *Seeing Off "Beidahuang"* (*Song Beidahuang*), this volume reflected the determination of officials to tame and transform the wilderness. They shared a deeply rooted expectation for local officials to "exhaust the land" (*jin dili*), a belief that had served as a yardstick in measuring the merits of local administration in imperial China.[81] The ambition of the provincial Party secretary, Qiang Xiaochu (1918–2007), to build a museum to preserve the memories of the lost wilderness was achieved in less than fifty years. The Beidahuang Museum was established in 2005 on one of the main streets of the provincial capital, Harbin, and has been designated as a patriotic education site since its opening. Qiang would be pleased to know that organized student tours constitute the major flow of museum visitors. Five of the six chambers of the museum display and document the CCP-led human efforts to conquer nature. Only the first hall contains elements from nature as the background of a 50-square-metre diorama of the 114 state farms and ranches in the Heilongjiang land reclamation area, with images from the Naoli River National Nature Reserve, complete with sound effects: the cry of wild geese and the howl of a wolf to "show a beautiful picture of the harmony between human and nature."[82] The conflicting message clearly reflects the changing perception of human/nature relationships in the past half-century, from a relationship of confrontation towards one of reconciliation.

When the war against the Earth was launched in the 1950s, it was carried out in its most literal sense, invigorated by combat language and validated through military ritual. Farm leaders delivered pep talks to teams of veterans before starting a particular land-clearing project. Fired up by waving flags and loud drums, veterans would eagerly request battle assignments (*qingzhan*), shouting slogans to demonstrate their determination. Land-opening squads then charged into the wilderness, singing the "Military Anthem of the Eighth Route Army."[83] The marshes

FIGURE 10.1 Diorama of state farms.
*Source:* Author's collection

in Sanjiang and elsewhere, which had served as militarized landscapes since imperial times,[84] became new battlegrounds for veterans-turned-farmworkers. With simple farming tools such as sickles and hoes in place of weapons, ex-soldiers regained masculinity through slashing their new "enemy": tall grass and tree trunks. The vegetation could hardly fight back, but mosquitoes in summer and extreme cold in winter were the most notorious and vicious enemies, testing veterans' spirit and endurance. Those who lost their lives in work-related accidents received honorary posthumous recognition as martyrs and were buried in revolutionary martyrs' cemeteries.

The most efficient, yet hazardous, practice for clearing land was burning – in which veterans literally opened fire on nature. Burning vegetation to prepare land for farming has long been part of agricultural routine in China. However, the inherent danger of using fire compelled state authorities – from imperial times to the PRC – to impose laws to ensure strict control over fire usage.[85] The hero narrative of Beidahuang credits Wang Zhen with setting the first fire to clear vegetation in the spring of 1956, with the sanction of the Ministry of Forestry and the Heilongjiang provincial government.[86] Immediately after, burning became common practice. Tieli Farm, which was founded by the Fourth Battalion of the Second Agricultural Construction Division in 1954, launched its first "burning wilderness battle" (*shaohuang zhandou*) in May 1956, clearing a thousand hectares in seven days and nights.[87] People in neighbouring towns, who "had never seen such a scale of blaze and smog," called the farm in panic.[88] Although air pollution was not a concern in the mid-1950s, the potential damage of fire was clearly evident.[89] Despite strict rules like the "Six Don'ts with Fire" (*liu bu shao*), in some cases land clearance fires did accidentally spread, causing death and state property destruction, and threatening residential areas.[90]

The emphasis on land opening and the optimism in the infinite size of the wilderness in the mid-1950s often led to ill-advised policies and reckless practices, often with destructive consequences. The *People's Daily* announced in March 1956 that "opening and cultivating wasteland" (*kaiken huangdi*) was a "century-lasting grand scheme" (*bainian daji*) for agricultural development.[91] In this political climate, the Railway Soldiers started to build Farm 852 in the same month. In response to the 1958 national guideline for building socialism "greater, faster, better and more cost-effective" (*duo kuai hao sheng*), Farm 852 devised its own approach, "simultaneously acting upon reclamation, production, construction, accumulation, and expansion," summed up as the "five-simultaneity strategy" (*wubian fangzhen*).[92] This strategy was soon adopted as local policy, and Farm 852 was praised by Vice Chairman Zhu De as the most cost-effective farm in the country, having cut its investment in land reclamation from 60 yuan per *mu* to a little over 14 yuan per *mu* (1 *mu* is about 0.16 acre).[93] The farm leader, a veteran from Nanniwan, was invited by the Chinese Academy of Agricultural Sciences to speak to scholars in August 1958. The "five-simultaneity strategy" amazed those who believed that production could only happen following a sequence of surveying, planning, infrastructure construction, hydraulic engineering, equipment purchase, personnel training, and then land clearance.[94]

The negative consequences of the "five-simultaneity strategy" started to become apparent after the haze of the Great Leap Forward. Assisted by vegetation burning, farms did significantly expand land clearance. However, because farms lacked sufficient manpower, machines, and drainage, most newly claimed land was left fallow or flooded. For example, in the seven years from 1957 to 1963, Farm 859 opened a total of 930,000 *mu*, but only 280,000 *mu* was seeded in 1963, leaving fallow about 70 percent of reclaimed land.[95] Beidahuang underwent further reclamation and more intensive farming in the next twenty years, under the Production and Construction Corps and then as part of the economic reform since the 1980s. This process of "exhausting the land" resulted in land degradation and salinization. Most farmlands were reclaimed from permanent wetlands; soil organic matter and total nitrogen decreased by 55.5 percent and 59.6 percent, respectively, in dry farmland, and 63.6 percent and 67.6 percent, respectively, in paddy land.[96] The layer of black soil, the most fertile soil on Earth and the pride and signature of Beidahuang, has diminished at an alarming rate. The original deposit of black soil in the region was usually 60 to 80 centimetres – as much as 100 centimetres in some places. By the late 2000s, this layer was being reduced by 0.4 to 0.7 centimetres annually due to erosion. It is worth remembering that 1 centimetre of black soil takes about four hundred years to form.[97]

In addition to the permanent loss of black soil, another negative consequence of land reclamation in the region is the permanent loss of large wetlands. Wetlands play a vital role in flood control, drought prevention, water quality improvement, and climate change mitigation. They also provide unique habitats for flora and fauna.[98] However, "land reclamation and drainage because of high human population density" are some of the primary causes of wetland destruction around the world.[99] The development and settlement of Beidahuang dealt a severe blow to the ecosystem of the Sanjiang Plain, China's largest area of freshwater wetlands, often referred to as the "kidney of Asia" (*Yazhou zhi shen*).[100] Chinese environmental scientists generally agree that cropland expansion from the mid-1950s to the mid-1970s, intensified by military farming, led to the most rapid wetland loss in the area.[101] Other land uses that accompany agricultural production and human settlement – such as irrigation canals, dykes, roads, and residential development – further reduced and fragmented wetlands. From 1954 to 1987, the number of wetlands decreased by 69.2 percent; total wetland area decreased by 60.6 percent.[102] When a contiguous string of wetland areas becomes fragmented, its ability to function properly – as drought prevention, floodwater storage, and wildlife habitat – is severely compromised if not destroyed.[103] Both droughts and floods have increased dramatically in number and severity since the 1970s. The average interval between extreme droughts declined from 4.2 years during 1950–70 to 1.8 years during 1971–90. The incidence of floods increased by 15 percent from 1949–69 to 1970–90.[104] Wetland destruction and the resulting habitat loss have led to an alarming decrease in species populations of rare and endangered waterfowl, including Red-Crowned Cranes, Great Swans, and Oriental White Storks. The Oriental White Stork population was observed to be in the thousands in the early 1950s; by 1984, the observed population was about two hundred. In 2000, this number had decreased to a few dozen.[105]

While agricultural productivity continues to take priority in the Heilongjiang land reclamation area, increasing environmental concerns have encouraged some change in the relationship between human and nature after the domination of military farming. The most prominent effort to reconcile with nature was the establishment of seventeen national and provincial wetland reserves in the Sanjiang Plain from 1985 to 2003.[106] In the Beidahuang Museum, however, reconciliation efforts feature only as background to human achievement; regional enterprises and their products, on the other hand, occupy an entire exhibition hall. Businesses were quick to take advantage of increasing environmental consciousness to tap into Beidahuang's new market value. Beidahuang has become a popular brand; now rid of its connotation of uncivilized desolation, the name conveys pure nature – green, organic, and uncontaminated. Contrary to former provincial Party head Qiang Xiaochu's 1959 revolutionary wish to eradicate Beidahuang, the name of the region is now the subject of corporate celebration, courtesy of the Beidahuang Group (*Beidahuang jituan*).

FIGURE 10.2 Beidahuang Group building.
*Source:* Collection of N. Smith

While the idea of a history museum served as a trope in the 1950s to represent the revolutionary aspirations to convert wasteland into a socialist agricultural modernity, the Beidahuang Museum itself provided the physical space to display the trophies of human victory over the non-human world. The mid-twentieth-century promises of Beidahuang have manifested themselves in increased grain production, a growing population, and endangered wetlands that depend on human conservation efforts for their survival. For the tens of thousands of veterans and female migrants who developed the border region, endured the hardships, and put down roots in Beidahuang, the myth can ultimately be summed up in the popular saying, "sacrificed my youth and then sacrifice my life (*xianle qingchun xian zhongshen*); sacrificed my life and then sacrifice my children (*xianle zhongshen xian zisun*)." Since the 1990s, their heroic struggle, endurance, and sacrifice have been captured in a new state propaganda device termed "Beidahuang Spirit" (*Beidahuang jingshen*), intended to accelerate the construction of the grand agricultural modernity and imbue it with socialist characteristics.

## Notes

*Acknowledgments:* I would like to thank Alan Christy for inspiring my research interest in post-colonial Northeast China, and Norman Smith for opening the door to Manchuria Studies by accepting my first inquiries about Maoist land reclamation in the 2013 UBC workshop. I am also deeply grateful to Bill Sewell and John Munro for their constructive comments and meticulous editing. Research for this chapter was generously funded by Saint Mary's University.

*Epigraphs:* Guo Moruo, *Xiang diqiu kaizhan – Beidahuang shiji* [Waging war against the Earth – an anthology of Beidahuang poetry] (Beijing: Nongken chubanshe, 1958), 1–4. All poetry translations are by the author.

1 Qi Changfa, ed., *Fuzhuan guanbing kaifa jianshe Beidahuang* [Demobilized and transferred officers and soldiers opening and building Beidahuang] (Harbin: Heilongjiang Nongken ribao chubanshe, 2000), 235.

2 Chinese literature boasts a magnitude of *shiwan fuzhuan guanbing* (100,000 demobilized and transferred officers and soldiers) to show mightiness; in fact, however, this figure includes about 20,000 non-veterans, such as family members of veterans and some "enthusiastic students." Ibid., 27.

3 Ibid., 14.

4 Zhen Ge, "Zhufu nin, Beidahuang" [Bless you, Beidahuang], in *Xiang diqiu kaizhan,* 24; Wang Zhuyu, "Xiang Beidahuang jinjun" [Marching into Beidahuang], in *Xiang diqiu kaizhan,* 29.

5 Feng Jianhua, "Feeding the Nation: China's Painful Memories of Famine Will Hopefully Act as Deterrent to Another Such Catastrophe," *Beijing Review,* June 14, 2007, 21.

6 For pioneer analysis of the impact of PRC political campaigns on the Dianchi Wetlands in Yunnan, see Judith Shapiro, *Mao's War against Nature: Politics and the Environment in Revolutionary China,* Studies in Environment and History (Cambridge, UK: Cambridge University Press, 2001).

7 For example, see Louise Young, *Japan's Total Empire: Manchuria and the Culture of Wartime Imperialism* (Berkeley: University of California Press, 1998).

8 For studies on political exiles, see Wang Ning, "Border Banishment: Rightists in the Army Farms of Beidahuang," in *The Chinese State at the Borders,* ed. Diana Lary (Vancouver: UBC Press, 2007), 211–33; and Willy Wang, "Discovering Xingkaihu: Political Inmates in a PRC Labor Camp," *East Asia* 25, no. 3 (2008): 267–92. For studies on urban development, see Jay Carter, "A Tale of Two Temples: Nation, Region, and Religious Architecture in Harbin, 1928–1998," *South Atlantic Quarterly* 99, no. 1 (2000): 97–115; and "The Future of Harbin's Past," *Itinerario* 35, no. 3 (2011): 73–85.

9 Guo Xiaochuan, "Ke zai Beidahuang de tudi shang" [Engraved onto the land of Beidahuang], in *Zhongguo dangdai zhuming zuojia wenku: Guo Xiaochuan daibiaozuo* [Collections of prominent writers in modern China: Guo Xiaochuan's masterpieces], ed. Li Lizhong (Zhengzhou: Huanghe wenyi chubanshe, 1986), 157–60. The poem was first published in *People's Daily* on January 27, 1963.

10 Zhang Xiangling, "Guanyu Heilongjiang lishi de jige wenti" [A few questions about the history of Heilongjiang], *Lilun tantao* [*Theoretical Investigation*], 5 (2003): 117.

11 Ibid.

12 James Reardon-Anderson, *Reluctant Pioneers: China's Expansion Northward, 1644–1937* (Stanford, CA: Stanford University Press, 2005), 143.

13 Ibid., 101.

14 Ibid., 107.

15 Ronald Suleski, "Northeast China under Japanese Control: The Role of the Manchurian Youth Corps, 1934–1945," *Modern China* 7, no. 3 (1981): 352.

16 Ibid., 367.

17 Liu Xingtu and Ma Xuehui, "Sanjiang pingyuan damianji kaihuang dui ziran huanjing yingxiang ji quyu shengtai huanjing baohu" [Influence of large-scale reclamation on natural environment and regional environmental protection in the Sanjiang Plain], *Dili kexue* (*Scientia Geographica Sinica*) 20, no. 1 (2000): 14.

18 Thomas R. Gottschang and Diana Lary, *Swallows and Settlers: The Great Migration from North China to Manchuria* (Ann Arbor: Center for Chinese Studies, University of Michigan, 2000), 136.

19 Ibid.

20 J. David Wood, "'The Last Frontier': Rationalizing the Spread of Farming into the Boreal Woods of Canada, c. 1910–1940," *Canadian Geographer* 50, no. 1 (2006): 38–55.

21 Wang Zhen, "Guanyu kaifa beidahuang wenti xiang zhongyang de baogao, jielu" [Report on developing Beidahuang to the central government, excerpt], in *Beidahuang* [Opening and building Beidahuang], ed. CCP Heilongjiang Provincial Branch Party History Research Center and the Heilongjiang Bureau of Agricultural Reclamation (Beijing: Zhonggong dangshi chubanshe, 1998), 146.

22 Xu Xianguo, "Yong bu fangxia qiang" [Never put down the gun], in *Xiang diqiu kaizhan*, 9.

23 Wang Zhen, "Qianwanren de xinsheng – gei Xu Xianguo tongzhi de yifengxin" [The heartfelt voice of hundreds of thousands of people – A letter to Comrade Xu Xianguo], *Renmin* ribao [*People's Daily*], May 26, 1958, 8.

24 Liu Peizhi, *Tiegu chunqiu* [Iron bones through springs and autumns] (Changsha: Hunan wenyi chubanshe, 1989), 136.

25 Hao Guangnong, "Rongjun nongchang jianchang ji" [Establishing the Rongjun Farm], in *Tuohuangzhe de huiyi* [Recollections of pioneers], ed. Zhonggong Heilongjiang sheng-

wei dangshi yanjiushi and Heilongjiang sheng guoying nongchang zongju (Harbin: Heilongjiang renmin chubanshe, 1989), 30–36.

26 Heilongjiang sheng difangzhi bianzhuan weiyuanhui, ed., *Heilongjiang sheng zhi: guoying nongchang zhi* [Gazetteers of Heilongjiang Province: State farms] (Harbin: Heilongjiang renmin chubanshe, 1992), 72.

27 Heilongjiang sheng difangzhi bianzhuan weiyuanhui, ed., *Heilongjiang sheng zhi: minzheng zhi* [Gazetteers of Heilongjiang Province: Civil affairs] (Harbin: Heilongjiang renmin chubanshe, 1993), 240.

28 Heilongjiang sheng difangzhi bianzhuan weiyuanhui, *Guoying nongchang zhi,* 74–75.

29 Liu, *Tiegu chunqiu,* 140.

30 Zhongyang renmin zhengfu renmin geming junshi weiyuanhui, "Guanyu 1950 nian jundui canjia shengchan jianshe gongzuo de zhishi" [Directions for the PLA to participate in production and construction in 1950], in Qi, *Kaifa jianshe Beidahuang,* 57.

31 Mao Zedong, "Renmin geming junshi weiyuanhui mingling, yijiuwuer nian er yue" [Order from the People's Revolutionary Military Commission, February 1952], in *Nongken gongzuo wenjian ziliao xuanbian* [Collections of documents on agricultural reclamation], ed. Nongkenbu zhengce yanjiushi (Beijing: Nongye chubanshe, 1983), 46–47.

32 Yitzhak Shichor, "Demobilization: The Dialectics of PLA Troop Reduction," *China Quarterly,* no. 146 (June 1996): 337.

33 Neil J. Diamant, "Conspicuous Silence: Veterans and the Depoliticization of War Memory in China," *Modern Asian Studies* 45, no. 2 (2011): 431–46.

34 Heilongjiang sheng difangzhi bianzhuan weiyuanhui, *Guoying nongchang zhi,* 82–83.

35 Zhongyang nongyebu dangzu, "Guanyu kaifa dongbei huangdi he nongjian ershi yiken dongbei wenti de baogao" [Report on opening wasteland and relocating the Second Agricultural Construction Division for farming in the northeast], in Qi, *Kaifa jianshe Beidahuang,* 133.

36 The Railway Soldiers, mainly responsible for railway construction, existed from 1950 to 1984 as a division of the People's Liberation Army.

37 Heilongjiang sheng difangzhi bianzhuan weiyuanhui, *Guoying nongchang zhi,* 88.

38 Wang, "Guanyu kaifa Beidahuang," 146.

39 Heilongjiang sheng difangzhi bianzhuan weiyuanhui, *Guoying nongchang zhi,* 91.

40 Xu, "Xin Zhongguo tunken shubian shilu" [Records of military farming and border development in New China], *Wuhan wenshi ziliao* [*Wuhan Literary and Historical Materials*], no. 8 (2013): 9.

41 "Zhonggong zhongyang guanyu fazhan junken nongchang de yijian" [Directions from the Central Committee of the CCP on developing military farms], in Qi, *Kaifa jianshe Beidahuang,* 183.

42 Zhao, *Huangye lingyin: mingren zai Beidahuang* [Spirited voices in the wilderness: Famous people in Beidahuang] (Harbin: Beifang wenyi chubanshe, 2000), 110–11.

43 Wang, "Qianwanren de xinsheng," 8.

44 Zhao, *Huangye lingyin,* 114.

45 Ibid., 117.

46 Ibid.

47 Ibid., 114.

48 Ibid., 116.

49 *Xiang diqiu kaizhan,* 101.

50 "Guojia jihua weiyuanhui, wuzibu junshi guanzhi weiyuanhui, caizhengbu junshi guanzhi weiyuanhui, nongkenbu junshi daibiao guanyu xinjiang junxu shengchan jianshe

bingtuang xiafang hou ruogan wenti de lianhe tongzhi" [Notice on the issues of Xinjiang Production and Construction Corps from the State Planning Commission, the military control committees of the Ministry of Materials and Equipment and the Ministry of Finance, and military representatives of the Ministry of Agricultural Reclamation], in Nongkenbu zhengce yanjiushi, *Nongken gongzuo wenjian ziliao xuanbian*, 778.

51 Heilongjiang sheng 859 nongchang zhi bianji bangongshi, ed., *859 nongchang zhi, 1956–1984* [Gazetteer of Farm 859, 1956–1984] (Harbin: Heilongjiang renmin chubanshe, 1985), http://210.76.63.176/trsweb/Search.wct?ChannelID=8081.

52 Heilongjiang sheng difangzhi bianzhuan weiyuanhui, *Guoying nongchang zhi*, 95.

53 For the Xinjiang Production and Construction Corps, see Donald H. McMillen, "Xinjiang and the Production and Construction Corps: A Han Organisation in a Non-Han Region," *Australian Journal of Chinese Affairs*, no. 6 (1981): 65–96; Thomas Matthew James Cliff, "Neo Oasis: The Xinjiang Bingtuan in the Twenty-First Century," *Asian Studies Review* 33, no. 1 (2009): 83–106. For rubber plantation in Yunnan, see Mette Halskov Hansen, *Frontier People: Han Settlers in Minority Areas of China* (Vancouver: UBC Press, 2005). I would like to express my gratitude to one of the reviewers for recommending Hansen's work on Yunnan.

54 Luo Lihua, "Zhongguo ge junqu Shengchan jianshe bingtuan shi yanjiu, 1968–1975" [History of the production and construction corps of Chinese military zones, 1968–1975], *Bingtuan dangxiao xuebao* [*Journal of Production and Construction Corps Party Academy*] 3 (2013): 75.

55 The Xinjiang PCC was reinstated in 1981 due to ethnic conflicts and border tensions, and remains the only PCC now.

56 Chen Di, "Chengshi guniang aishang le Beidahuang" [City girls fell in love with Beidahuang], in *Xiang diqiu kaizhan*,19.

57 James Z. Gao, "The Call of the Oases: The 'Peaceful Liberation' of Xinjiang, 1949–53," in *Dilemmas of Victory: The Early Years of the People's Republic of China*, ed. Jeremy Brown and Paul G. Pickowicz (Cambridge, MA: Harvard University Press, 2007), 198.

58 "Wang Zhen shuaishi kaifa Beidahuang de lishi liucheng jiqi moulue" [Historical sequences of Wang Zhen leading troops opening Beidahuang and his strategies] (hereafter cited as "Wang Zhen shuaishi"), in Qi, *Fuzhuan guanbing kaifa jianshe Beidahuang*, 259–60.

59 He Libo, "Ershi shiji zui weida de hongniang" [The greatest marriage broker of the twentieth century], *Zhongguo gongchandang xinwen* [*News of the Communist Party of China*], http://cpc.people.com.cn/GB/85037/85038/7706265.html.

60 Gong Huanzhi, "Buzhang yu an xintiexin" [Heart-to-heart with the minister], in *Jiangjun yu Beidahuangren* [The general and Beidahuang people], ed. Zheng Jiazhen and Qi Changfa (Shanghai: Shanghai shehui kexuyuan chubanshe, 1991), 141.

61 Zheng Jiazhen, "Beidahuang yimin lu" [Records of Beidahuang migrants], in Zheng and Qi, *Jiangjun yu Beidahuangren*, 232.

62 Wang Guilin, "Heitudi de zuji" [Footsteps on the black soil], in Qi, *Fuzhuan guanbing kaifa jianshe Beidahuang*, 133.

63 852 nongchang zhi bianshen weiyuanhui, ed., *852 nongchang zhi, 1956–1984* [Gazetteer of Farm 852, 1956-1984], http://210.76.63.176/trsweb/Search.wct?ChannelID=2654.

64 "Tiedaobing nongken zongju guanyu jianchang yilai de gongzuo zongjie" [Review of establishing farms by the railway soldiers General Bureau of Agricultural Reclamation], in Qi, *Fuzhuan guanbing kaifa jianshe Beidahuang*, 168.

65 852 nongchang zhi bianshen weiyuanhui, *852 nongchang zhi*.

66 Ibid.

67 Ibid.

68 Ibid.

69 Tie Xun, "Chang nügong" [Ode to women workers], in *Xiang diqiu kaizhan,* 55.

70 Ibid., 220.

71 Ba Wushan, "Women huainian de nübing" [A woman soldier in our memory], in Zhonggong and Heilongjiang, *Tuohuangzhe de huiyi,* 219.

72 Ibid., 222.

73 Heilongjiang sheng nongken zongju, ed., *Beidahuang quanshu: jianshi juan* [A complete collection on Beidahuang: A concise history] (Harbin: Heilongjiang renmin chubanshe, 2010), http://210.76.63.176/trsweb/Search.wct?ChannelID=6454.

74 "Wang Zhen shuaishi," 240.

75 Heilongjiang sheng 859 nongchang zhi bianji bangongshi, *859 nongchang zhi.*

76 Fang Pei'en, "Zuichu de huiyi" [The early days], in Zhonggong and Heilongjiang, *Tuohuangzhe de huiyi,* 89.

77 Heilongjiang sheng 859 nongchang zhi bianji bangongshi, *859 nongchang zhi.*

78 "Heilongjiang sheng weiwen gongzuotuan guanyu weiwen Hejiang diqu shengchan jianshe bingtuan qingkuang de baogao" [Report on supporting the Production and Construction Corps in Hejiang District by the Heilongjiang Provincial Service Team], in Qi, *Fuzhuan guanbing kaifa jianshe Beidahuang,* 74.

79 "Wang Zhen shuaishi," 242.

80 Qiang Xiaochu, "Song 'Beidahuang'" [Seeing off "Beidahuang"], in *Song "Beidahuang"* [Seeing off "Beidahuang"], ed. Zhongguo zuojia xiehui Heilongjiang fenhui (Harbin: Beifang wenyi chubanshe, 1959), 16.

81 Peter C. Perdue, *Exhausting the Earth: State and Peasant in Hunan, 1500–1850* (Cambridge, MA: Harvard University Press, 1987).

82 Zhao Guochun, *Zoujin Beidahuang Bowuguan* [Entering the Beidahuang Museum] (Harbin: Heilongjiang remin chubanshe, 2009), 5.

83 Ding Lüshu, "Nanwang de suiyue" [Unforgettable years], in Zhonggong and Heilongjiang, *Tuohuangzhe de huiyi,* 43.

84 Jack Patrick Hayes, "From Great Green Walls to Deadly Mires: Wetlands as Military Environments and Ecosystems in Chinese History," *Water History* 5, no. 1 (2013): 7–26.

85 Jack Patrick Hayes, "Fire and Society in Modern China: Fire Disasters and Natural Landscapes in East Asian Environmental History (1820–1965)," *ASIA Network Exchange* 20, no. 1 (2012): 23–35.

86 "Wang Zhen shuaishi," 244.

87 Zhang Tiankong, "Nongjian situan zhan Tieli" [The Fourth Agricultural Construction Battalion fighting Tieli], in Zhonggong and Heilongjiang, *Tuohuangzhe de huiyi,* 84.

88 Ibid.

89 Recent studies have demonstrated that open crop residue burning can cause severe air pollution. See K. Yamaji, J. Li, I. Uno, Y. Kanaya, H. Irie, M. Takigawa, Y. Komazaki, P. Pochanart, Y. Liu, H. Tanimoto, T. Ohara, X. Yan, Z. Wang, and H. Akimoto, "Impact of Open Crop Residual Burning on Air Quality over Central Eastern China during the Mount Tai Experiment 2006," *Atmospheric Chemistry and Physics* 10 (2010): 7353–68; Zhu Bin, Su Ji-feng, Kang Han-qing, and Cai Yuan, "The Impact of Crop Residue Burning on Air Quality over Yangtze River Delta, China: Observation and Simulation," *IEEE International Geoscience and Remote Sensing Symposium* (July 22–27, 2012): 3642–45. While this chapter was being prepared, a smog emergency of PM2.5, caused by crop residue burning in rural Jiangsu in December 2013, hit big cities like Nanjing and led to closures

of schools, highways, ferries, and the airport. See Cang Wei and Song Wenwei, "Smog Disrupts Daily Life in Nanjing," *China Daily,* December 6, 2013: http://usa.chinadaily.com.cn/china/2013-12/06/content_17155333.htm.

90 Hayes, "Fire and Society in Modern China," 31.

91 "Kaiken huangdi" [Opening and cultivating wasteland], *Renmin ribao* [*People's Daily*], March 6, 1956, 4.

92 Li Guilian, "Cong Nanniwan dao Beidahuang" [From Nanniwan to Beidahuang], in Qi, *Fuzhuan guanbing kaifa jianshe Beidahuang,* 164.

93 Ibid.

94 Ibid., 165.

95 Heilongjiang sheng 859 nongchang zhi bianji bangongshi, *859 nongchang zhi.*

96 Wei Ouyang, Yinming Xu, Fanghua Hao, Xuelei Wang, Chen Siyang, and Chunye Lin, "Effect of Long-Term Agricultural Cultivation and Land Use Conversion on Soil Nutrient Contents in the Sanjiang Plain," *Catena* 104 (2013): 243.

97 Huili Gong, Meng Dan, Li Xiaojuan, and Zhu Feng, "Soil Degradation and Food Security Coupled with Global Climate Change in Northeastern China," *Chinese Geographical Science* 23, no. 5 (2013): 568.

98 William J. Mitsch and James G. Gosselink, "The Value of Wetlands: Importance of Scale and Landscape Setting," *Ecological Economics* 35, no. 1 (2000): 25–33.

99 Wolfgang J. Junk, Shuqing An, C.M. Finlayson, Brij Gopal, Jan Kvet, Stephen A. Mitchell, William J. Mitsch, and Richard D. Robarts, "Current State of Knowledge Regarding the World's Wetlands and Their Future under Global Climate Change: A Synthesis," *Aquatic Sciences* 75, no. 1 (2013): 157.

100 Zhao, *Zoujin Beidahuang Bowugan*, 2009, 7.

101 Zongming Wang, Kaishan Song, Wenhong Ma, Chunying Ren, Bai Zhang, Dianwei Liu, Jing Chen, and Changchun Song, "Loss and Fragmentation of Marshes in the Sanjiang Plain, Northeast China, 1954–2005," *Wetlands* 31, no. 5 (2011): 952; Hongyu Liu and Qing Yang, "Impacts on Wetlands of Large-Scale Land-Use Changes by Agricultural Development: The Small Sanjiang Plain, China," *AMBIO: A Journal of the Human Environment* 33, no. 6 (2004): 307.

102 Wang et al., ibid., 945.

103 Y.M. Xu, Y. Li, W. Ouyang, F.H. Hao, Z.L. Ding, and D.L. Wang, "The Impact of Long-Term Agricultural Development on the Wetlands Landscape Pattern in Sanjiang Plain," *Procedia Environmental Sciences* 13 (2012): 1930.

104 Xiaohui Liu, Guihua Dong, Xigang Wang, Zhenshan Xue, Ming Jiang, Xianguo Lu, and Yuan Zhang, "Characterizing the Spatial Pattern of Marshlands in the Sanjiang Plain, Northeast China," *Ecological Engineering* 53 (April 2013): 340; Liu and Ma, "Sanjiang pingyuan," 18.

105 Liu and Yang, "Impacts on Wetlands," 308.

106 Wang et al., "Loss and Fragmentation," 952.

# Glossary

| | | |
|---|---|---|
| Ai Ren | 靄人 | |
| Amakasu Masahiko | 甘粕正彥 | |
| *Anda* | 安達 | official supervising tribute collection |
| *Anda* | 安達 | name of a town |
| *anjiafei* | 安家費 | money to settle the family |
| Araragi Shinzō | 蘭信三 | |
| Bahai | 巴海 | |
| *bai xiong pai* | 白熊牌 | Polar Bear brand |
| *bainian daji* | 百年大计 | century-lasting grand scheme |
| *ban zhe ju zhi shenghuo* | 半蟄居之生活 | half hibernation life |
| *banbiantian* | 半边天 | Half the Sky |
| *batou* | 巴头 | labour contractor |
| *Beidacang* | 北大仓 | Great Northern Granary |
| *Beidahuang* | 北大荒 | Great Northern Wilderness |
| Bi Kegang | 畢克剛 | |
| *Bingtian shishe* | 冰天詩社 | Poetry Club of the Ice World |
| *bolaipin* | 舶來品 | exotic |
| *buleili* | 布累理 | prairie |
| *busheng* | 補牲 | hunters |
| *Buteha qi* | 布特哈旗 | Butha (hunting) banners |
| *caibu* | 採補 | gathering and trapping |
| *caigong shanchang* | 采貢山場 | mountain foraging reserves |

| | | |
|---|---|---|
| *caizhu hekou* | 采珠河口 | mouths of rivers where pearls are gathered |
| Changbaishan / (Ma., Golmin Šanggiyan Alin) | 長白山 | Long White Mountains |
| *chao shi xing* | 潮濕性 | cold, damp nature |
| *chaogong* | 朝貢 | tribute |
| *Chaoxian ondol* | 朝鲜온돌 | (North) Korean *ondol* (underfloor heating) |
| Chen Di | 谌笛 | |
| Chen Fengxiao | 陈奉孝 | |
| Chen Jingyin | 陈敬尹 | |
| Chen Lu | 陳露 | |
| Chen Zhilin | 陈之遴 | |
| Cheng Dechang | 程德昌 | |
| Chengde | 成德 | |
| *cheng mu* | 乘木 | ride the wood |
| *cheng yingxiong* | 逞英雄 | heroism |
| Chun Fang | 春芳 | |
| Cisan | 奇三 | |
| *culgan* | 楚勒罕 | official market where tribute pelts are collected |
| *cunzhu* | 蠢猪 | stupid pigs |
| *Daban* | 鞑板 | Tartar boards /skis |
| Dagūr | 達斡爾 | |
| Dai Huang | 戴煌 | |
| Daikan'en | 大觀園 | Garden of Grand Vision |
| *dan'ganhu* | 单干户 | single-person household |
| Dasheng Wula | 打牲烏拉 | |
| Deoelgiya | 都爾嘉 | |
| *detianduhou* | 得天獨厚 | richly endowed by nature |
| *dianlu* | 電爐 | electric stoves or hotplates |
| *dihan shenghuo* | 低寒生活 | low cold life |
| *diwen shenghuo* | 低溫生活 | low-temperature lifestyle |
| *dong chang de Manzhou* | 冬長的滿洲 | long winter Manchuria |
| Dong Xueqin | 董雪芹 | |
| Dorja | 都爾嘉 | |
| *duo kuai hao sheng* | 多快好省 | greater, faster, better, and more cost-effective |

| | | |
|---|---|---|
| Enje | 恩澤 | |
| Fang Gongqian | 方拱乾 | |
| Feng Daren | 馮大人 / 馮打人 | |
| Feng Dengdiao | 馮登鈞 | |
| *fengche xing* | 風車性 | centrifugal fans |
| Fiyaka | 費雅喀 | |
| Fu Yu | 傅玉 | |
| Fuchikami Hakuyō | 淵上白楊 | |
| *fudutong* | 副都統 | lieutenant general |
| Fuheng | 福恆 | |
| Fujiadian | 傅家甸 / 傅家店 | |
| Fujin jie | 富錦街 | Fujin Street |
| Fujiyama Kazuo | 藤山一雄 | |
| Fuk'anggan | 福康安 | |
| *fundoshi* | ふんどし/褌 | loincloth |
| Galdan | 噶爾丹 | |
| *gaoji shechi pin* | 高级奢侈品 | luxury item |
| *gen'gu huangyuan* | 亘古荒原 | eternal wasteland |
| *gongye xue Daqing* | 工業學大慶 | industry learn from Daqing |
| Gotō Reiji | 後藤冷次 | |
| Guan Zhong | 管仲 | |
| Guanboo | 官保 | |
| *guanggun* | 光棍 | bare sticks / unmarried men |
| *guangrong shengli de junzhuang* | 光荣胜利的军装 | glorious and victorious uniforms |
| *guanzhuang* | 官庄 | government farms |
| Guanzi | 管子 | [Writings of] Master Guan |
| *guniang* | 姑娘 | girls |
| Guo Songgen | 郭松根 | |
| Guo Xiaochuan | 郭小川 | |
| *guochi* | 國恥 | national humiliation |
| *guohao* | 国号 | name of a country |
| *guoqu wei geming liuxue, xianzai wei jianshe liuhan* | 过去为革命流血，现在为建设流汗 | bled for the revolution in the past and sweat for the construction in the present |
| *hafan* | 哈番 | officials |
| Hafungga | 哈豐阿 | |
| *haiwai cun* | 海外村 | village across the sea |
| *han gui* | 寒鬼 | cold ghost |

| | | |
|---|---|---|
| *Han minzu de danchun xing* | 漢民族的單純性 | Chinese people's simple character |
| *hanbing* | 寒病 | disease of the cold |
| Hanke | 函可 | |
| *haomai qigai* | 豪迈气概 | imposing manners |
| *haori* | 羽織 | jacket to go over a kimono |
| *haramaki* | 腹巻き | woolen sashes wrapped around the stomach |
| Hejen | 赫哲 | |
| *hengxin* | 恆心 | perseverance |
| *hezuohua* | 合作化 | collectivize |
| *hongniang* | 红娘 | marriage broker |
| Hongshan | 紅山 | |
| Huang Fujun | 黄富俊 | |
| Huang Zhan | 黄湛 | |
| *huaxue de xiang* | 滑雪的鄉 | home of skiing |
| *huidian* | 會典 | statutes |
| *huishang* | 回賞 | gifts in return |
| *huolu* | 火爐 | heating stoves |
| *huopan* | 火盘 | brazier |
| Hūrha | 呼爾哈 | |
| *huwai yundong* | 戶外運動 | outdoor sports |
| Iida, Nagano | 長野県 飯田市 | Iida city in Nagano Prefecture |
| Ishiwara Kanji | 石原莞爾 | |
| Ji Kaisheng | 季开生 | |
| *jiankang di'yi, yao dao waibian qu* | 健康第一 要到外邊去 | for good health, go outside |
| *jianqiang youli de bibang* | 坚强有力的臂膀 | strong and powerful arms |
| *jiashu* | 家属 | dependents |
| *jiating tuolei* | 家庭拖累 | family burdens |
| *jibian kuhan* | 極邊苦寒 | furthest frontiers, with bitter cold |
| Jilin jiangjun yamen hu si | 吉林將軍衙門戶司 | Jilin general yamen's revenue division |
| *jin dili* | 尽地力 | exhaust the land |
| Jin Jian | 金簡 | |
| Jingqili | 精奇哩 | Zeia River (M. Jingkiri) |
| *jiujiu xiaohan* | 九九消寒 | Double-Nine Expulsions of the Cold |
| *juhiyan* | 珠軒 | foraging detachment |

| | | |
|---|---|---|
| *juhiyan da* | 珠軒達 | foraging detachment chief |
| *kaiken huangdi* | 开垦荒地 | opening and cultivating wasteland |
| *kaitaku-dan* | 開拓団 | settlement group |
| *kaitaku senshi* | 開拓戦士 | rural development warriors |
| Kantōgun | 關東軍 | Kantō Army |
| Katō Toyotaka | 加藤豊隆 | |
| Kawamura Yasuo | 河村泰男 | |
| Kawano Village | 河野村 | |
| *kefu ziran de tiaojian* | 克服自然的條件 | conquer nature's conditions |
| *kefu ziran li* | 克服自然力 | conquering nature power |
| *kenkoku* | 建國 | nation building |
| Kiler-Ewenki | 奇勒爾-鄂倫春 | |
| Kitaiskaia Street | キタイスカヤ通り | |
| Kitaiskaya Street | キタイエスカヤ | (Ch., 中央大街) |
| *komekura Manshū* | 米倉満洲 | grain storehouse Manchuria |
| Konoe Fumimaro | 近衛文麿 | |
| Kubota Isamu | 久保田諫 | |
| *laogai* | 勞改 | labour reform camps |
| *laojiao* | 勞教 | labour re-education camps |
| *laomianyang* | 老绵羊 | old sheep |
| Li Jianru | 李兼汝 | |
| Li Tieyan | 李鐵言 | |
| Li Xiangchen | 李相臣 | |
| *liji zhuyi* | 利己主義 | egoism |
| Lin Jitang | 林幾唐 | |
| Liu Baokun | 刘寶坤 | |
| *liu bu shao* | 六不烧 | Six Don'ts with Fire |
| *long xing zhi di* | 龍興之地 | birthplace of the dragon |
| Lü Liuliang | 呂留良 | |
| Lunjišan | 倫吉善 | |
| *majia* | 马架 | saddle-shaped shelters |
| Mandarhan | 滿達爾漢 | |
| Man-Mō kaitaku o katari-tsuzuku kai | 満蒙開拓を語り繼く会 | Society for the Propagation of Narratives about the Colonization of Manchuria and Mongolia |
| Manshū shashin sakka kyôkai | 満州写真作家協会 | Manchurian Photographic Artists Association |

| | | |
|---|---|---|
| Manshukoku | 満洲国 | Manchukuo/Manchoukuo/ Manzhouguo |
| Mantie liang cheng ke | 滿鐵練成科 | (South) Manchuria Railway Company Excel Division |
| *Manzhou difang bixu jinxing de naihan de duanlian* | 滿州地方必須進行的耐寒的鍛錬 | conduct essential Manchurian cold-resistant exercises |
| Manzhouguo | 满洲国 | Manchukuo/Manchoukuo/ Manzhouguo |
| *mao dong* | 猫冬 | hiding from the winter |
| Menguhutulingga | 蒙庫瑚圖靈阿 | |
| Minami Manshū tetsudō kabushiki geisha | 南滿州鉄道株式会社 | South Manchuria Railway Company/SMRC |
| *minglang* | 明朗 | bright and clear |
| Mingliang | 明亮 | |
| *minzu xing* | 民族性 | national character |
| Miura Unichi | 三浦運一 | |
| *monpe* | モンペ | drawstring pants fashioned out of the bottom portion of the kimono |
| *moringga Orončon* | 摩凌阿鄂倫春 | Equestrian Orochen |
| *moxing* | 鬼星 | devil star |
| Mudanjiang | 牡丹江 | Hūrha River |
| Nakagawa Kōichi | 中川好一 | |
| Nakaii Hisaji | 中井久兒 | |
| Nayanceng | 那彥成 | |
| Nayantai | 那彥泰 | |
| *neibu wenjian* | 内部文件 | internal document |
| Neiwufu | 内務府 | Imperial Household Department |
| *neng shui liang bu shui chao* | 能睡涼不睡潮 | can sleep cold, don't sleep damp |
| Nie Er | 聂耳 | |
| Nogi Maresuke | 乃木希典 | |
| *nōgyō imin* | 農業移民 | agricultural immigrants |
| *nōhonshugi* | 農本主義 | agrarianism |
| Noji Kaeko | 野地香惠子 | |
| Nongjian ershi | 农建二师 | Second Agricultural Construction Division |

| | | |
|---|---|---|
| Nongkenbu | 农垦部 | Ministry of Agricultural Reclamation |
| Onitsuka Hiroshi | 鬼塚博 | |
| Orončon | 鄂倫春 | |
| *pa leng* | 怕冷 | fear of the cold |
| *panbozhe* | 盤剝者 | usurer |
| *pantsu* | パンツ | pants (underwear, boxer shorts) |
| *pech* | печь | (Russian) stove |
| *peiqika* | 培氣卡 | (Russian) stove |
| *piku* | 皮庫 | fur storehouse (of the Imperial Household Department) |
| *qi mu er xing* | 骑木而行 | ride the wood and go |
| Qiang Xiaochu | 强晓初 | |
| *qiangshou* | 抢收 | shock harvesting |
| *qingxu wending* | 情绪稳定 | stabilized veterans' moods |
| *qingzhan* | 请战 | request battle assignments |
| *Qizhe zhihui* | 七謫之会 | Society of Seven Exiles |
| *qun bu ping sha / jieban buxing* | 群步平沙 结伴步行 | group walk |
| *renshao diduo* | 人少地多 | few people, much land |
| *risōkyō* | 理想郷 | utopian homeland |
| *rongjun* | 荣军 | honoured soldiers |
| Saboo | 薩保 | |
| Saitō Toshie | 齊藤俊江 | |
| Šarhūda | 沙爾虎達 | |
| Satō Shin'ichirō | 佐藤慎一郎 | |
| Seishōnen giyūgun | 青少年義勇軍 | Volunteer Youth Corps |
| *senbei* | せんべい/ 煎餅 | salty cracker |
| *senden* | 宣伝 | propaganda/advertising |
| *senshi-sha* | 戦死者 | a fallen warrior |
| Šertu | 舍爾圖 | |
| *shaohuang zhandou* | 烧荒战斗 | burning wilderness battle |
| Shengchan jianshe bingtuan | 生产建设兵团 | Production and Construction Corps |
| Shi Songqiao | 世嵩乔 | |
| *shibian* | 實邊 | frontier colonization |
| Shibuya Saburō | 澁谷三郎 | |
| *shifan nongmin* | 示范农民 | modelling collective farming for peasants |

| | | |
|---|---|---|
| Shilu | 使鹿 | Reindeer Herder (tribe) |
| Shimoina no naka no Manshū | 下伊那のなかの滿洲 | Shimoina and Manchuria |
| Shinmai shō | 信濃賞 | Shinmai prize |
| Shiquan | 使犬 | Dog Keeper (tribe) |
| *shiwan fuzhuan guanbing* | 十万复转官兵 | 100,000 demobilized and transferred officers and soldiers |
| Šindali | 辛達里 | |
| *sixiang jiancha* | 思想檢查 | thought examination |
| Solon | 索倫 | |
| Su Xipo | 蘇西坡 | |
| Šuliang | 舒亮 | |
| Sun Wenliang | 孙文良 | |
| Sun Yang | 孙旸 | |
| Suolun | 索倫 | |
| Šutongga | 舒通阿 | |
| Tachibana Kōzaburō | 橘孝三郎 | |
| Tai Yuzhen | 邰玉鎮 | |
| *tan shui de exi xing* | 貪睡的惡習性 | greedy for sleep evil habit |
| Tang Zunwen | 唐尊文 | |
| *Tennō heika no tame ni* | 天皇階下の為に | service for the august Emperor |
| Tian Han | 田漢 | |
| *tiaojie qihou* | 調節氣候 | regulate the climate |
| *Tiedaobing* | 铁道兵 | railway soldiers |
| *tieli* | 铁力 | iron force |
| Tieli/Tetsurei | 鉄驪 | |
| *timian* | 体眠 | dormant |
| Tokunaga Sunao | 德永直 | |
| Tsutsui Aikichi | 筒井愛吉 | |
| *tufei* | 土匪 | bandits |
| *tugong* | 土貢 | tribute |
| Tuhuru | 圖呼勒河 | Tuhuru River |
| *Ula* | 烏拉 | Ula banner/river/grass |
| *ume yo, fuyase yo* | 生めよ、増やせよ | give birth and reproduce! |
| *wakaku takumashiki chikara* | 若く逞しき力 | young, robust power |
| *Wang da jiao lu lü* | 王大叫艪驢 | Braying Mule Wang |
| Wang Luo | 王洛 | |
| Wang Tinglin | 王亭霖 | |
| Wang Zhen | 王震 | |

| | | |
|---|---|---|
| Warka | 瓦爾喀 | |
| Wen Nailang | 温乃朗 | |
| *wenziyu* | 文字獄 | literary inquisition |
| Wu Zhaoqian | 吳兆騫 | |
| Wu Zhenchen | 吳桭臣 | |
| *wubian fangzhen* | 五边方针 | five-simultaneity strategy |
| Wu Liande | 伍連德 | |
| *wu wei hanleng* | 勿畏寒冷 | do not fear cold |
| *xianle qingchun xian zhongshen* | 献了青春献终身 | sacrificed my youth and then sacrifice my life |
| *xianle zhongshen xian zisun* | 献了终身献子孙 | sacrificed my life and then sacrifice my children |
| *xilin yu* | 細鱗魚 | lenok |
| *xin jianiang* | 新嫁娘 | new bride |
| *Xin Manzhou (Ma., Ice Manju)* | 新滿洲 | New Manchus |
| Xu Renjun | 许人俊 | |
| Xu Shichang | 徐世昌 | |
| Xu Xianguo | 徐先国 | |
| *xuanchuan zhuliyuan* | 宣传助理员 | propaganda assistant |
| *xue di zou* | 雪地走 | snow walking |
| *xueji jinran de junzhuang* | 血迹浸染的军装 | blood-stained uniforms |
| *xunhua* | 訓話 | rebuking |
| *yafahan Orončon* | 雅發罕鄂倫春 | Foot Orochen |
| Yagi | 柳 | Nicolai Nicolaivich Yagi |
| Yaksa (Ya-ke-sa) | 雅克薩 | Albazin (in Russian) |
| *Yamato hatarake* | 大和働け | serve Japan |
| Yanmoo | 延茂 | |
| Yang Ancheng | 杨安城 | |
| Yang Bin | 杨宾 | |
| Yang Congdao | 杨崇道 | |
| *yaran, hadaki ni natchaou* | やらん、はだかになっちゃう | but then, we'll be naked |
| *Yazhou zhi shen* | 亚洲之肾 | kidney of Asia |
| *yeshou yang de* | 野兽樣的 | beast-like |
| *yi liang wei gang* | 以粮为纲 | grain first policy |
| *yiku* | 衣庫 | imperial wardrobe [in the Imperial Household Department] |

| | | |
|---|---|---|
| *yin se yun* | 銀色運 | silver-coloured movement |
| Yin Yi | 殷毅 | |
| Yu Minzhong | 于敏中 | |
| *zanryū koji* | 残留孤児 | left-behind orphans |
| *zapaijun* | 杂牌军 | Motley crew |
| Zha Siting | 查嗣庭 | |
| *zhagen* | 扎根 | take root |
| Zhang Huangyan | 張煌言 | |
| Zhang Jiyou | 張繼有 | |
| Zhang Runan | 张汝楠 | |
| Zhang Tangong | 张坦公 | |
| *zhengzhibu* | 政治部 | Ideological Department |
| *zhiqi bi tian gao de yingxiong haohan* | 志气比天高的英雄好汉 | heroes with ambitions higher than the sky |
| Zhou Deqing | 周得慶 | |
| *zhuangding* | 壯丁 | able-bodied adult males |
| *zhuju weisheng* | 住居衛生 | housing hygiene |
| *zhulong* | 豬龍 | pig-dragon |
| *zidan chuantou de shangba* | 子弹穿透的伤疤 | bullet-penetrated scars |
| *zucheng* | 族称 | name of an ethnic group |
| Zunghar | 準噶爾 | |
| Zuo Mu | 佐木 | |
| *zuo zhan* | 作戰 | do battle |
| *zuoling* | 佐領 | company captain |

# Contributors

**David A. BELLO** is a professor of East Asian history at Washington and Lee University. He holds a PhD from the University of Southern California, Los Angeles. His current research interests focus on Qing borderland environmental history. His book *Across Forest, Steppe and Mountain: Environment, Identity and Empire in Qing China's Borderlands* has been published in the Studies in Environment and History series from Cambridge University Press (2016).

**Blaine CHIASSON** is an associate professor of history at Wilfrid Laurier University. He holds a PhD from the University of Toronto and is the author of *Administering the Colonizer: Manchuria's Russians under Chinese Rule, 1919–1929*. His current projects focus on the Boxer Rebellion and the multinational occupation of Beijing, and the work of Russian scientists in Manchuria under both Chinese and Japanese rule.

**Annika A. CULVER** is an associate professor of East Asian history at Florida State University. A graduate of Harvard University and the University of Chicago, she is the author of *Glorify the Empire: Japanese Avant-Garde Propaganda in Manchukuo*. She serves as a scholar in Cohort II of the US-Japan Network for the Future, and received the 2013 William F. Sibley Memorial Translation Prize for the English translation of Japanese proletarian writer Nakano Shigeharu's 1928 short story "Heading for Moscow," about two Korean independence activists journeying through Manchuria.

**Loretta KIM** is an assistant professor and director of China Studies at the University of Hong Kong. She is a graduate of Harvard University and was formerly an assistant professor at the State University of New York (Albany) and Hong Kong Baptist University. Her current writing projects are about the Orochen people of northern Heilongjiang, social and cultural interactions in the Amur River region, and career trajectories of Qing frontier officials.

**Diana LARY** is a professor emerita in history at the University of British Columbia, and grandmother of three. Her monographs include *China's Republic; Chinese Migration from Antiquity to the Present*; *Region and Nation: The Kwangsi Clique in Chinese Politics, 1925–1937*; *Swallows and Settlers: The Great Migration From North China to Manchuria*; *The Chinese People at War: Human Suffering and Social Transformation, 1937–1945*; *The Chinese State at the Borders*; *Warlord Soldiers: Chinese Common Soldiers, 1911–1937*; and *China's Civil War: A Social History, 1945–1949*.

**Kathryn MEYER** is a professor of history at Wright State University. She obtained her PhD from Temple University and formerly studied for two years at the Stanford Center in Taipei. Her books include *Life and Death in the Garden: Sex, Drugs, Cops and Robbers in Wartime China* and *Webs of Smoke: Smugglers, Warlords, Spies and the History of the International Drug Trade*. Her current research interests focus on the Manchuria Aviation Company.

**Norman SMITH** is a professor of history at the University of Guelph. He holds a PhD from the University of British Columbia. His books include *Beyond Suffering: Recounting War in Modern China*; *Intoxicating Manchuria: Alcohol, Opium, and Culture in China's Northeast*; and *Resisting Manchukuo: Chinese Women Writers and the Japanese Occupation*.

**Ronald SULESKI** is a professor of history and director of the Rosenberg Institute for East Asian Studies at Suffolk University, Boston. He holds a PhD from the University of Michigan. He was formerly president of the Asiatic Society of Japan and assistant director of the Fairbank Center for Chinese Studies at Harvard University. His books include *Civil Government in Warlord China: Tradition, Modernity and Manchuria*; *Manshū no seishōnen zō* (*Images of Japanese Youth in Manchuria*); and *The Modernization of Manchuria: An Annotated Bibliography*.

**SUN Xiaoping** is an assistant professor of history at St. Mary's University. She holds a PhD from the University of California, Santa Cruz, and an MA in comparative literature from Liaoning University. She previously taught at Beijing Jiaotong University and McMaster University. Her research interests include gender, war and society, and state-society relations in twentieth-century China.

**WANG Ning** is an associate professor of history at Brock University. His research interests are modern China, especially Chinese intellectuals, foreign-trained Chinese students, border banishment, and Chinese social history. His book, *Banished to the Great Northern Wilderness: Political Exile and Re-education in Mao's China*, is scheduled to be published in 2016.

# Index

bold = illustration

**Contemporary Chinese Studies**

Glen Peterson, *The Power of Words: Literacy and Revolution in South China, 1949–95*

Wing Chung Ng, *The Chinese in Vancouver, 1945–80: The Pursuit of Identity and Power*

Yijiang Ding, *Chinese Democracy after Tiananmen*

Diana Lary and Stephen MacKinnon, eds., *Scars of War: The Impact of Warfare on Modern China*

Eliza W.Y. Lee, ed., *Gender and Change in Hong Kong: Globalization, Postcolonialism, and Chinese Patriarchy*

Christopher A. Reed, *Gutenberg in Shanghai: Chinese Print Capitalism, 1876–1937*

James A. Flath, *The Cult of Happiness: Nianhua, Art, and History in Rural North China*

Erika E.S. Evasdottir, *Obedient Autonomy: Chinese Intellectuals and the Achievement of Orderly Life*

Hsiao-ting Lin, *Tibet and Nationalist China's Frontier: Intrigues and Ethnopolitics, 1928–49*

Xiaoping Cong, *Teachers' Schools and the Making of the Modern Chinese Nation-State, 1897–1937*

Diana Lary, ed., *The Chinese State at the Borders*

Norman Smith, *Resisting Manchukuo: Chinese Women Writers and the Japanese Occupation*

Hasan H. Karrar, *The New Silk Road Diplomacy: China's Central Asian Foreign Policy since the Cold War*

Richard King, ed., *Art in Turmoil: The Chinese Cultural Revolution, 1966–76*

Blaine R. Chiasson, *Administering the Colonizer: Manchuria's Russians under Chinese Rule, 1918–29*

Emily M. Hill, *Smokeless Sugar: The Death of a Provincial Bureaucrat and the Construction of China's National Economy*

Kimberley Ens Manning and Felix Wemheuer, eds., *Eating Bitterness: New Perspectives on China's Great Leap Forward and Famine*

Helen M. Schneider, *Keeping the Nation's House: Domestic Management and the Making of Modern China*

James A. Flath and Norman Smith, eds., *Beyond Suffering: Recounting War in Modern China*

Elizabeth R. VanderVen, *A School in Every Village: Educational Reform in a Northeast China County, 1904–31*

Norman Smith, *Intoxicating Manchuria: Alcohol, Opium, and Culture in China's Northeast*

Juan Wang, *Merry Laughter and Angry Curses: The Shanghai Tabloid Press, 1897–1911*

Richard King, *Milestones on a Golden Road: Writing for Chinese Socialism, 1945–80*

David Faure and Ho Ts'ui-P'ing, eds., *Chieftains into Ancestors: Imperial Expansion and Indigenous Society in Southwest China*

Yunxiang Gao, *Sporting Gender: Women Athletes and Celebrity-Making during China's National Crisis, 1931–45*

Peipei Qiu, with Su Zhiliang and Chen Lifei, *Chinese Comfort Women: Testimonies from Imperial Japan's Sex Slaves*

Julia Kuehn, Kam Louie, and David M. Pomfret, eds., *Diasporic Chineseness after the Rise of China: Communities and Cultural Production*

Bridie Andrews, *The Making of Modern Chinese Medicine, 1850–1960*

Kelvin E.Y. Low, *Remembering the Samsui Women: Migration and Social Memory in Singapore and China*

Jiayan Zhang, *Coping with Calamity: Environmental Change and Peasant Response in Central China, 1736–1949*

Alison R. Marshall, *Cultivating Connections: The Making of Chinese Prairie Canada*

Ruoyun Bai, *Staging Corruption: Chinese Television and Politics*

Christopher G. Rea and Nicolai Volland, eds., *The Business of Culture: Cultural Entrepreneurs in China and Southeast Asia, 1900–65*

Eric Hyer, *The Pragmatic Dragon: China's Grand Strategy and Boundary Settlements*

Ning Wang, *Banished to the Great Northern Wilderness: Political Exile and Re-education in Mao's China*

Lloyd L. Wong, ed., *Trans-Pacific Mobilities: The Chinese in Canada*

Printed and bound in Canada by Friesens
Set in Garamond by Apex CoVantage, LLC.
Copy editor: Jillian Shoichet
Proofreader: Frank Chow
Indexer: Dianne Tiefensee
Cartographer: Eric Leinberger
Cover designer: David Drummond